Refractory Fundamentals in Metallurgical Practice

Jürgen Pötschke

Refractory Fundamentals in Metallurgical Practice

 Springer

Jürgen Pötschke
Essen, Germany

ISBN 978-3-031-63711-7   ISBN 978-3-031-63709-4   (eBook)
https://doi.org/10.1007/978-3-031-63709-4

This Springer imprint is published by the registered company Springer Nature Switzerland AG
The registered company address is: Gewerbestrasse 11, 6330 Cham, Switzerland

If disposing of this product, please recycle the paper.

Foreword by H. Harmuth

Since it was issued in 2015, the German edition of the present book was eagerly embraced by experts in the field of refractories and service applications for refractory materials. There were many requests for an English edition. Thanks are due to the publisher and the author for making this wish finally come true. The success and the appreciation of this technical book in recent years prove that it bridges a gap. This may be due to the fundamental concept the author already revealed in the preface of the first edition and what might also be the motto of his own scientific research: To describe the interaction of refractories with the respective environment of their application by methods of physical chemistry, heat and mass transfer, thermomechanics and mineralogy. For this purpose, the author further remarks, experiences and laboratory results have to be traced back to these fundamentals. This is not always an easy task. This basic idea clearly conveys the goal to leave the traditional, mainly phenomenological approach to refractory science and know-how behind to a greater extent. Even for still justified and applied phenomenological testing methods, understanding of scientific fundamentals increases the scope of conclusions and offers promising alternative investigation methods. A reader will learn from the book which relations to basic laws of various scientific disciplines constitute refractory "behavior." This obviously increases the problem-solving skills which is an attribute of powerful education. In view of the use of simulation methods in refractory material science, the author's approach is particularly contemporary. Thanks to the clear structure, the explicit presentation and the excellent graphical material, the book is easy to read despite its scientific approach. With this English edition, the book now fulfills all requirements to promote a contemporary approach to refractory science and technology for the worldwide community of both refractory experts and students. It will be met with great interest!

Leoben, Austria
2023

Prof. Dr. H. Harmuth

Foreword by Gangolf Stegh

The second, updated edition of the reference book *Service Life of Refractory Materials*, compiled by Prof. Dr.-Ing. Jürgen Pötschke once again conveys the extensive level of knowledge that has to be taken into account in the selection of refractory materials depending on the type of plant and its operation specifics. The German Association for Refractory and Chimney Construction (dgfs) sees its main task in setting quality standards for the benefit of plant operators and in carrying out research and development work as well as training and further education measures in the field of refractory engineering. The second, updated new edition of the technical book with the extensive data on refractory materials is a supplement to the dgfs publications and, thus, also fulfills the requirements under which the specialist companies conduct their activities as members of the dgfs family.

We can only recommend to the entrepreneurs active in the refractory industry to carefully read all chapters of this technical reference to gain additional know-how leading to longer service lives of the materials.

We wish Prof. Dr.-Ing. Jürgen Pötschke a successful publication of the updated new edition and hope that he will continue to be available to us with his expertise for a very long time, specifically regarding his activities within the dgfs.

2017

Dipl.-Ing. Gangolf Stegh
Chairman
German Association for Refractory
and Chimney Construction
Bonn, Germany

Preface

The service life of a refractory material is the result of its interaction with its environment, for example, liquid metal, slag and the gas atmosphere. This complex process can basically be described by methods of physical chemistry, mass and heat transfer, thermomechanics and mineralogy. For this purpose, practical experience and laboratory results must be traced back to these fundamentals. This is not always easy.

The present technical reference book also offers an approach to this in the second, updated edition. On the basis of numerous examples, experimental results are traced back to the above-mentioned basic principles and quantitatively calculated. In this way, the reader is enabled to solve his own problems without much effort and to improve the service life of his refractory material. Furthermore, the knowledge gained will improve the methodology used and broaden the range of applications. New and expanded subchapters supplement the updated edition.

I am indebted to a large number of gentlemen: Profs. Dr. K. Löhberg and Dr. M. G. Frohberg, Berlin. Dr. O. Knacke, Aachen, taught me the enthusiasm to seek answers to scientific questions. Prof. Dr. A. Eschner, Osterode, Prof. Dr. H. Harmuth, Leoben and Dr. Hopf, Langewiesen, gave me suggestions and discussed results with me. Dr. E. Brochen and Dr. C. Brüggmann provided me with excerpts of their dissertations prepared at DIFK, Bonn, for which I thank them. Professor Dr. O. Krause and the staff of the DIFK, Hörgrenzhausen, supported me with numerous useful hints. For further technical advice, which was also a prerequisite for the present work, I would like to thank the following gentlemen: Prof. Dr. E. Baake, Dr. Th. Deinet, Dr. E. Dötsch, T. Dünnes, Dr. D. Fünders, Dr. K. E. Granitzki, Prof. Dr. K. Hack, Prof. Dr. G. Henning, Dr. K. Junge, R. Krebs, Dr. D. Lenzner, J. Mendheim, G. Mittler, Dr. S. Postrach, Dr. P. Röhl, G. Routschka and Prof. Dr. P. Scheller.

I would like to express my gratitude to Mr. Loren Mark Hamersley for his translation into English and the required English language refractory terminology expertise. Mr. Jan Niklas Trier is also thanked for his assistance regarding IT-related questions to enable translation of this book.

My special thanks go to my wife Erika for her patience and to my two sons Stephan and Axel for their encouragement to finish the work and to publish an updated edition.

Essen, Germany
2023

Prof. Dr.-Ing. habil. Jürgen Pötschke

Contents

Chapter 1
Introduction

If in metallurgical practice there were a thermochemical equilibrium of the phases, e.g. liquid steel, slag, refractory material and atmosphere, there would be no chemical wear of the refractory material. Then, however, there would be no metallurgical work. On the contrary, in order to increase the material turnover, the distance to equilibrium is kept as large as possible during the entire process, to the disadvantage of the refractory lining. The lining dissolves and reacts chemically, forming or changing a slag. Furthermore, forming inclusions in the steel and escaping gaseous products. The physical and thermomechanical behavior of the lining also gives rise to wear. These are all undesirable processes.

What is desired is an inert refractory material. But this does not exist. Consequently, we are searching for ways to get closer to this ideal state. To this end, industry utilizes a large number of varying laboratory tests and careful observation of daily practice. Proposals for improvement are derived from the conclusions.

The present work sets itself the task and objective of assisting this procedure. In addition to empirical experience, theoretical knowledge is very helpful in these reconditioning processes and will lower costs. For this reason, the theoretical fundamentals of refractory material wear are briefly presented on the basis of practical calculation examples. Basic derivations of the formal correlations used are largely omitted in favor of their practical application, and reference is made to literature references instead.

This work is not a textbook but the compilation of the methods used and developed in many years of practical activity to calculate the wear of refractory materials and effects on the quality of the product, especially the steel. This work does not take the place of a thorough study of the individual fields, such as thermodynamics, mass and heat transfer, thermomechanics, but shows a path toward understanding their interaction on various corrosion phenomena as quantitatively as possible without using numerical calculation models that are often obscure to the user. By means of systematically selected and calculated examples, the reader is given the opportunity to adapt his problem with little effort and to calculate himself in order to derive enhancements

© The Author(s), under exclusive license to Springer Nature Switzerland AG 2024

J. Pötschke, *Refractory Fundamentals in Metallurgical Practice*,

https://doi.org/10.1007/978-3-031-63709-4_1

for the selection of materials, manufacture and use of refractory materials. Several of my own published works serve as a basis. Above all, however, this work is intended as an incentive to deal quantitatively with the corrosion of refractory materials and, thus, to further improve the solutions proposed here.

The presentation is given in complete chapters which means that occasional repetitions cannot be avoided.

Chapter 2
Basics

2.1 Introduction to Thermodynamics

The fundamentals presented here concentrate solely on the part of thermodynamics required in this work. Much more comprehensive and in-depth descriptions or illustration can be found in various textbooks and tables, e.g. by Frohberg [1] and Knacke et al. [2].

2.1.1 Gibbs' Phase Law

A system in a completely uniform state (pressure P, temperature T, chemical composition X) is designated as a homogeneous phase P. A heterogeneous system consists of several phases and their phase boundaries. A phase can be composed of different components K. Reactions R can occur in a system which are constrained by conditions B, such as stoichiometric conditions. The total number of variables required for the unambiguous description of the system (so-called freedoms F: P, T, X) according to Gibbs is:

$$F = K + 2 - P - R - B. \qquad (2.1.1)$$

2.1.2 State Variables

Each heat exchange is equal to the change of the internal energy plus the external work (1st law):

© The Author(s), under exclusive license to Springer Nature Switzerland AG 2024
J. Pötschke, *Refractory Fundamentals in Metallurgical Practice*,
https://doi.org/10.1007/978-3-031-63709-4_2

$$dQ_V = dU + PdV \ [\text{J}] \tag{2.1.2.1}$$

$$dQ_P = dH + VdP \ [\text{J}]. \tag{2.1.2.2}$$

With unchanged volume, the change in internal energy dU equals the exchanged heat dQ_V. At constant pressure, the same is valid for the change in enthalpy dH. The change of both state variables depends on the change of temperature:

$$dU = C_V \, dT \ [\text{J/mol}] \tag{2.1.3.1}$$

$$dH = C_P dT \ [\text{J/mol}]. \tag{2.1.3.2}$$

The specific heat at constant volume C_V or rather constant pressure C_P should be unchanged in the temperature range looked at here. In the ideal gas,

$$C_P - C_V = R \ [\text{J/K mol}]. \tag{2.1.4}$$

The ideal gas constant has the commonly used values (counts) $\mathbf{R} = 8.31$ [J/K mol], 0.8479 [kp m/K mol] and 84.79 [atm cm^3/mol K].

Without external influence, all irreversible processes run in only one direction. Thereby, the entropy increases, $dS > 0$. If the entropy S reaches its maximum, equilibrium prevails. For example, a hot body always heats a colder one until temperature equilibrium is reached. No heat flows in the opposite direction.

In thermal equilibrium, the heat exchange is reversible. The definition of entropy is then:

$$dS = dQ_{\text{rev}}/T \ [\text{J/K mol}]. \tag{2.1.5}$$

At equilibrium, $dQ = 0$ and, consequently, $dS = 0$.

For isobaric processes, $dQ_P = dH$ (Eq. 2.1.2.2) and, therefore,

$$dS_P = dH/T = C_P dT/T = C_P d \ln T \ [\text{J/K mol}]. \tag{2.1.6}$$

For constant volume, the same applies for dS_V (refer to Eq. 2.1.2.1).

In most cases, metallurgical processes take place at constant pressure.

The connection between free enthalpy and enthalpy is established by entropy and temperature. According to Helmholtz, the following applies:

$$dG = dH - TdS \ [\text{J/mol}]. \tag{2.1.7}$$

G is the free enthalpy (Gibbs enthalpy). It is the part of the enthalpy which can be converted into external work. The entropy part (share) TdS remains in the system. Heat cannot be converted quantitatively into work (2nd law). Equilibrium exists when the free enthalpy reaches its minimum. Then the entropy is maximum. In a closed system, in equilibrium, $dG = 0$ and $dS = 0$.

G, H and S are state variables that fully describe the thermodynamic state of the system.

Basically, Gibbs's "fundamental equation" describes the change in free enthalpy:

$$dG = VdP - SdT + \Sigma \mu_i dn_i + \sigma dF_O + \Sigma z_j F \Phi_j dn_j \ [\text{J}]. \qquad (2.1.8)$$

σdF_O is the work required to create the surface F_O of the system. σ [J/m^2] is the specific interfacial free energy (refer to Sect. 2.1.3.4). The last part is the electric work (refer to Sect. 2.1.4.2) with valence z_j of component j in solution, its electric potential Φ_j and Faraday constant $F = 9.56 \times 10^4$ J/mol Volt.

If the number n of particle species i changes in an open system, $\sum \mu_i dn_i$ is the associated change in free enthalpy. μ_i is designated the chemical potential of component i. It is defined as

$$\mu_i = (\partial G / \partial n_i)_{T,P,F,nk} (i \neq k) \ [\text{J/mol}]. \qquad (2.1.9)$$

$$(\partial G / \partial n_i)_{T,P,F,nk} = \overline{G}_i \ [\text{J/mol}] \qquad (2.1.10)$$

is the partial molar-free enthalpy of component i in a mixture (solution). G [J] $= \overline{G}_1 \cdot n_1 + \overline{G}_2 \cdot n_2 + \dots$ gives the total free enthalpy of the mixture. The comparison of both Eqs. 2.1.9 and 2.1.10 with each other gives:

$$\overline{G}_i \equiv \mu_i \ [\text{J/mol}]. \qquad (2.1.11)$$

μ_i is, therefore, also a state variable. In the standard state, "pure substance"

$$\mu_i^\circ \equiv G_i^\circ.$$

By definition, at 298 K, the enthalpy $H^\circ = 0$ applies for all stable substances.

Example What is the free enthalpy of formation of corundum at 2000 K?

$$4/3 \ \text{Al} + O_2 = 2/3 \ \text{Al}_2O_3 \ [\text{FU} \equiv \text{formula conversion}]. \qquad (2.1.12)$$

The values at 2000 K are taken from the tables of Knacke et al. [2]:

$$\mu^\circ_{\text{Al}_2O_3} = -2{,}009{,}667 \ [\text{J/mol}], \quad \mu^\circ_{\text{Al}} = -128{,}289 \ [\text{J/mol}],$$
$$\mu^\circ_{O_2} = -478{,}259 \ [\text{J/mol}].$$

Using the stoichiometric scheme "sum of product substances minus sum of starting (base) substances," we obtain

$$\Delta\mu^{\mathrm{o}}_{\mathrm{Al_2O_3}} = 2/3\,\mu^{\mathrm{o}}_{\mathrm{Al_2O_3}} - \left(4/3\mu^{\mathrm{o}}_{\mathrm{Al}} + \mu^{\mathrm{o}}_{\mathrm{O_2}}\right) = -690{,}467 \ [\mathrm{J/mol}]. \qquad (2.1.13)$$

The same scheme applies to the calculation of ΔH^{o} and ΔS^{o}. (At 298 K, by definition, both are pure substances, H_{Al} and $H_{\mathrm{O_2}} = 0$ [J/mol]).

If the heat content of the starting (base) materials (R) and the products (P) of a chemical reaction remains unchanged, i.e. if $\Delta H = 0$, this is called an adiabatic process: the temperature changes. The temperature T of the products (P) is calculated at which their enthalpy is equal to that of the starting (base) materials (R):

$$\sum n_i H_i(R) - \sum n_i H_i(P) = 0 \ [\mathrm{J}]. \qquad (2.1.14)$$

Example How much does molten steel heat up as result of the addition of metallic aluminum. The enthalpy of solution of 0.02 mol aluminum in 1 mol liquid iron at 1600 °C is $\overline{\Delta H}_{[\mathrm{Al}]_{\mathrm{Fe}}} = -1500$ J/mol [3]. The specific heat of liquid iron is $C_{P_{\mathrm{Fe}}} = 46$ J/mol K [4]. From Eq. 2.1.6, it follows for the heating of the entire melt $\Delta T = -\ \overline{\Delta H}_{[\mathrm{Al}]_{\mathrm{Fe}}}/n \cdot C_{P_{\mathrm{Fe}}} = 1500/1.02 \cdot 46 = 32$ K. The addition of about 1% by weight aluminum to a steel melt increases the bath temperature by about 30 °C. If the enthalpy of fusion of the aluminum and its heating starting at 25 °C are also taken into account, this effect drops to 11 °C [5] which is why the addition of molten aluminum is recommended in practice.

2.1.3 Equilibria

The equilibrium conditions discussed in Sect. 2.1.2 are applied to various cases seen in metallurgy.

2.1.3.1 Phase Transformation

For the isothermal evaporation of a mole at temperature T, the Clausius–Clapeyron equation follows from Eqs. (2.1.6–2.1.8) in the form:

$$\frac{dP}{dT} = \frac{\Delta S_{\mathrm{li/gas}}}{\Delta V_{\mathrm{li/gas}}} = \frac{\Delta H_{\mathrm{li/gas}}}{T\left(V_{\mathrm{li}} - V_{\mathrm{gas}}\right)} \ [\mathrm{atm/K}]. \qquad (2.1.15)$$

$\Delta H_{\mathrm{li/gas}}$ is the molar enthalpy of vaporization at constant values (counts) of T and P. If the ideal gas law $V = RT/P$ applies, then $V_{\mathrm{li}} \ll V_{\mathrm{gas}}$ gives the steam pressure formula

$$\frac{\mathrm{d}\ln P}{\mathrm{d}T} = \frac{\Delta H_{\mathrm{li/gas}}}{RT^2} \ \left[\mathrm{K}^{-1}\right]. \qquad (2.1.16)$$

From the phase law of Gibbs (Sect. 2.1) follows with $K = 1, P = 2$ and $R = B = 0$ as variable $F = 1$. Temperature and pressure determine each other. Consequently, the boiling temperature T_S depends on the external pressure.

From the indefinite integration of Eq. (2.1.16) follows:

$$\ln P = \frac{\Delta H_{\text{li/gas}}}{RT} + C. \tag{2.1.17}$$

Figure 2.1 shows the steam pressures of some metals as $\log P\ (1/T)$. Integration of Eq. (2.1.16) in the limits $P(T_1)$ to $P(T_2)$ gives

$$\ln \frac{P_1}{P_2} = \frac{\Delta H_{\text{li/gas}}}{R} \left(\frac{1}{T_1} - \frac{1}{T_2} \right). \tag{2.1.18}$$

For example, $P_2\ (T_2)$ can be the boiling point at $P_o = 1$ atm.

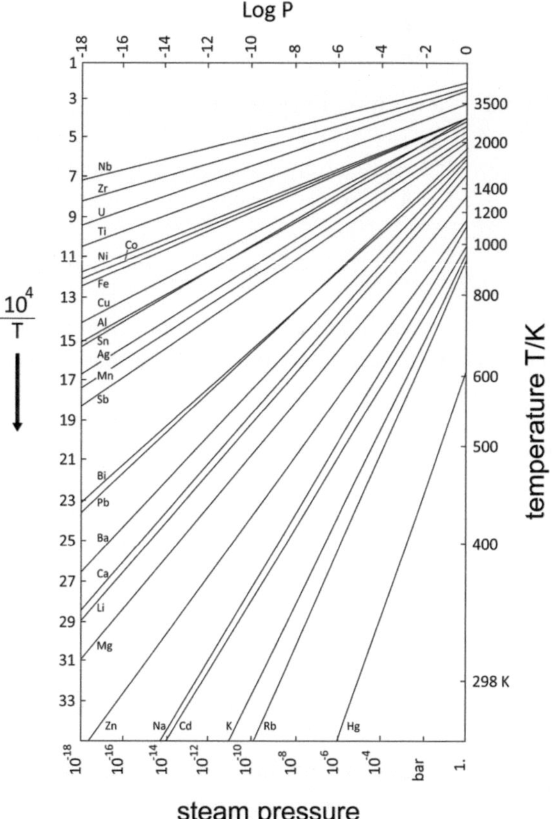

Fig. 2.1 Steam pressure of some metals [1]

Since the conditions for Sect. (2.1.18) apply quite generally, the equation of Clausius–Clapeyron (2.1.15) also describes the melting equilibrium between solid and liquid phases. Then $\Delta H_{li/gas}$ is to be substituted by the melting enthalpy $\Delta H_{sol/li}$.

An **example** of the application of the equation of Clausius–Clapeyron (2.1.15) as difference equation $\Delta T / \Delta P$ is the calculation of the change of the melting temperature of ice $T_m = 273$ K (1atm) by external pressure P. With the material values $\Delta H_{(sol \to li)} = 6$ kJ/mol, $V_{li} = 18.18$ cm^3/mol, $V_{sol} = 19.82$ cm^3/mol, as well as the conversion 1 cm$^3 \times$ atm $= 10^{-4}$ kJ, we get

$$\frac{\Delta T}{\Delta P} = \frac{T_m(V_{li} - V_{sol})}{\Delta H_{(sol \to li)}} = \frac{273(-1.64)}{6 \times 10^{+3}} = -7.5 \times 10^{-2} \text{ K/atm}. \qquad (2.1.19)$$

As a result of a pressure increase of 135 atm, the melting point drops by 1 K so that the slightly undercooled ice melts back into water. Glaciers glide down in direction of the valley on this water film. However, the water film required for skates to glide is additionally explained by the considerable frictional heat that occurs during sliding (dynamic) friction.

2.1.3.2 New Phase Formation

If a new phase arises from the isothermal chemical reaction.

$A + 2B = AB_2$, so the free enthalpy of reaction is

$$\Delta G = \Delta G^o + RT \cdot \ln K_{AB_2} [\text{J/mol}]. \qquad (2.1.20)$$

ΔG^o is the free enthalpy in the standard state and K is the equilibrium constant of the reaction. If the molar volume of the final phase is greater than the sum of the consumed ones, the volume increase is $\Delta V = V_{AB_2} - (V + 2V_{AB})$. Consequently, the external mechanical work $P \cdot \Delta V$ has to be applied. It is equal to the change in free enthalpy ΔG. It follows from Eq. (2.1.20)

$$P \Delta V + \Delta G^o + RT \cdot \ln K = 0 \text{ [J/mol]}. \qquad (2.1.21)$$

The mechanical, so-called crystallization pressure π, is calculated from the equation

$$\pi - 1 = \frac{-\Delta G^o - R_1 T \cdot \ln K_{AB_2}}{(V_{AB2} - V_A - 2V_B) \cdot R_1/R_2} \text{ [atm]}. \qquad (2.1.22)$$

The ideal gas constants are $R_1 = 8.31$ J/mol K and $R_2 = 84.8$ atm cm^3/mol K. $R_1/R_2 = 0.1$ J/atm cm^3. Without external work $\pi = 1$.

Example The formation of calcium sulfate in basic refractory material at 1200 °C: $\langle CaO \rangle + \{SO_3\} = \langle CaSO_4 \rangle$, $\Delta G^o = -141.7 \times 10^3$ J/mol, $V_{CaO} = 16.7$ ccm/mol, $V_{CaSO_4} = 46$ ccm/mol. The gas has partial pressure $P_{SO_3} = 1$ atm (assumption). However, it does not exert any mechanical pressure on its surroundings. Therefore, the crystallization pressure of the calcium sulfate, e.g. in a waste incinerator, can be

$$\pi - 1 = \frac{141,700}{(46 - 16.7) \cdot 0.1} = 48 \times 10^3 \text{ [atm]}. \tag{2.1.23}$$

It destroys the lining.

2.1.3.3 Chemical Equilibrium

The chemical reaction

$$\alpha A + \beta B = \gamma C + \delta D \tag{2.1.24}$$

is in (kinetic) equilibrium if the there and back reactions proceed at the same rate (speed): $\overrightarrow{k} = \overleftarrow{k}$ [s^{-1}]. It follows

$$K = \frac{\overrightarrow{k}}{\overleftarrow{k}} = \frac{a_A^\alpha \cdot a_B^\beta}{a_C^\gamma \cdot a_D^\delta}. \tag{2.1.25}$$

K is the equilibrium constant of the reaction. The chemical activity of component i is denoted by a_i. Pure substances have the activity 1.

The definition of the activity depends on the selected standard state:

$$a_i = P_i/P_o \tag{2.1.26}$$

describes the steam pressure of the dissolved component i compared to that of its pure substance. For example, the oxygen partial pressure of the formation of corundum (Eq. 2.1.12) at 2000 K, $P_{O_2} \approx 10^{-18}$ atm, versus pure oxygen at 1 atm [refer to (2.1.34)].

Raoult's law is applied is $a_i \rightarrow x_i$. $x_i \rightarrow 1$ is the mole fraction of component i in the highly concentrated multicomponent solution.

In a highly dilute solution, Henry's law applies:

$$a_i = \gamma_i \cdot x_i. \tag{2.1.27}$$

x_i is the mole fraction of component i and γ_i is the corresponding activity coefficient in the infinitely diluted solution. It is determined by applying the tangent line to the experimentally determined course of activity $a(x_i)$ at the point $x_i = 0$ and reading its value a_i at the point $x_i = 1$ (Fig. 2.2). $\gamma_i < 1$, as long as the activity is below

Raoult's straight line, and $\gamma_i > 1$ if it is above. In the molten iron-carbon system, $\gamma_c < 1$ because the activities for $x_C < 0.05$ (1% by weight) are below the Raoult straight line and only later rise steeply above the Raoult straight line (refer to Fig. 2.3 and [6]).

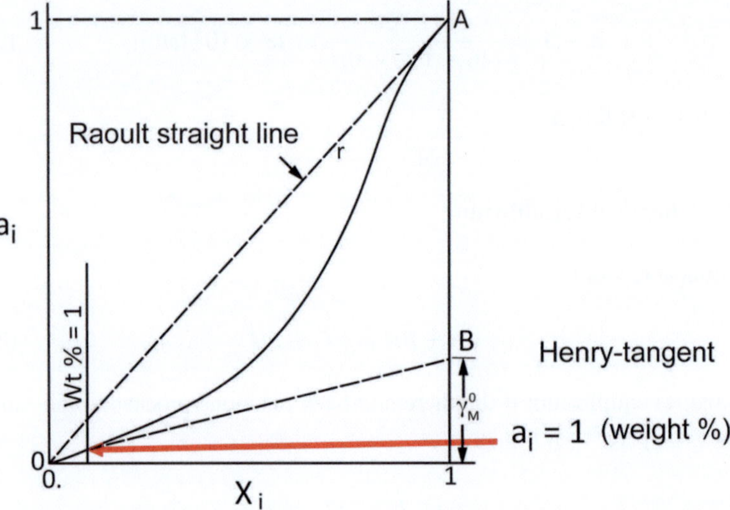

Fig. 2.2 Activity progression explaining the standard states [15] (Raoult: 0-A, Henry: 0-B, 1% solution: 0-C)

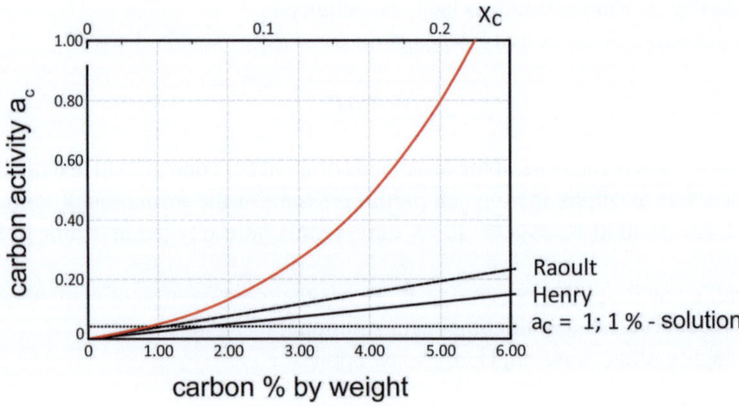

Fig. 2.3 Carbon activity in iron at 1600 °C [5]

Metallurgical systems very often show satisfactory conformity between experience and theoretical calculation for the standard condition of one percent solution introduced by Chipman [6]. According to this, the activity is defined to one once the dissolved concentration of component i just reaches 1% by weight:

$$[a_i/(\%)_i]_{(\%)i \to 1} = f_i = 1. \tag{2.1.28}$$

If, for example, $3 \times 10^{-3}\%$ by weight of oxygen is dissolved in the steel, its activity is $a_{[O]} = 0.003$.

Figure 2.2 shows an activity progression schematically and Fig. 2.3 the example iron-carbon at 1600 °C. The dissolved carbon activity, calculated according to Fact-Sage [5], deviates strongly positively from the Raoult straight line above 1% by weight carbon and already reaches the value $a_C = 1$ at $[C] = 5.4\%$.

The value (count) of an activity coefficient depends on the solvent and the composition of the solution. The effect of components (0), (1), (2), (3), (4)...(j) on component (2) is the product of the individual so-called activity coefficients:

$$f_2 = f_2^{(0)} \cdot f_2^{(2)} \cdot f_2^{(3)} \cdot f_2^{(4)} \cdots . \tag{2.1.29}$$

$f_2^{(0)} = 1$ is the activity coefficient of component (2) in the not contaminated, **pure** solvent 1. $f_2^{(2)}$ is the influence of the observed solute element (2) on itself, i.e. $f_2 \equiv f_2^{(0)} \times f_2^{(2)}$ is identical with the activity coefficient f_2 in the two-substance solution. The $f_2^{(3\cdots j)}$ describe the influence of all other additional elements on 2 and are multiplied member (part) by member (part) (refer to 2.3.6). The conversion of different activity coefficients into each other is illustrated in detail in [1, 6, 7].

Wagner [1] calculated the interaction of several components j dissolved in one phase on component i by the partial derivative of the logarithms of their activity coefficients as Taylor series at the position $\log f_2^{(0)} = 0$. If the composition of the solution is constant, the following holds for the first member (part):

$$e_i^j = \left(\frac{\partial(\log f_i)}{\partial(\%_j)} \right)_{(\%j) \to 0} . \tag{2.1.30.1}$$

At constant activity, $\partial(\log f_i) = -\partial(\log \%_i)$ so that using the mole fraction, x can be written analogously to Eq. (2.1.30.1):

$$\omega_i^j = -\left(\frac{\partial(\ln x_i)}{\partial(x_j)} \right)_{(x_j) \to 0} . \tag{2.1.30.2}$$

These so-called interaction coefficient $e_i^{(j)}$ (% by weight = const.) and $\omega_i^j (a_i = $ const) describe the mutual influence of components j and i in dilute solution. The values (counts) are tabulated for liquid iron [1, 6, 7]. The graphical illustration in Fig. 2.4 [1, 7] suggests a connection with the atomic number of the additional element **j** in the periodic table.

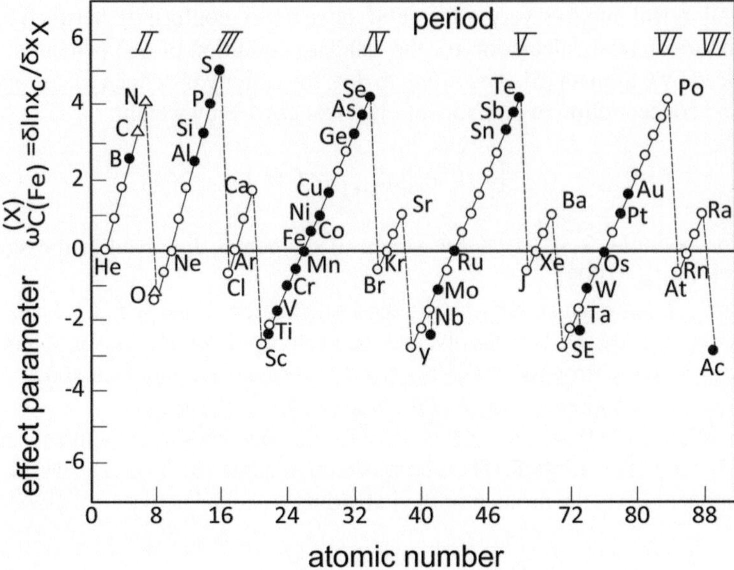

Fig. 2.4 Dependence of the Interaction Coefficient in Fe–C–X melts on the atomic number of the elements dissolved in the iron

Elements whose interaction coefficient $\omega_i^j > 0$ increase the activity of carbon, i.e. reduce its solubility in iron. A well-known example seen in foundries is the formation of simmering foam graphite as a result of a silicon addition. The opposite is true for $\omega_i^j < 0$. For example, chromium decreases the activity of carbon and, therefore, increases its solubility in liquid iron.

With the interaction coefficient of all additional elements, the logarithm of the activity coefficient of component (2) is calculated as follows:

$$\log f_2 = e_2^{(2)} \cdot (\%)_2 + e_2^{(3)} \cdot (\%)_3 + e_2^{(4)} \cdot (\%)_4 + \cdots . \tag{2.1.31}$$

In highly diluted solution: $e_i^{(j)} = e_j^{(i)}$.

Molar fraction x_i and (% by weight)$_i$ can be converted into each other; for example, in iron, the atomic weights M_i and $M_{Fe} = 56$ g/mol:

$$(x_i)_{Fe} = \left[\left(\frac{100}{(\%)_i} - 1 \right) \cdot \frac{M_i}{M_{Fe}} + 1 \right]^{-1} . \tag{2.1.32}$$

Knowing the activities, the free enthalpy is calculated with the help of the law of mass action (refer to 2.1.21 and 2.1.25):

$$\Delta G = \Delta G^{\circ} + RT \cdot \ln K \ [\text{J/mol}]. \tag{2.1.32}$$

ΔG^o is the tabulated basic potential μ_o [2] and $RT \ln K$ is the so-called residual potential.

The Richardson-Ellingham diagram, frequently used for the calculation of metallurgical and refractory equilibria, is derived from this context (Fig. 2.5) [1.5]. For easier understanding, the following example is given:

What is the oxygen partial pressure of the reaction (2.1.12) at equilibrium at 2000 K?

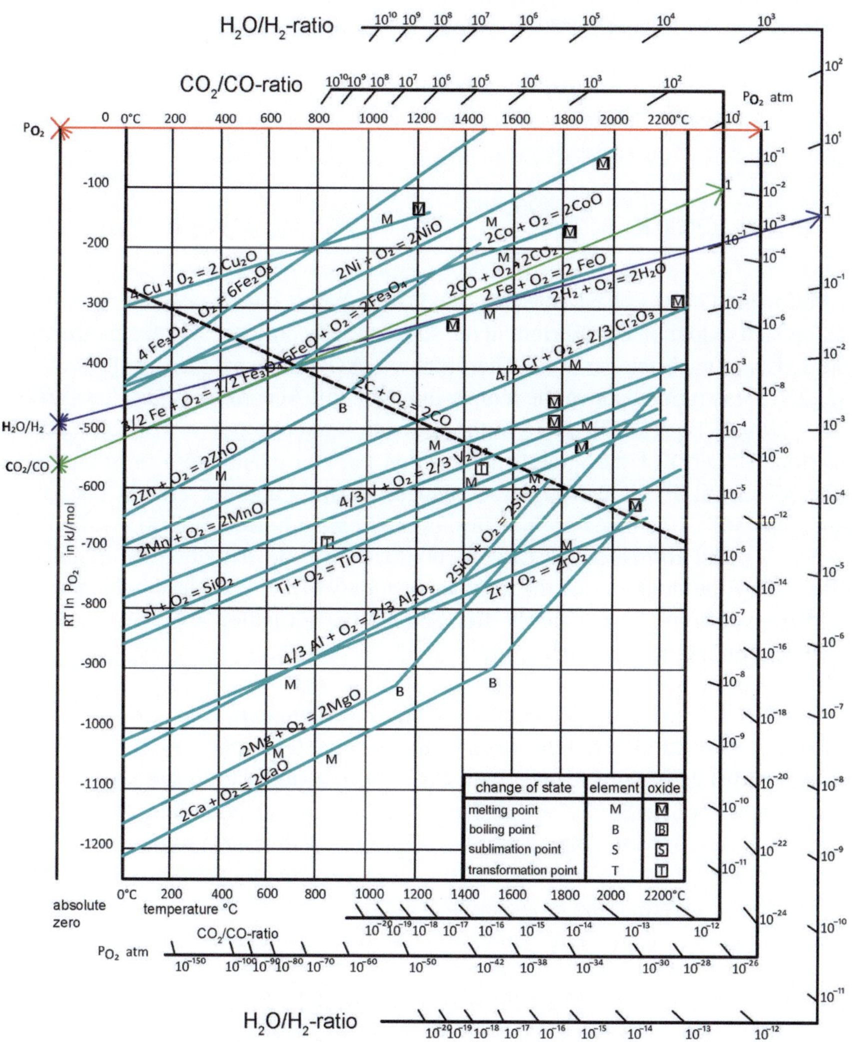

Fig. 2.5 Richardson–Ellingham diagram for comparison of thermochemical stability of various oxides. *Note* This figure could be used as a fold-out cover for the reader to work with repeatedly

$$4/3\,Al + O_2 = 2/3\,Al_2O_3\ [mol], \quad \Delta G^\circ(2000\,K) = -690{,}467\ [J/mol]. \quad (2.1.12)$$

Since Al_2O_3 and Al are pure substances, their activity is 1:

$$K = \frac{a_{Al_2O_3}^{2/3}}{a_{Al}^{4/3} \cdot P_{O_2}} = \frac{1}{P_{O_2}}. \tag{2.1.33}$$

The standard state of pure oxygen is $P_{O_2}^O = 1$ atm.
From Eqs. 2.1.32 and 2.1.33, it follows in equilibrium ($\Delta G = 0$):

$$\Delta G^\circ = +RT \ln P_{O_2}\ [J/mol], \tag{2.1.34}$$

i.e.

$$\ln P_{O_2}(2000\,K) = \frac{-690{,}467}{8.31 \cdot 2{,}000} = -41.54, \quad P_{O_2} = 9 \times 10^{-19}\,atm.$$

Following Gibbs's phase rule (Eq. 2.1.1), the system has one degree of freedom, i.e. for each oxidation equilibrium in question, exactly only one oxygen partial pressure exists at each temperature. Therefore, $-\Delta G^\circ = -RT \ln P_{O_2}$ [J/mol O_2] Eq. (2.1.34) is plotted above the temperature T [°C]. According to $\Delta G^\circ = \Delta H^o - T\Delta S^o$ [J/mol] Eq. (2.1.7), ΔH^o is the ordinate intercept at $T = 0$ K and ΔS^o is the slope (ascent) of the straight line. Upon passing evaporation or sublimation points of metals **B**, the entropy ΔS^o changes abruptly and the straight line gets a bend "upward". Examples are magnesium and calcium according to Eq. (2.1.37). Conversely, the formation of carbon monoxide by the oxidation of graphite is followed by the doubling of the number of moles of the gas phase which is why the entropy ΔS° increases and the free enthalpy of formation ΔG° becomes more negative with increasing temperature.

To calculate graphically the reaction of aluminum with oxygen to corundum (refer to 2.1.12), using the **Richardson-Ellingham diagram** (Fig. 2.6), at 1727 °C (2000 K), a straight line (red) is drawn from the upper left corner (**P_{O_2}**) starting from the standard state of pure oxygen with **P_{O_2}** $= 1$ atm (red arrow) at $T = -273$ °C. This straight line is extended over the intersection of the reaction lines (4/3 Al + O_2 = 2/3 Al_2O_3 (red)) with temperature $T = 1727$ °C. On the right ordinate, the oxygen partial pressure of the reaction $P_{O_2} = 10^{-18}$ atm. (Practically, turn the red arrow **P_{O_2}** in Fig. 2.5 downward to the intersection at 1727 °C.)

In addition, the diagram takes into account that the reactions

$$2\{CO\} + \{O_2\} = 2\{CO_2\} \tag{2.1.35.1}$$

and

$$2\{H_2\} + \{O_2\} = 2\{H_2O\}, \tag{2.1.35.2}$$

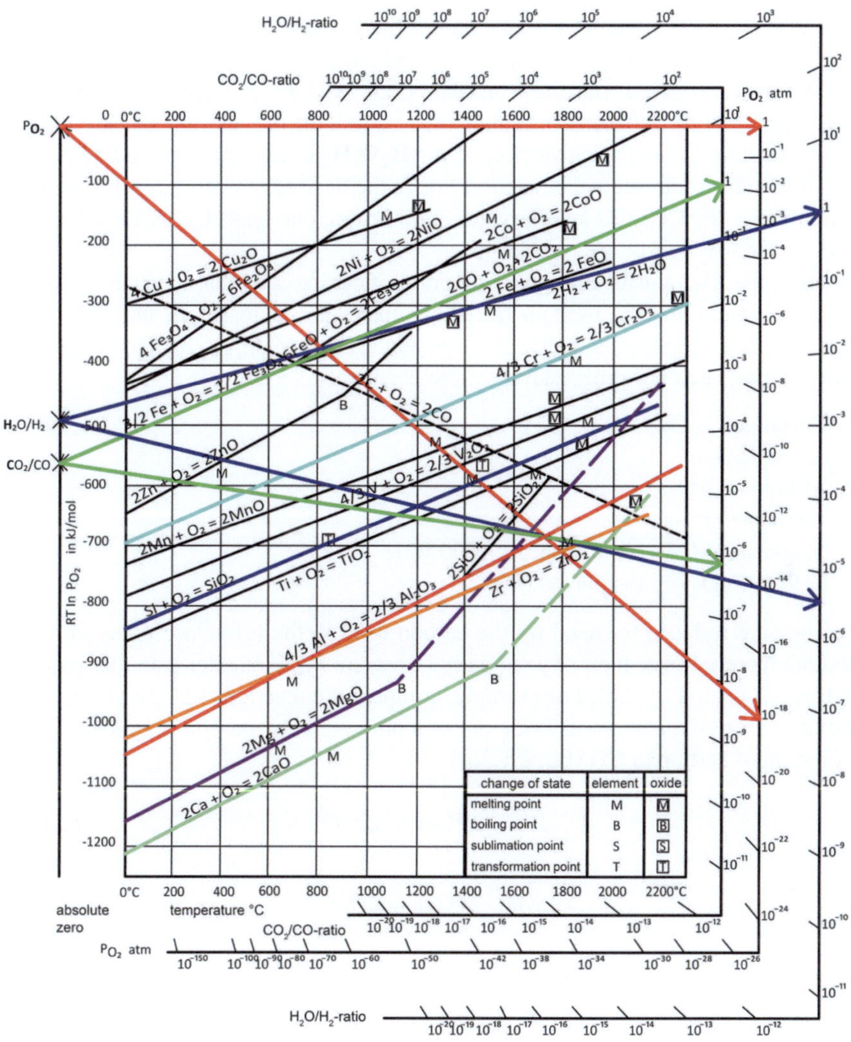

Fig. 2.6 Richardson–Ellingham diagram. Example: Oxidation of aluminum to corundum at 1727 °C

due to their strict linkage with the oxygen partial pressure offer the possibility of describing, as auxiliary gas equilibria independent of the total pressure, the redox reactions shown in the Richardson-Ellingham diagram:

$$P_{O_2} = \frac{1}{K_1} \cdot \frac{p_{CO_2}^2}{p_{CO}^2} = \frac{1}{K_2} \cdot \frac{p_{H_2O}^2}{p_{H_2}^2}. \qquad (2.1.35.3)$$

These ratios are read off by establishing a linear connection between the "reaction line"/temperature interface and the right-hand edge scale **CO_2/CO** or **H_2O/H_2** (on the left-hand ordinate), starting from the points "CO_2/CO" (carbon) or "H_2O/H_2" (hydrogen) (on the left-hand ordinate), and reading off the relevant gas equilibrium there. (Practically, one turns the blue arrow **H_2O/H_2** or rather the green **CO_2/CO** in Fig. 2.6, until it intersects the reaction line at the desired temperature). In the given example, for the CO_2/CO auxiliary gas equilibrium, one reads $CO_2/CO = 9 \times 10^{-7}$ and for $H_2O/H_2 = 4 \times 10^{-6}$.

From the Richardson-Ellingham diagram, it can be seen that the stability of the six pure oxides mainly used in refractory materials increases in the order indicated because their free enthalpy of formation $\Delta G° = RT\ln p_{O2}$, related to the same temperature, becomes progressively more negative.

- Chromia (Cr_2O_3).
- Silica (SiO_2).
- Alumina Al_2O_3).
- Zirconia (ZrO_2).
- Magnesia (MgO) (up to approx. 1500 °C).
- Calcia (CaO) (up to approx. 2000 °C).

They are reduced to metal by the carbon used in the refractory material as the seventh material once their oxygen partial pressure P_{O_2} is greater than that required to form CO gas (Fig. 2.6). For example, compare the reactions (2.1.37 to 2.1.39) of the formation of a pure solid oxide out of a pure metal by gaseous oxygen with the oxidation of carbon to CO (Eq. 2.1.36):

$$2\langle C\rangle + \{O_2\} = 2\{CO\} \tag{2.1.36}$$

$$2\langle Me\rangle + \{O_2\} = 2\langle MeO\rangle \tag{2.1.37}$$

$$\frac{4}{3}\langle Me\rangle + \{O_2\} = \frac{2}{3}\langle Me_2O_3\rangle \tag{2.1.38}$$

$$\langle Me\rangle + \{O_2\} = \langle MeO_2\rangle. \tag{2.1.39}$$

As long as the $\Delta G°$ values ($\Delta G°$ = free standard enthalpy of formation) of the oxides, which increase with temperature, do not intersect the descending straight line of carbon oxidation (black), the oxides are stable in 1 bar CO atmosphere; chromium oxide (light blue) up to 1200 °C, silica (blue) up to 1600 °C, magnesia (violet) up to 1830 °C, alumina (red) up to 2000 °C, zirconia (orange) up to 2100 °C and calcia (green) up to 2100 °C.

As an example, the reduction of Al_2O_3 by graphite is calculated:

$$2/3 Al_2O_3 \rightleftarrows \{O_2\} + 4/3 Al, \quad \Delta G^\circ > 0 \quad \text{for all temperatures (2.1.38, resp.2.1.12)}$$

$$2\langle C \rangle + \{O_2\} \rightleftarrows 2\{CO\}, \quad \Delta \mathbf{G}^\circ < 0 \quad \text{for all temperatures} \qquad (2.1.36)$$

$$\overline{2/3 Al_2O_3 + 2C \rightleftarrows 2\{CO\} + 4/3 Al, \ \Delta \mathbf{G}^\circ \overset{?}{=} 0} : \text{ at } T = 1960\,°C. \qquad (2.1.40)$$

Below 1960 °C, ΔG° of the sum reaction (Eq. (2.1.40)) is > 0, i.e. graphite is not able to reduce corundum. As soon as ΔG° (Eq. (2.1.40)) < 0 is reached, i.e. starting at 1960 °C, this reaction runs to the right and corundum is no longer stable at P_{CO} = 1 atm. This can be clearly seen in the Richardson-Ellingham diagram at 1960 °C. It can also be seen that at 1960 °C for both individual reactions (2.1.12 and 2.1.36), $\Delta G^\circ = \pm\ 640$ kJ.

With FactSage [5], this result is obtained as a "Table Reaction" from Eq. (2.1.40) by inserting $P_{CO} = 1$ under "non-standard state" and $\Delta G^\circ = 0$ before doing the calculation.

From the Richardson-Ellingham diagram (Fig. 2.7), by comparing the two straight lines for the formation of SiO_2 (blue) and Al_2O_3 (red), it can be seen that pure aluminum always reduces silica because Al_2O_3 is thermodynamically more stable than SiO_2.

The question arises whether or not the possibly free silica, using a ladle in a steel foundry which has been lined with bauxite or fireclay, is reduced at 1600 °C by the deoxidation of the steel with aluminum. The steel would contain 10 ppm of aluminum dissolved. For 1600 °C, one reads from Fig. 2.7 at the "Al_2O_3 -line" (red) on the left ordinate $\Delta G^\circ = -\ 710$ kJ/mol. In the one percent solution, by definition, 10 ppm of aluminum corresponds to the activity 10^{-3}, so it follows from Eqs. (2.1.38) and (2.1.32) with $K = a_{[Al]} = 10^{-3}$ that $\Delta G = -\ 710{,}000 - \frac{4}{3} 8.31 \cdot 1873 - \ln 10^{-3} = -\ 567{,}000$ J/mol.

This value slightly exceeds the free enthalpy value seen at 1600 °C on the "SiO_2 line" (blue). The "Al_2O_3-straight line" turns toward lower values of the negative free enthalpy, i.e. upwards (dotted red) as a result of the decreasing aluminum activity. The reducing effect of the aluminum dissolved in the steel has comparatively decreased. The intersection of the two reaction lines is at 1510 °C. Thus, with less than 10 ppm aluminum in the molten steel, the refractory material consisting of "silica" would not be affected above 1510 °C. The oxygen partial pressure at 1510 °C is 10^{-17} atm.

Figure 2.5 does not show the stability of oxide compounds. Their stability lies between those of the oxide components forming them.

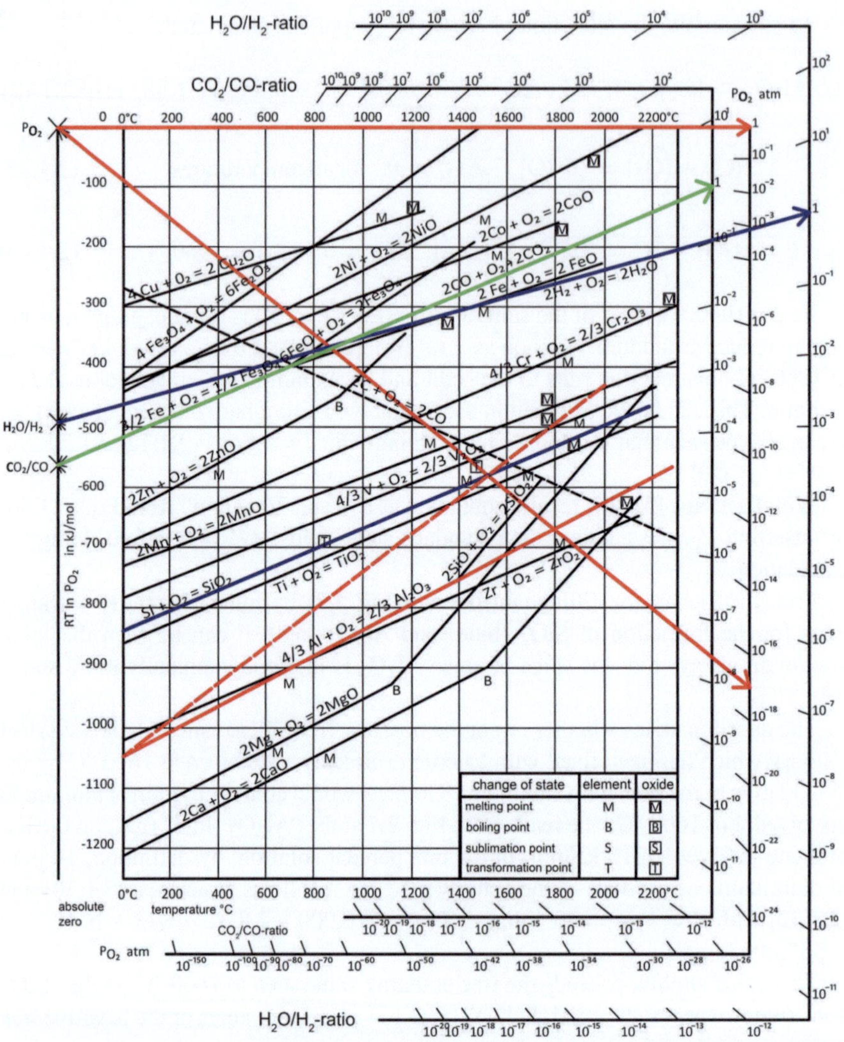

Fig. 2.7 Richardson–Ellingham diagram [1]. Example: Reduction of silica by aluminum dissolved in liquid iron at 1600 °C

Figure 2.8 shows a section of the Richardson-Ellingham diagram (Fig. 2.5). As is the case there, for the sake of comparability, the calculation of the complex oxides is done from the elements and not from the simple oxides which may already exist in practice. Examples are: Chromium spinel ($MgO \cdot Cr_2O_3$) is formed from the less stable chromia (Cr_2O_3) and the very stable magnesia (MgO). It is less stable than silica.

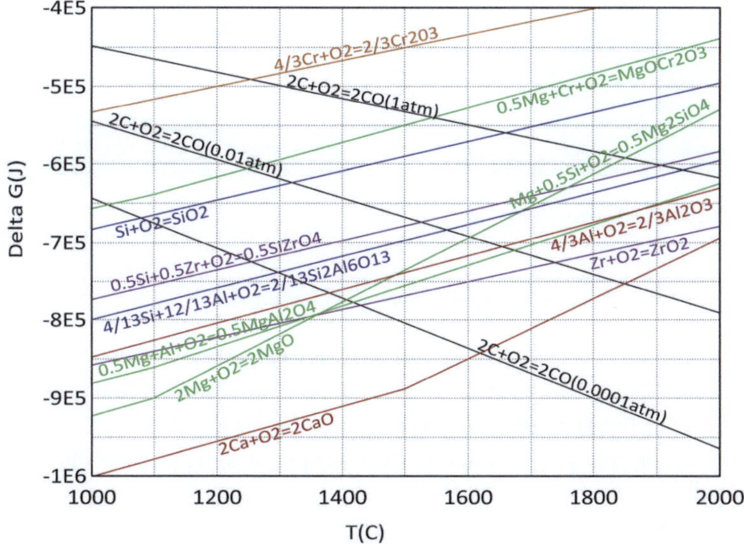

Fig. 2.8 Temperature dependence of the free enthalpy of various oxide compounds (calculated with FactSage [5])

Forsterite (2MgO·SiO$_2$), on the other hand, is almost as stable as alumina, i.e. more stable than chromium spinel (MgO·Cr$_2$O$_3$) and silica (SiO$_2$). This explains, among other things, its use in regenerators of glass tanks.

Mullite (2SiO$_2$·3Al$_2$O$_3$) is more stable than silica but less stable than alumina which experience has shown to limit its application possibilities.

Mg spinel (MgO·Al$_2$O$_3$) is thermodynamically about equivalent to corundum (Al$_2$O$_3$) and above 1350 °C much more stable than periclase (MgO).

Figure 2.8 also shows that, with decreasing CO partial pressure, the reduction of the oxides takes place at increasingly lower temperatures. The reduction of Al$_2$O$_3$ occurs at $P_{CO} = 1$ atm above 2000 °C, at $P_{CO} = 10^{-1}$ atm above 1650 °C and at $P_{CO} = 10^{-2}$ atm already above 1350 °C (also refer to Sect. 13.1.4.2, Fig. 13.11).

2.1.3.4 Interfacial Phenomena

The "Fundamental Equation of Thermodynamics" (2.1.8) also describes the contribution of an interface F to the free enthalpy of the total system:

$$\left(\frac{\partial G}{\partial F}\right)_{T,P,ni} = \sigma\left[J/m^2\right]. \tag{2.1.41}$$

σ is the specific interfacial free energy of two adjacent phases. In the case of "liquid/gas," this is referred to as the surface tension of the liquid.

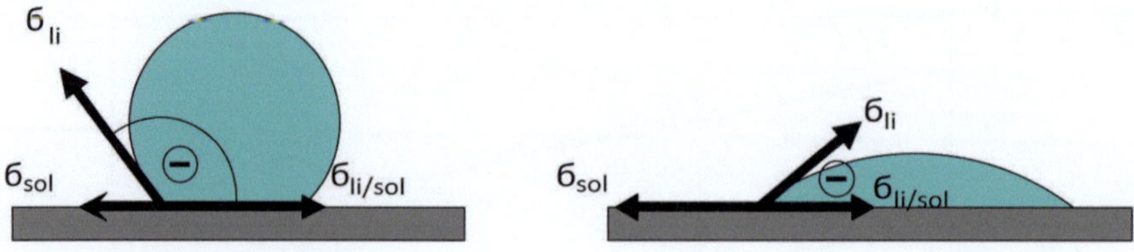

Fig. 2.9 Wetting and non-wetting

The temperature dependence describes the equation:

$$\sigma_1 - \sigma_2 = -k(T_1 - T_2)\ \left[J/m^2\right]. \tag{2.1.42}$$

Typically for liquid metals and slags, $k \approx 0{,}1$ J/m^2 x K.

According to Young (1804), the mutual wetting of three phases is given by the vector equilibrium:

$$\sigma_{SG} - \sigma_{LS} = \sigma_{LG} - \cos\theta\ \left[J/m^2\right]. \tag{2.1.43}$$

Figure 2.9 illustrates this equilibrium graphically. In the left part, the melt does not wet the solid support because the wetting angle θ is $> 90°$. In the right part of the figure, $\theta < 90°$ and is referred to as wetting. $(\sigma_{SG} - \sigma_{LS})$ is called adhesive stress. If $(\sigma_{SG} - \sigma_{LS}) > \sigma_{LS}$, there is no wetting angle but the fluid creeps completely along the given wall.

The work of adhesion is given by

$$W_A = \sigma + \sigma_{LGSG} - \sigma_{LS}\left[J/m^2\right]. \tag{2.1.44}$$

Capillarity refers to the behavior of liquids in narrow tubes/pipes. Upon wetting, the liquid rises against gravity g [N] by the height h [m] until capillary pressure and hydrostatic pressure compensate each other. This is referred to as capillary ascension:

$$2\sigma - \cos\theta/R = (\rho_1 - \rho_2) \cdot g \cdot h,\ \left[N/m^2\right],\quad (\rho_1 > \rho_2). \tag{2.1.45}$$

In case of non-wetting, it drives the liquid from the tube/pipe of radius R [m] downwards, capillary depression (refer to Fig. 2.10). $\Delta\rho$ [kg/m^3] is the density difference of the adjacent liquid phases.

The pressure over the liquid is in equilibrium with the capillary pressure. At the surface of the liquid, the tangential force and the normal force are equal to each other:

$$2\pi r\sigma = 4\pi r^2 P_K,\quad P_K = 2\sigma/r = 2\sigma \cdot \cos\theta/R\ [\text{atm}]. \tag{2.1.46}$$

The radius of curvature of the liquid is r and the radius of the capillary is R, i.e. $R = r \cdot \cos\theta$. Equation (2.1.46) is also valid for bubbles with radius r.

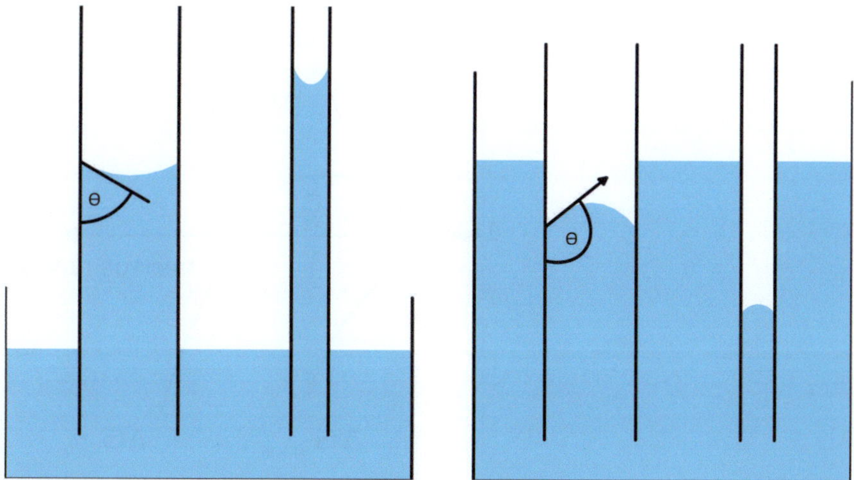

Fig. 2.10 Liquid level in capillaries: $\theta < 90°$ (left) $\theta > 90°$ (right)

The steam pressure of small droplets is calculated from the equation of *W*. Gibbs and Thomson (Lord Kelvin) [1]:

$$\ln \frac{P_r}{P_\infty} = \frac{2\sigma V}{RTr}.$$ (2.1.47)

V is the molar volume of the liquid. Over a convex curved surface, the steam pressure P_r is greater than over the flat surface P_∞. Over a concave surface, however, it is smaller. Therefore, a liquid cannot boil with homogeneous bubble nucleation because the steam pressure in the bubble is always lower than the pressure over the extended phase. Heterogeneous nucleation is required, e.g. at scratches (scrapes) and crevices (refer to Sect. 2.1.3.5).

Accordingly, a convex surface melts at lower temperature than its infinitely extended phase. According to Thomson [8] applies:

$$\frac{T_\infty - T_r}{T_\infty} = \frac{2\sigma V}{\Delta H_{\text{sol/li}}\, r}.$$ (2.1.48)

The more convexly a phase is curved, the lower its melting point.

2.1.3.5 Nucleation

To form a new interface in a homogeneous phase, homogeneous nucleation work ΔG_K is required. This is the difference between the work done to create the new surface $\Delta G_F = \sigma \cdot \Delta F$ and the volume work gained $\Delta G_V = \Delta V_K \cdot P_K$ (refer to 2.1.8). The spherical nucleus has radius r, so it follows with Eq. (2.1.46):

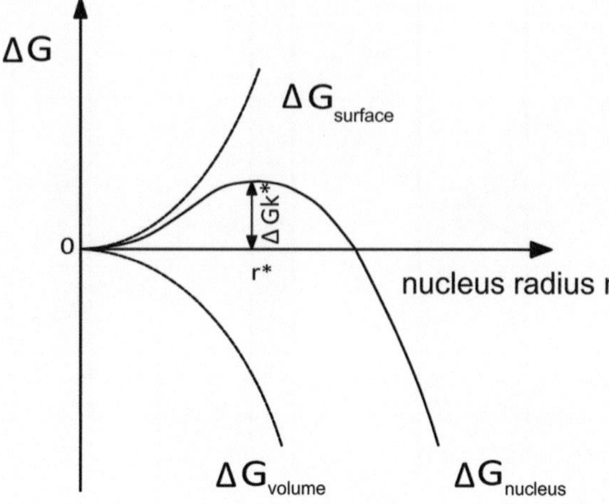

Fig. 2.11 Surface work ΔG_O, volume work ΔG_V and nucleation work ΔG_K [1]

$$\Delta G_K = \left(4\pi r^2 - 8\pi r^3/3r\right)\sigma = 1/3\left(4\pi r^2\right)\sigma = F \cdot \sigma/3 \; [\text{J/nucleus}]. \quad (2.1.49)$$

The nucleation work is the third part of the total work to form the nucleation surface F (Gibbs). With increasing radius, the nucleation work initially increases. After exceeding the size of the critical nucleus r_K^*, it then decreases; the nucleus is now stable. The new phase can continue to grow while gaining energy. The condition is (refer to Fig. 2.11):

$$\left(\frac{\partial \Delta G_K}{\partial r_K}\right)_{r_K=r_K^*} = 0. \quad (2.1.50)$$

For nucleation, the mother phase I must be supersaturated at the new phase II to be formed:

$$\Delta G_K = \frac{16}{3}\pi\left(\frac{V^{\mathrm{II}}}{RT}\right)^2 \frac{\sigma^3}{\ln^2\left(\frac{a^{\mathrm{II}}}{a_\infty^{\mathrm{I}}}\right)} f_H(\theta) \; [\text{J/nucleus}]. \quad (2.1.51)$$

The work involving heterogeneous nucleation ΔG_K depends on the wetting angle θ between the phase of nucleus II and its support. ΔG_K is multiplied by the factor [8]:

$$f_H(\theta) = 0.25(2 + \cos\theta) \cdot (1 - \cos\theta)^2. \quad (2.1.52)$$

With complete wetting, $\theta = 0$, no nucleation work is required. If $\theta = 180°$, the entire nucleation work must be expended.

The frequency of nucleation is calculated from this equation:

$$J = J_0 \exp\left(-\frac{\Delta G_K}{kT}\right) \left[\text{nuclei/cm}^3 \text{ s}\right].$$ (2.1.53)

The frequency factor can be calculated [8]. In liquid metals $J_0 \approx 10^{30}$–10^{35} nuclei/cm^3 s [9]. J should, therefore, be significantly greater than 1 nucleus/cm^3 s in melts in order to speak of successful nucleation.

2.1.3.6 Boiling

If a liquid is in equilibrium with its steam, homogeneous nucleation of steam bubbles is impossible because above the concave liquid surface of the (imaginary) bubble, the steam pressure is always lower than above the flat liquid surface (Gibbs–Thomson Eq. 2.1.48). The bubble would be crushed and its steam would dissipate. Nucleation occurs heterogeneously at surface roughness areas, as shown in Fig. 2.12.

The criterion for heterogeneous nucleation is:

$$\beta > 2\theta - 180 \text{ nucleation work} > 0$$
$$\boldsymbol{\beta^* = 2\theta - 180 \text{ nucleation work} = 0}$$
$$\beta^* < 2\theta - 180 \text{ nucleation work} < 0.$$ (2.1.54)

* There is already a gas phase in the gap.

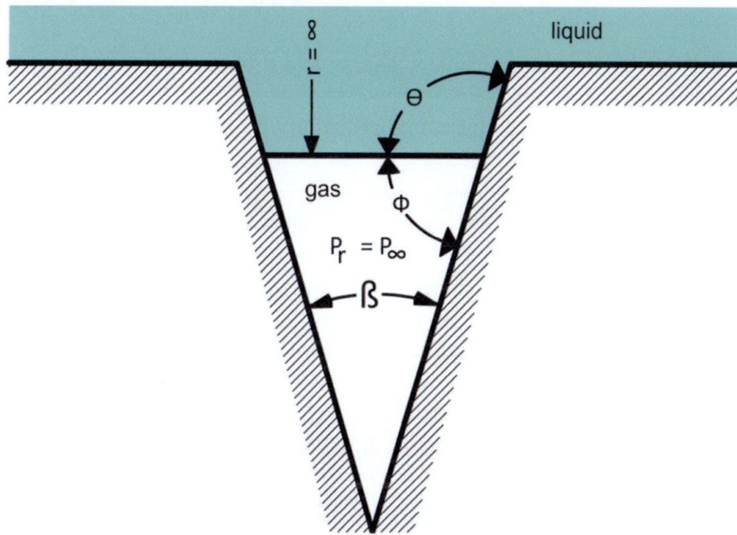

Fig. 2.12 Nucleation work of a steam bubble

It should be noted that a bubble is not stable until the pressure in the bubble P_B balances the sum of the external pressure, liquid pressure and capillary pressure:

$$P_B = P_{\text{external}} + (\rho g\, h + 4\sigma/d)\ [\text{N/m}^2] \tag{2.1.55}$$

2.1.4 Electrochemistry

2.1.4.1 Electrolytic Conduction

An electric field E [V/m] acts on an electrolyte. As a result, an electric current of density [10] flows:

$$j = \sigma \cdot E\ \left[\text{A/m}^2\right]. \tag{2.1.56}$$

The specific conductivity of the electrolyte σ [A/V m] is the sum of the cation and anion partial conductivities of all the ions present.

The ion mobility is

$$\mu_{\text{total}} = (\mu_+ + \mu_-) = \frac{\sigma}{e \cdot n}\ \left[\text{m}^2/\text{Vs}\right]. \tag{2.1.57}$$

Their migration speed is

$$U_{\text{total}} = (u_+ + u_-) = (\mu_+ + \mu)E\ [\text{m/s}]. \tag{2.1.58}$$

$e = 1.6 \times 10^{-19}$ A s is the elementary charge and n is the total number of ions in $1\ \text{m}^3$ of electrolyte.

A prerequisite, however, is that no polarization occurs at the two electrodes, i.e. the field is not weakened or collapses. If, for example, the ions at the electrodes are discharged, their surface changes, reducing the applied field strength. (excretion polarization). The continuation of the excretion then requires the overcoming of an often considerable overvoltage.

If depletion of a component occurs in the immediate vicinity of the electrode as a result of excretion, diffusive replenishment of the minority component from the electrolyte determines the excretion rate. This so-called concentration polarization also reduces the current density.

In all cases, the "counter EMF" can be derived from Nernst's equation (Sect. 2.1.4.2) if the concentration ratios at the electrode are known. However, this is only very rarely the case.

The electrochemically deposited (dissolved) mass m_i of component i with current I in time t is:

$$m_i = \frac{M_i}{z_i \cdot F} \cdot I \cdot t \; [g].$$

(2.1.59)

Dissolution of copper anodes in aqueous electrolyte solution, for example, occurs at a bath voltage of 0.2–0.3 V and a current density of about 250 A/m^2 [11]. The molecular weight of copper is $M = 64$ g/mol and its valence $z = 2+$. The Faraday constant is $F = 9.56 \times 10^4$ J/mol V. The amount of copper dissolved in one hour according to Eq. (2.1.59) is, thus, $m = 300$ g/m^2.

2.1.4.2 Electrochemical Potential

W. Nernst showed the equivalence of electric potential and chemical potential [10] (refer to Sect. 2.1.8):

$$z_i F \left(\Phi_\beta \times \Phi_\alpha \right) = \Sigma v_i \cdot \mu_i \; [J/mol].$$

(2.1.60)

z_i is the valence of component i in the solution. $(\Phi_\beta - \Phi_\alpha)$ is the electric potential (electromotive force, EMF) between the β and α electrodes. $F = 9.56 \times 10^4$ J/mol V is the Faraday constant. v_i is the stoichiometric factor of component i in the chemical reaction and μ_i is its chemical potential (refer to Eqs. 2.1.8 and 2.1.32):

$$\mu_i = \mu_i^o + RT \cdot \ln a_i \; [J/mol].$$

(2.1.61)

A well-known example is fused-salt electrolysis for the reduction of alumina to metallic aluminum at 1000 °C in cryolite (Na$_3$AlF$_6$). If one converts the values (counts) of the free enthalpy ΔG^o [J/mol] shown in Fig. 2.6 into the electromotive force EMF [volts] using Eq. (2.1.60) ($\Delta G^o = - z \cdot F \cdot E$), one reads from Fig. 2.13 that at least $E = 2.2$ V is required as the applied cell voltage. In practice [11], it is 4–5 V because of various polarization effects and high line resistances.

2.1.4.3 Electrocapillarity

Lippmann [12] described the correlation between a DC electrical voltage U [V] and the interfacial voltage between two liquid phases, e.g. mercury and the aqueous electrolyte potassium sulfate:

$$(d\sigma/dU)_{P,T,\mu} = -q_M \; [A \, s/m^2].$$

(2.1.62)

q_M is the excess electrical charge on the metal surface. It changes with the applied voltage:

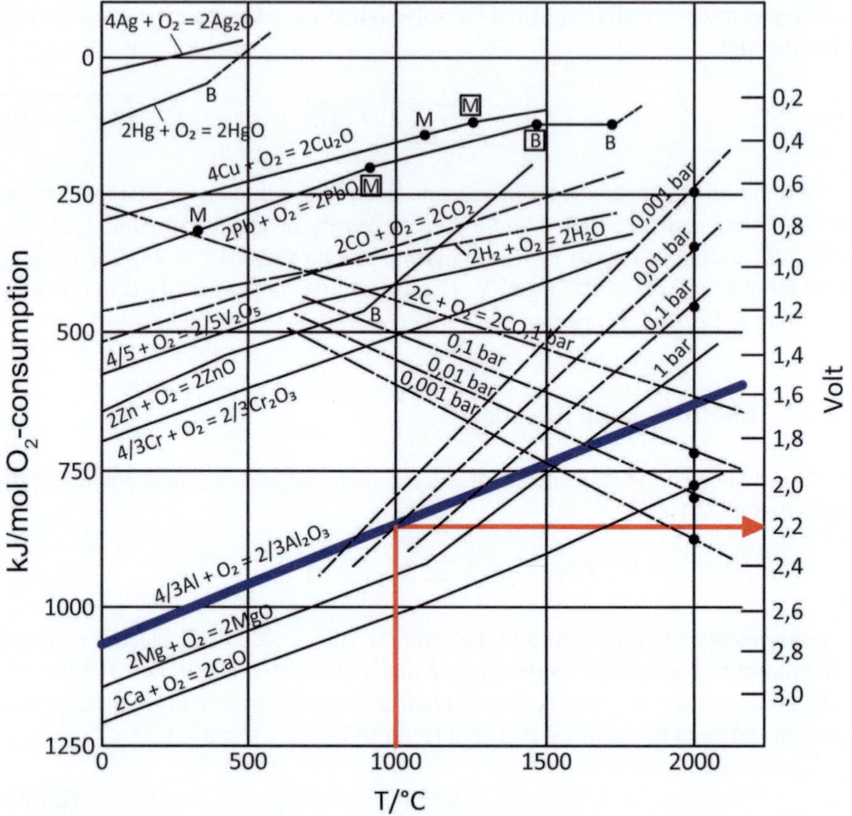

Fig. 2.13 Richardson–Ellingham diagram. Example: Melt flow electrolysis of Al_2O_3 to aluminum at 1000 °C [11]

$$dq_M = C \cdot dU \ [\mathrm{A\ s/m^2}]. \tag{2.1.63}$$

Substituting in Eq. (2.1.62) and integrated we get:

$$\sigma - \sigma_o = -0.5 \cdot C \cdot U^2 \ [\mathrm{V\ A\ s/m^2}]. \tag{2.1.64}$$

The interfacial voltage thus changes parabolically with U. At its maximum, $dq_M/dU = 0$, i.e. the charge of the capacitor (condenser) C is zero.

Figure 2.14 shows a principle illustration. Since the voltage is squared, its sign is irrelevant, i.e. AC voltage also changes the interfacial tension.

Experimental evidence suggests that in practice this effect alters the infiltration of the refractory material and, therefore, influences its wear [13, 14].

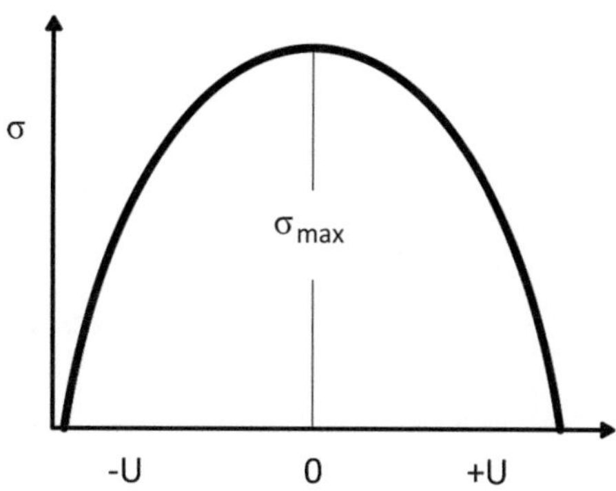

Fig. 2.14 Idealized illustration of a "Lippmann curve"

2.2 Introduction to Kinetics

2.2.1 Transport Sizes

Whereas the direction and end position of a process are thermodynamically determined, kinetics describes its rate.

The three transport quantities "thermal diffusivity" **a**, "diffusion coefficient" **D** and "kinematic viscosity" **v** not only have the same dimension $[m^2/s]$ but also formally occur in the same transport law. Their definition equations are [16]:

$$j_q = -\lambda \frac{dT}{dx} \left[J/m^2\ s \right] \text{ (Fourier I) (energy transport), with } a \equiv \frac{\lambda}{C_p \cdot \rho} \qquad (2.2.1)$$

$$j_m = -D \frac{dc}{dx} \left[mol/m^2\ s \right] \text{ (Fick I) (mass transport),} \qquad (2.2.2)$$

$$j_i = -\eta \frac{du}{dy} [(kg\ m/s)/(m^2\ s)] \text{ (Newton) (momentum transfer),} \quad \text{with } v \equiv /\eta\rho.$$

$$(2.2.3)$$

These three equations apply to the steady (stationary) state.

Let j be the flux density, T [K] the temperature, c [mol/m^3] the concentration, u [m/s] the shear rate perpendicular to the x-direction, λ [W/m K] the thermal conductivity, C_P [J/kg K] the specific heat at constant pressure, ρ [kg/m^3] the density, and η [kg/m s] the dynamic viscosity. a [m^2/s] is the thermal diffusivity, D [m^2/s] the diffusion coefficient, and v [m^2/s] the kinematic viscosity.

In the transient case, the following partial differential equations apply:

$$\frac{\partial T}{\partial t} = a \cdot \left(\frac{\partial^2 T}{\partial r^2} + \frac{n}{r} \cdot \frac{\partial T}{\partial r} \right) [K/s] \text{ (Fourier II)} \qquad (2.2.4)$$

$$\frac{\partial c}{\partial t} = D \cdot \left(\frac{\partial^2 c}{\partial r^2} + \frac{n}{r} \cdot \frac{\partial c}{\partial r} \right) \left[\text{mol/m}^3 \text{ s} \right] \text{ (Fick II)} \qquad (2.2.5)$$

Length, respectively radius is denoted by r [m]. For the plate $n = 0$, the cylinder $n = 1$ and the sphere $n = 2$. Because $v = \eta/\rho$ one obtains analogously:

$$\rho \cdot \frac{\partial u}{\partial t} = \eta \cdot \left(\frac{\partial^2 u}{\partial y^2} + \frac{1}{y} \cdot \frac{\partial u}{\partial y} \right) \left[(\text{kg/m}^3) \cdot (\text{m/s}^2) \right] \text{ (Navier–Stokes)} \qquad (2.2.6)$$

This is the Navier–Stokes equation without consideration of an additional external volume force density such as gravity, for example. It describes the motion of fluids. According to this, the following force densities are in equilibrium with each other [10]:

$$\text{Acceleration} = \text{friction} + \text{pressure}. \qquad (2.2.6.1)$$

In the steady state, the acceleration force density is zero. Consequently, friction and pressure are in equilibrium, resulting in the well-known equation from Hagen–Poiseuille for the mean flow speed (velocity) $\bar{u}$ [m/s] in a pipe having a radius R [m] and length L [m] as result of the pressure gradient $\Delta P/L$ [kg/m^{2}s^2)] [10]:

$$\bar{u} = \frac{\Delta P \cdot R^2}{8\eta \cdot L} \text{ [m/s]}. \qquad (2.2.7)$$

The following similarity indices are derived from the transport constants [16]:

$$\text{Heat: Pr} = v/a \text{ (Prandtl) and Le} = a/D \text{ (Lewis)} \qquad (2.2.8)$$

$$\text{Substance (material): Sc} = v/D \text{ (Schmidt)} \qquad (2.2.9)$$

$$\text{Momentum: Re} = u \cdot x/v \text{ (Reynolds).} \qquad (2.2.10)$$

The similarity indices allow to compare model experiments with transport processes in practice in a simple way if the relevant indices are equal in both cases. For example, experience shows that the flow in a pipe is turbulent for Re > 2000 [10].

2.2.2 Solutions and Approximations for Transient Diffusion

These calculations are performed for mass (material) transfer but apply equally to heat transfer.

2.2.2.1 Differential Equation of Transient Diffusion

For the transient diffusion in plate, cylinder and sphere, the following differential equation is valid:

$$\frac{\partial c}{\partial t} = D \cdot \left(\frac{\partial^2 c}{\partial r^2} + \frac{n}{r} \cdot \frac{\partial c}{\partial r} \right). \tag{2.2.11}$$

The following applies for the plate $n = 0$, for the cylinder $n = 1$ and for the sphere $n = 2$. The solution depends on the initial and boundary conditions. The following specifications apply:

The reduced concentration φ is adapted to the problem. For example

$$\varphi \equiv \frac{c - c_0}{c_\infty - c_0} \tag{2.2.12}$$

with $c = c(r, t)$, $c = c_0$ (for all r, $t = 0$), $c = c_\infty$ ($r = 0$, $t > 0$).

For the reduced dimension

$$\rho \equiv \frac{r}{R} \tag{2.2.13}$$

with $r = r(t)$, $R =$ half the thickness of the infinitely thick plate or rather radius of the cylinder or sphere.

For the reduced time

$$\tau \equiv \frac{D \cdot t}{R^2}. \tag{2.2.14}$$

2.2.2.2 Diffusion into an Infinite Half-Space [21]

An infinitely thick plate ($R \to \infty$) is looked at. In its interior, the concentration $c_0 = c(r, t = 0)$ is present at all locations r at the beginning. At the surface ($r = 0$), the concentration drops abruptly to the constant value $c_\infty = c(r = 0, t)$. The resulting mass (material) flow is directed outward. The solution for the decrease of the concentration in the interior with $\varphi = \frac{c_0 - c}{c_0 - c_\infty}$ is:

$$\varphi = \mathrm{erf}\left(\frac{1}{2 \cdot \sqrt{\tau_P}} \right) \equiv \mathrm{erf}\left(\frac{x}{2 \cdot \sqrt{D \cdot t}} \right) \equiv \mathrm{erf}(A) \tag{2.2.15}$$

with

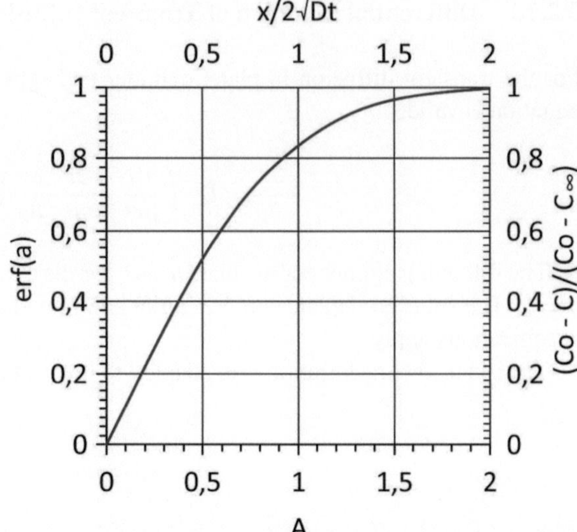

Fig. 2.15 Error function erf (A)

$$\mathrm{erf}(A) = \frac{2}{\sqrt{\pi}} \cdot \int_0^A e^{-z^2}\, \mathrm{d}z, \tag{2.2.16}$$

i.e. the error function. The progression is shown in Fig. 2.15. This solution is "classical" and quite common in tabulated form. A is the argument of the error function. For $A = 0$, $\mathrm{erf}(0) = 0$ and for $A \to \infty$, $\mathrm{erf}(\infty) = 1$. For $A = 1$, $\mathrm{erf}(1) = 0.84$ and for $A = 2$, $\mathrm{erf}(2) = 0.995 \approx 1$. Therefore, the plate can be considered infinitely thick once its dimension exceeds $R = 2 \cdot 2\sqrt{D \cdot t}$. If the concentration inside the body increases with time, the complementary error function $\mathrm{erfc}(A) \equiv (1 - \mathrm{erf}(A))$ must be used.

2.2.2.3 Diffusion in Confined Bodies [21]

R is the radius of a sphere, an infinitely long cylinder and $2R$ the thickness of a plate. Starting from these basic forms, Newman [22] have calculated the concentration distribution in various bodies. Here, the coordinate origin is at the center ($r = 0$) and the dimension ($r = R$) is at the surface. Initially, the body contains uniformly distributed the concentration $c_0 = c(0 \leq r \leq R, t = 0)$. The concentration at the surface changes abruptly at the beginning of diffusion to the final value $c_\infty = c(R, t)$ and remains constant. The reduced concentration is

$$\varphi_1 = \frac{C - C_0}{C_\infty - C_0} \quad \text{or} \quad \varphi_2 = \frac{C_\infty - C}{C_\infty - C_0}, \tag{2.2.17}$$

as the concentration increases or decreases over time. It is $\varphi_2 = 1 - \varphi_1$.

In the center of the body always has $\left.\frac{dc}{dr}\right|_{r=0,t} = 0$. Then this applies:

$$\varphi_1 = 0 \text{ for } t = 0 \text{ and respectively } \varphi_2 = 1 \text{ for } t = 0 \text{ and}$$

$$\varphi_1 = 1 \text{ for } t = \infty, \; \varphi_2 = 0 \text{ for } t = \infty. \tag{2.2.18}$$

As a result, one obtains series whose evaluation is complex. Their geometrical illustration with reduced quantities is done, among others, by Carslaw and Jaeger [23], Bird et al. [16], and Crank [24]. As a rough estimate, it is shown that for $\rho \equiv (r/R) \rightarrow 0$ and $\tau \equiv \left(\frac{D \cdot t}{R^2}\right) \rightarrow 1$, for a given φ_1 the approximation is valid:

$$\tau_P(\text{plate}) \approx 2 \cdot \tau_Z(\text{cylinder}) \quad \text{and} \quad \tau_P \approx 3 \cdot \tau_K(\text{sphere}). \tag{2.2.19}$$

In the cylinder and sphere, a given concentration is, thus, achieved in 2 to 3 times less time (also refer to Fig. 2.16 (1–3)):

$$\varphi_1 = 0.90; \; r/R < 0.3; \; \tau_P = 1; \; \tau_Z \approx 0.5; \; \tau_K \approx 0.33. \tag{2.2.20}$$

Another simplification for $\tau_P > 0.3$ is shown by an **example**: in order to estimate the rel. concentration $\varphi(\tau_P)$ for a plate ($n = 0$) with $\tau_P > 0.3$, the error function $\text{erf}(4\tau_P/\pi)$ is calculated for the magnitude $(4\tau_P/\pi)$. Let us consider a plate of thickness $2R = 0.02$ m and an infinitely long cylinder and sphere of equal radius $R = 0.01$ m: let the diffusion coefficient be $D = 10^{-6}$ m²/h, the time considered $t_P = 100$ h and, therefore, $\tau_P = 1$ and $\varphi = 0.9$ (refer to Fig. 2.16 (1)). One obtains $\text{erf}(4\tau_P/\pi)$ $= \text{erf}(1.27) = 0.9$ for the plate. For the cylinder, according to Eq. (2.2.20), $t_Z \approx 50$ h and for the sphere $t_K \approx 33$ h with $\varphi = 0.9$ (refer to Fig. 2.16 (2 and 3)).

If the thickness of the plate was $2R = 2 \cdot 2\sqrt{D \cdot t} = 0.08$ m, the solution (2.2.15), that is the error function $\text{erf}\left(x/(2\sqrt{(D \cdot t)})\right)$ could be used to calculate φ because with $R = 0.04$ the value of the error function practically reaches $\varphi = 1$.

2.2.2.4 Calculation of the Mean Concentration in Confined Bodies [21]

In many practical cases, it is not the concentration distribution in the body at the moment of termination of the experiment that is of interest but the mean concentration c_m, which would have been set in the entire body subsequently following a very long waiting period. This calculation was also performed and presented by Newman [22]. The integral mean value of the concentration distribution is given by

$$\varphi_m = \frac{1}{R^{n+1}} \cdot \int_0^R \varphi(\tau) \cdot r^n dr \tag{2.2.21}$$

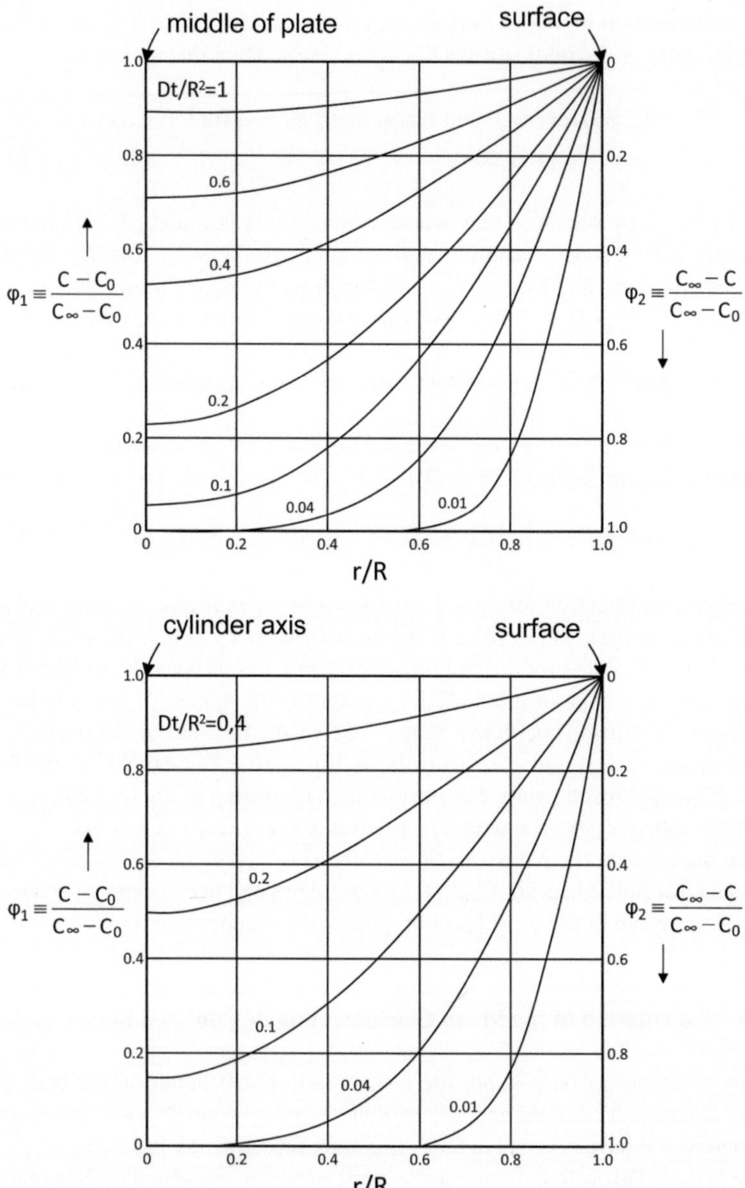

Fig. 2.16 1–3 Distribution of concentration (temperature) during transient diffusion (heat conduction), (1) plate, (2) cylinder, (3) sphere [16]

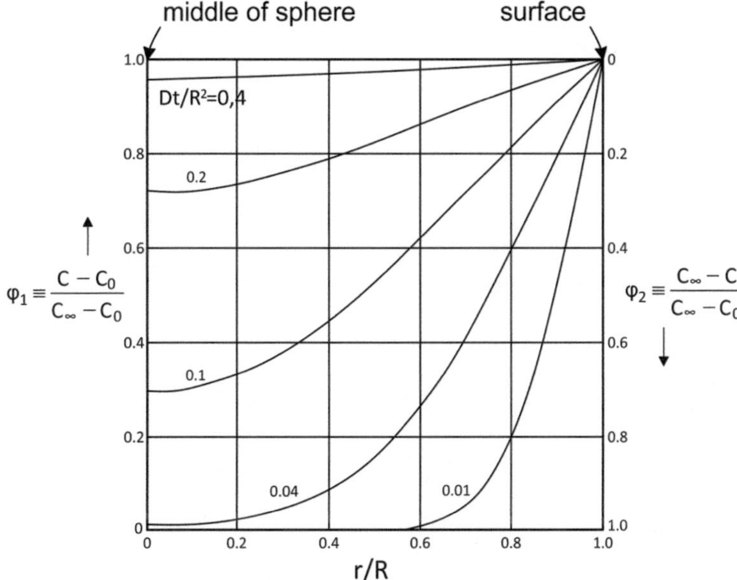

Fig. 2.16 (continued)

with n = 0 for the plate, n = 1 for the cylinder and n = 2 for the sphere.

The reduced mean concentration is

$$\varphi_m \equiv \frac{c_m - c_0}{c_s - c_0}.$$

(2.2.22)

The tabular and graphical illustration can be found, among others, in Darken and Gurry [15] (refer to Table 2.1 and Fig. 2.17). Here, c_0 is the initial concentration uniformly distributed in the body, c_s is the concentration set at the surface R at time $t = 0$.

The time progression of the mean concentration in a plate of a thickness 2R follows the equation for $\tau_P = D \cdot t/R^2 \leq 2$:

$$\varphi_m(P) = \left(-0.0757\tau_P^4 + 0.4169\tau_P^3 - 0.8199\tau_P^2 + 0.6133\tau_P + 0.9702\right) \cdot \mathrm{erf}\left(\sqrt{\tau_P}\right).$$

(2.2.23)

Comparing all τ_P in Table 2.1 with the corresponding values τ_Z for the cylinder and τ_K for the sphere, given a common value φ_m, the following applies for the cylinder:

$$\left(\frac{\tau_P}{\tau_Z}\right)^{\frac{1}{2}} = 2 - 0.4 \cdot \varphi_m$$

(2.2.24)

Table 2.1 Partial saturation (mean concentration) of plate, cylinder and sphere in transient diffusion according to [15]

Dt/R^2		$(C_m - C_0)/(C_s - C_0)$	
[−]	Plate	Cylinder	Sphere
0.005	0.078	0.157	0.226
0.01	0.110	0.216	0.310
0.02	0.161	0.302	0.421
0.03	0.195	0.360	0.500
0.04	0.227	0.412	0.560
0.05	0.251	0.452	0.604
0.06	0.275	0.488	0.648
0.08	0.320	0.550	0.720
0.10	0.357	0.606	0.774
0.15	0.438	0.708	0.861
0.20	0.503	0.781	0.916
0.25	0.560	0.832	0.948
0.30	0.612	0.875	0.969
0.40	0.702	0.9316	0.988
0.50	0.767	0.9616	0.9957
0.60	0.816	0.9785	0.9984
0.70	0.856	0.9879	0.9994
0.80	0.887	0.9932	0.9998
0.90	0.912	0.9960	0.9999
1.00	0.931	0.9979	
1.50	0.980	0.9999	
2.00	0.9942		
3.00	0.9995		

and for the sphere:

$$\left(\frac{\tau_P}{\tau_K}\right)^{\frac{1}{2}} = 3 - \varphi_m. \tag{2.2.25}$$

For example, $\varphi = 0.93$ for $\tau_P = 1$ and, consequently,

$$\tau_Z = \tau_P/(2 - 0.4\,\varphi_m)^2 = 0.38\,\tau_P \tag{2.2.26.1}$$

and

$$\tau_K = \tau_P/(3 - \varphi_m)^2 = 0.23 \cdot \tau_P. \tag{2.2.26.2}$$

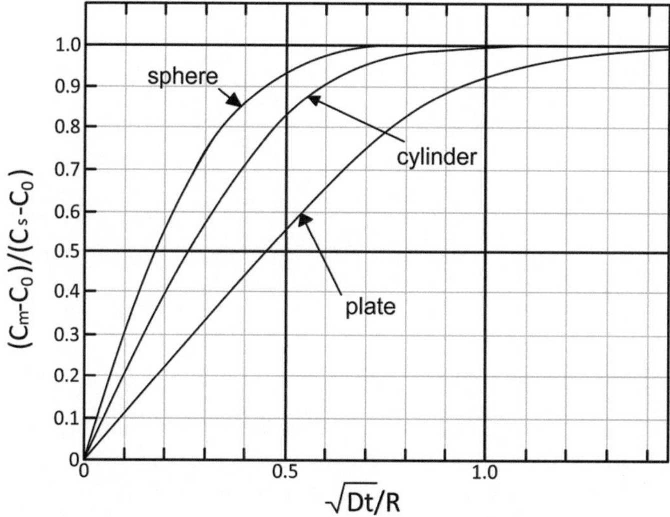

Fig. 2.17 Partial saturation (mean concentration) of plate, cylinder and sphere during transient diffusion according to [15]

According to Eq. 2.2.23), the average generalized saturation φ_m (P) for $\tau_P = 1$ follows the error function erf($\sqrt{\tau_P}$) by a factor of 0.93 (rounded up to 1). Therefore, approximately

$$A_P : A : A_{ZK} \approx 1 : 1/\sqrt{0.38} : 1/\sqrt{0.23} = 1 : 1.6 : 2.1 \tag{2.2.27}$$

can be expected once diffusion is almost complete.

For the special case $\tau \leq 0.25$ the following applies in good approximation:

$$\varphi_m = \frac{n}{\sqrt{\pi}} \cdot \sqrt{\tau} - m \cdot \tau \tag{2.2.28}$$

with $n = 2$ and $m = 0$ for the plate, $n = 4$ and $m = 1.2$ for the cylinder, and $n = 6$ and $m = 3$ for the sphere.

2.2.3 Boundary Layers

2.2.3.1 Flow Boundary Layer

Transport in moving media is the sum of diffusion and convection. The hydrodynamic adhesion condition requires that at a solid surface, the flow speed (velocity) becomes zero. At the contact surface of two fluid phases, e.g. liquid steel/slag, the relative velocity is also zero and one speaks of a frictionless flow.

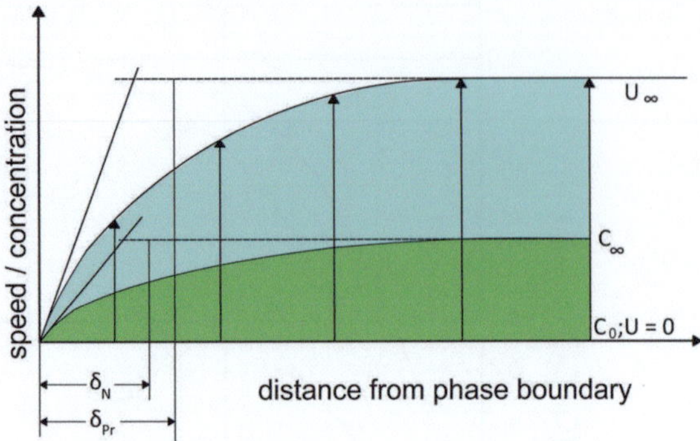

Fig. 2.18 Flow boundary layer (δ_{Pr}) and diffusion boundary layer ($\delta)_N$

Laminar flow is distinguished from turbulent flow. With a laminar flow, the flow lines run parallel to each other. In turbulent flow, convection balls are formed and the mass (material) transport is comparatively greater. The Reynolds number Re (Eq. 2.2.10) expresses the transition condition.

As a solid wall is approached, the so-called laminar boundary zone is formed in any case. In turbulent flow, this is very thin and is called the laminar sublayer. The thickness of this flow boundary layer δ_{Pr} mediating between resting flow ($u = 0$) and main flow ($u = u_\infty$) (Eq. 2.2.8) is named after Prandtl and is defined according to (refer to Fig. 2.18):

$$\left(\frac{du}{dx}\right)_{x=0} = \frac{u_\infty}{\delta_{Pr}}\left[\text{s}^{-1}\right]. \tag{2.2.29}$$

2.2.3.2   Diffusion Boundary Layer

The diffusion boundary layer is embedded in the flow boundary layer.

It is always less strong and describes the range in which mass (material) transport is to take place exclusively by diffusion. (It should be emphasized that the same applies to heat transport. Then formally the diffusion coefficient. D [m^2/s] is to be substituted by the thermal diffusivity a [m^2/s] and one speaks of the temperature boundary layer). Figure 2.18 shows that the diffusion boundary layer is defined as:

$$\left(\frac{dc}{dx}\right)_{x=0} = \frac{c_\infty - c_o}{\delta_N} \ [\text{mol/m}]. \tag{2.2.30}$$

It is proven that both boundary layers are connected via the Schmidt number (Eq. 2.2.9) [25]:

$$\frac{\delta_N}{\delta_{Pr}} = f \cdot \left(\frac{\nu}{D}\right)^{-\frac{1}{3}}. \tag{2.2.31}$$

With laminar flow $f = 1$, with turbulent $f = 0.72$ [25]. In liquid steel $\delta_N/\delta_{\text{Pr}} \approx 0.2$ and in slag $\delta_N/\delta_{\text{Pr}} \approx 0.005$.

2.2.4 Mass Transfer

2.2.4.1 Mass Transfer Coefficient

According to Eq. 2.2.2, the mass flow at the point $x = 0$ (Fig. 2.18) is

$$j_m = -D \left(\frac{dc}{dx}\right)_{x=0} = \beta \cdot (c_0 - c_\infty) \; [\text{mol/m}^2 \; \text{s}]. \tag{2.2.32}$$

Taken together, Eqs. 2.2.30 and 2.2.32 give the so-called mass transfer coefficient β [m/s]:

$$\beta = D/\delta \; [\text{m/s}]. \tag{2.2.33}$$

In liquid steel, $\beta \approx 10^{-4}$ m/s and in slag $\beta \approx 10^{-5}$ m/s.

The mean mass (material) transfer coefficient can be calculated directly from the flow speed (velocity) u and a "characteristic length," preferably the so-called inflow length y [16, 25, 26]:

$$\beta_y = \left(\frac{4 \cdot D \cdot u}{\pi \cdot y}\right)^{\frac{1}{2}} \cdot \left(\frac{D}{v}\right)^{\frac{1}{6}} \; [\text{m/s}]. \tag{2.2.34}$$

This equation is valid with wall friction; without it, the second bracket is dropped. For molten steel, $\left(\frac{D}{v}\right)^{\frac{1}{6}} \approx 0.3$ and for slag, $\left(\frac{D}{v}\right)^{\frac{1}{6}} \approx 0.1$, this means that wall friction noticeably reduces mass transfer.

In many cases, one uses dimensionless indicators, e.g. to compare processes in practice with laboratory experiments. From Eq. (2.2.34) and the two indicators $S_C = v/D$ (2.2.9) and Re $= u \cdot y/v$ (2.2.10), the Sherwood number Sh $\equiv \beta \cdot y/D$ is calculated to be

$$\text{Sh} = \frac{2}{\sqrt{\pi}} \cdot \text{Re}^{\frac{1}{2}} \cdot \text{Sc}^{\frac{1}{3}}. \tag{2.2.35}$$

It characterizes the mass (material) transfer.

2.2.4.2 Mass Transfer in Confined Vessels

Dissolution of a Plane Wall into a Confined Melt

The density of the mass flow from a plane wall having the area F [cm^2] into the melt of volume V [cm^3] and concentration $C(t)$ [g/cm^3] is

$$i = \beta \cdot (C_S - C(t)) \ [\text{g/cm}^2 \ \text{s}] \tag{2.2.36}$$

The increase in concentration in the melt over time is

$$i \cdot F = V \cdot \frac{dc}{dt} \ [\text{g/s}]. \tag{2.2.37}$$

In summary, the integral relation is

$$\int_{C_0}^{C} \frac{dC}{C_s - C_0} = -\beta \cdot \frac{F}{V} \cdot \int_{0}^{t} dt \tag{2.2.38}$$

to be solved with the boundary conditions $t = 0$: $C = C_0$ and $t = \infty$: $C = C_S$, $C_S > C_0$:

$$\frac{C(t) - C_0}{C_S - C_0} = 1 - \exp\left(-\beta \cdot \frac{F}{V} \cdot t\right). \tag{2.2.39}$$

The thickness of the wall layer removed by the substance transport is

$$x(t) - x_0 = -\frac{V \cdot \rho_{li}}{100 \cdot F \cdot \rho_R} \cdot (C(t) - C_0) \ [\text{cm}]. \tag{2.2.40}$$

The over time decay of the initial wall thickness x_0 is obtained by (taking into account the initial condition $t = 0$, $x(t) = x_0$) having the concentration difference $C(t) - C_0$ in Eq. 2.2.40 substituted by Eq. 2.2.39:

$$x(t) - x_0 = -\frac{V \cdot \rho_{Sl}}{100 \cdot F \cdot \rho_R} \cdot (C_s - C_0) \cdot \left(1 - \exp\left(-\beta_{Sl} \cdot \frac{F}{V} \cdot t\right)\right) \ [\text{cm}]. \tag{2.2.41}$$

The corrosion rate $u_{corr}(t) = dx/dt$ is obtained by deriving Eq. (2.2.41) with respect to time:

$$u_{\text{corr}}(t) = 360 \cdot \beta \cdot \frac{\rho_{Sl}}{\rho_R} \cdot (C_s - C_0) \cdot \exp\left(-\beta_{Sl} \cdot \frac{F}{V} \cdot t\right) \ [\text{mm/h}]. \tag{2.2.42}$$

The concentration in the melt $C(t)$ and the wear of the wall $(x(t) - x_0)$ increase with time, whereas the corrosion rate $u_{\text{corr}}(t)$ decreases over time.

Dissolution of a Rod in a Geometrically Constrained Melt

The starting point is the mass (material) flow density i of the diffusion-controlled dissolution of the cylindrical, densely sintered specimen into the slag, refer to (Eq. (2.2.36) and (2.2.37)):

$$i = \beta_{Sl} \cdot (C_S - C(t)) \ [\text{g/cm}^2 \ \text{s}]. \tag{2.2.43}$$

The increase in concentration in the slag over time is

$$i \cdot F = V \cdot \frac{dC}{dt} \ [\text{g/s}]. \tag{2.2.44}$$

The volume V [cm^3] of the melt remains approximately unchanged. The surface area F [cm^2] changes with time.

The substance balance applies:

$$V \cdot (C - C_0) = m_0 \cdot \left(1 - \frac{R^2}{R_0^2}\right). \tag{2.2.45}$$

The front surface of the specimen is not considered, only the immersed shell area $F = 2\pi R \cdot h$ [cm^2]. The thickness of the diffusion boundary layer is much smaller than the specimen radius: $\delta \ll R$ [cm]. To simplify the calculation, the following dimensionless quantities are formed [17]:

$$\text{Generalized time: } \tau \equiv \frac{D \cdot F_0}{\delta \cdot V} \cdot t \tag{2.2.46.1}$$

$$\text{Reduced saturation ratio: } \gamma \equiv \frac{C - C_0}{C_S - C_0} \tag{2.2.46.2}$$

$$\text{Related radii ratio: } \varphi \equiv \left(\frac{R_0}{R_S}\right)^2. \tag{2.2.46.3}$$

The index "0" denotes the initial state and "S" the saturation. R_S is the radius whose resolution just leads to the saturation of the solution. Consequently, φ is the ratio of the initial mass m_0 to the mass $m_S = V \cdot (C_S - C_0)$. This applies:

- $(\varphi = 1)$—the dissolved initial mass (material) just saturates the melt.
- $(\varphi < 1)$—The initial mass (material) is dissolved before saturation is accomplished.
- $(\varphi > 1)$—The initial mass (material) is not completely dissolved once saturation is accomplished.

The solutions of the differential equations resulting for the cylinder

$$d\tau = \sqrt{\varphi} \cdot \frac{d\gamma}{(1-\gamma) \cdot \sqrt{\varphi - \gamma}} \tag{2.2.47}$$

are [17]:

$$\tau = \frac{2}{\sqrt{1-\gamma}} - 2 \quad (\text{for } \varphi = 1; \ 0 \le \gamma \le 1), \tag{2.2.48.1}$$

$$\tau = 2 \cdot \sqrt{\frac{\varphi}{1-\varphi}} \cdot \left(\arctan \sqrt{\frac{\varphi}{1-\varphi}} - \arctan \sqrt{\frac{\varphi - \gamma}{1-\varphi}} \right) \quad (\text{for } \varphi < 1; \ 0 \le \gamma \le \varphi)$$

$$\tau = 2 \cdot \sqrt{\frac{\varphi}{1-\varphi}} \cdot \left(\arctan \sqrt{\frac{\varphi}{1-\varphi}} - \arctan \sqrt{\frac{\varphi - \gamma}{1-\varphi}} \right) \quad (\text{for } \varphi < 1; \ 0 \le \gamma \le \varphi),$$

$$\tag{2.2.48.2}$$

$$\tau = \frac{\sqrt{\varphi}}{\sqrt{\varphi - 1}} \left(\ln \frac{\sqrt{\varphi} - \sqrt{\varphi - 1}}{\sqrt{\varphi} + \sqrt{\varphi - 1}} - \ln \frac{\sqrt{\varphi - \gamma} - \sqrt{\varphi - 1}}{\sqrt{\varphi - \gamma} + \sqrt{\varphi - 1}} \right) (\text{for } \varphi > 1; 0 \le \gamma \le 1).$$

$$\tag{2.2.48.3}$$

Corresponding solutions are also available for the sphere [17]. For the plate there is only one solution in all cases: $\tau = -\ln(1-\gamma)$ (refer to Eq. (2.2.39)). Figure 2.19 (1–4) show the calculated results. In many cases, the quick comparison of the results with different boundary conditions is already of valuable assistance in making decisions.

2.2.4.3 Evaporation and Vaporization

These processes play a particularly important role in vacuum metallurgy. A liquid evaporates below its boiling temperature T_m (1 atm) and, upon reaching it, the liquid evaporates. The Clausius–Clapeyron equation describes the steam pressure as a function of temperature (Eq. 2.1.15).

If the negative pressure reaches the value 10^{-2} atm, the mean free path length of the steam molecules is approximately in the dimension of standard vacuum apparatus. It is given by the equation [27]

$$\lambda = \frac{kT}{\sqrt{2} \cdot \pi d^2 \cdot p} \ [\text{m}]. \tag{2.2.49}$$

$k = 1.38 \times 10^{-23}$ J/K is the Boltzmann constant and d [m] the dimension typical for the system. We then speak of a Knudsen flow in which the collision of the molecules

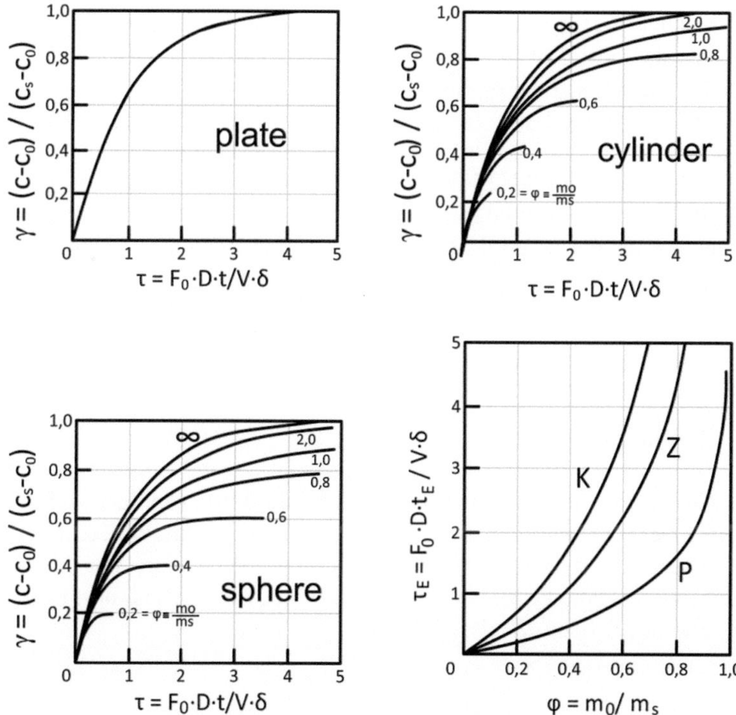

Fig. 2.19 1–3 Reduced saturation ratio γ as a function of generalized time τ for plate, cylinder and sphere [17]. 4 Generalized dissolution time τ as a function of mass (material) ratio φ for plate, cylinder and sphere [17]

with each other no longer determines the flow velocity but only that with the walls. It is significantly larger than that calculated according to the Hagen–Poiseuille Eq. (2.2.7).

The Hertz-Knudsen equation for practice is [28]:

$$j = 45 \cdot \alpha \cdot P_s \cdot \sqrt{\frac{M}{T}}\ [\text{g/cm}^2\,\text{s}]. \tag{2.2.50}$$

If there is thermodynamic equilibrium between the melt and the atmosphere (steam phase), evaporation and condensation occur at the same rate. If the vacuum is increased further and, in addition, the steam condenses on the cold furnace wall, the above equation describes the evaporation process quite well.

The condensation coefficient of liquid metals is $\alpha \le 1$ and is usually between 0.2 and 0.9. It is heavily dependent on the type and quantity of capillary-active solution partners. Oxygen and sulfur, for example, accumulate on the surface of the bath and impair mass transfer [3].

If it turns out that the evaporation rate becomes independent of the pressure, the mass (material) transfer of the evaporating species within the melt determines the rate and (Eq. 2.2.32) is to be applied.

If the majority of the evaporating molecules reach the free surface, the so-called surface renewal theory [16, 25, 26] states that the mass (material) transfer coefficient for flow without friction (Eq. 2.2.34) is to be applied.

The selection of the method for calculating the evaporation rate should be made taking into account the experimental result in order to avoid making some errors.

2.2.5 Interfacial Convection

2.2.5.1 Capillary Flow

Section 2.1.3.4 describes the climb height of a liquid in a capillary with radius R (Fig. 2.10). Here, the question is answered how long the liquid need to reach its equilibrium position starting from the immersion of the capillary. For this purpose, in the Hagen–Poiseuille Eq. (2.2.7), the external pressure (ΔP) is substituted by the capillary pressure $2\sigma \cdot \cos \theta / R$ (Eq. 2.1.45):

$$\frac{\mathrm{d}L}{\mathrm{d}t} = \frac{\sigma_{\mathrm{LS}} \cdot \cos \theta \cdot R}{4\eta \cdot L} \ [\mathrm{m/s}] \tag{2.2.51}$$

After separation of the variables and integration, one obtains for the time duration:

$$t = \left(\frac{2 \cdot \eta \cdot L^2}{R \cdot \sigma \cdot \cos \theta} \right) \ [\mathrm{s}]. \tag{2.2.52}$$

η [kg/m· s] is the dynamic viscosity of the liquid. This equation describes, among other things, the infiltration of steel and slag into the refractory material.

2.2.5.2 Marangoni Convection

A pressure difference ΔP [kg/m s^2] moves a liquid film of thickness d [m] and the viscosity η [kg/m s] with the mean speed (velocity) [26]:

$$u = \Delta P \cdot \frac{\mathrm{d}}{4\eta} \ [\mathrm{m/s}]. \tag{2.2.53}$$

The pressure difference can be caused, among other things, by the gradient of dissolved capillary-active components *gradc*$_i$ [mol/m] or a temperature gradient grad T [K/m], if the surface tension changes due to this influence:

$$\Delta P = \frac{d\sigma}{dc_i} \cdot \mathrm{grad}c_i \quad \text{or} \quad \Delta P = \frac{d\sigma}{dT} \cdot \mathrm{grad}T \; \left[\mathrm{N/m^2}\right]. \tag{2.2.54}$$

The combination of both gives the equation of Marangoni convection [26]:

$$u = \frac{d}{4\eta} \cdot \frac{d\sigma}{dc} \cdot \mathrm{grad}c \quad \text{or} \quad u = \frac{d}{4\eta} \cdot \frac{d\sigma}{dT} \cdot \mathrm{grad}T \; [\mathrm{m/s}]. \tag{2.2.55}$$

Clearly visible examples are the lightning-fast spreading of gasoline to water and the premature corrosion of the refractory lining by molten glass or rather slag and steel.

2.2.6 Heat Transfer

2.2.6.1 Stationary Heat Transfer

In Sect. 2.2.1, it was already pointed out that the transport processes for material and heat can be treated formally in the same way. Some peculiarities of the stationary heat flow are discussed here [16, 27].

The heat transfer within a phase is determined by its thermal conductivity λ [W/mK] and the thickness of the conductive layer δ [m] and the temperature difference ΔT [K].

$$\dot{q} = \lambda \cdot \frac{\Delta T}{\delta} \; \left[\mathrm{W/m^2}\right]. \tag{2.2.56}$$

The heat transfer to the body is controlled by the heat transfer coefficient h [W/m^2 K]:

$$\dot{q} = h \cdot \Delta T \; \left[\mathrm{W/m^2}\right]. \tag{2.2.57}$$

ΔT [K] is the temperature difference between the two phases in contact with each other. In practice, a temperature jump is often observed, the amount of which cannot be considered here.

If several layers are in contact, λ, h and δ together determine the heat flux:

$$\dot{q} = k \cdot \Delta T \; \left[\mathrm{W/m^2}\right]. \tag{2.2.58}$$

For the coefficient of heat transfer, the following applies:

$$k = \frac{1}{\frac{\delta_1}{\lambda_1} + \frac{1}{h_1} + \frac{\delta_2}{\lambda_2} + \frac{1}{h_2} + \cdots} \; \left[\mathrm{W/m^2 \, K}\right] \tag{2.2.59}$$

Formally, the heat loss through the wall of a metallurgical vessel is calculated in this way.

2.2.6.2 Transient Heat Transfer

Pure heat conduction within a phase has already been described in Sect. 2.2.2 as a diffusion process. Here, we are concerned with the case where a non-negligible heat transfer to the workpiece occurs during temperature equalization. At the beginning, a semi-infinite plate is considered.

The differential equation to be solved is (refer to **3.1.5**)

$$\frac{\partial T}{\partial t} = a \cdot \frac{\partial^2 T}{\partial x^2}.$$
(2.2.60)

As dimensionless temperatures we enter

$$\theta = \frac{T - T_0}{T_U - T_0}$$
(2.2.61.1)

for heating and

$$\theta = \frac{T - T_U}{T_0 - T_U}$$
(2.2.61.2)

for cooling.

For very intensive heat transfer (Bi $\to \infty$) with the boundary conditions $T_U = T(x = 0, t)$ and $T_0 = T(x, t = 0)$, $(T_U > T_0)$, one obtains for the temperature rise in the body the known solution

$$\theta = 1 - \mathrm{erf}\left(\frac{1}{\sqrt{4 \cdot Fo}}\right) \quad (\mathrm{Bi} \geq 1),$$
(2.2.62)

which corresponds to Eq. 2.2.15. The Biot number Bi describes the ratio of heat transfer to heat conduction and is defined in Eq. 2.2.66. The Fourier number is defined as:

$$Fo \equiv \frac{a \cdot t}{L^2}.$$
(2.2.63)

The thermal diffusivity a [m^2/s] is linked to the thermal conductivity λ [W/m K], the specific heat C_P [J/kg K] and the density ρ [kg/m^3]:

$$a = \frac{\lambda}{C_p \cdot \rho} \ [\mathrm{m}^2/\mathrm{s}].$$
(2.2.64)

L [m] is a length characterizing the shape of the body which can be estimated by the ratio of the volume to the surface area.

If the surface temperature of the body changes with the constant speed (velocity) $\mathbf{k} \equiv dT/dt$, the following applies according to Carlslaw and Jaeger [23]

$$T(x, t) = k \cdot t \cdot \left(\left(1 + \frac{x^2}{2 \cdot a \cdot t}\right) \cdot \left(1 - \text{erf}\left(\frac{x}{2\sqrt{a \cdot t}}\right)\right) - \frac{x}{\sqrt{\pi \cdot a \cdot t}} \cdot \exp\left(-\frac{x^2}{4 \cdot a \cdot t}\right)\right). \tag{2.2.65}$$

There is no statement about the heat transfer between environment and body. This is described by the so-called Biot number:

$$\text{Bi} = \frac{h \cdot L}{\lambda} \tag{2.2.66}$$

h [W m^{-2} K^{-1}] is the heat transfer coefficient, λ [W m^{-1} K^{-1}] is the thermal conductivity, and L [m] is a characteristic length. The larger Bi is, the stronger a thermal shock will be, for example.

If the heat transfer occurs only very hesitantly (Bi $\to$ 0), one obtains [18, 29].

$$\theta = 1 - \exp(-\text{Bi} \cdot \text{Fo}) \cdot (\text{Bi} \leq 1) \tag{2.2.67}$$

If now a semi-infinite plate with a time-variable surface temperature is considered and both heat transfer and temperature conduction have a remarkable influence, the surface of the body is formally shifted outward by the fictitious length $x^* = \lambda/h$ [m] while maintaining its physical properties. It is possible to describe the time progression of the temperature as a whole, i.e. also at the surface ($x = 0$), as a consequence of the temperature jump of the environment to the value (count) $T = T_U$ approximately [30] (refer to Fig. 2.20).

The temperature $\theta(x, t)$ of the plate changes in the example shown in Fig. 2.20 with x^* and the suited corrected time t^*:

$$x^* = \lambda/h \text{ [m]}, \quad t^* = \left(\frac{\lambda}{h}\right)^2 \cdot \frac{1}{4 \cdot a} \text{ [s]}. \tag{2.2.68}$$

like [29]:

$$\theta = 1 - \text{erf}\left(\frac{x + x^*}{2 \cdot \sqrt{a \cdot (t + t^*)}}\right) = 1 - \text{erf}\left(\frac{1 + 1/\text{Bi}}{\sqrt{4\text{Fo} + (1/\text{Bi})^2}}\right). \tag{2.2.69}$$

The change in temperature over time is obtained by deriving Eqs. 2.2.67 and 2.2.69 with respect to time:

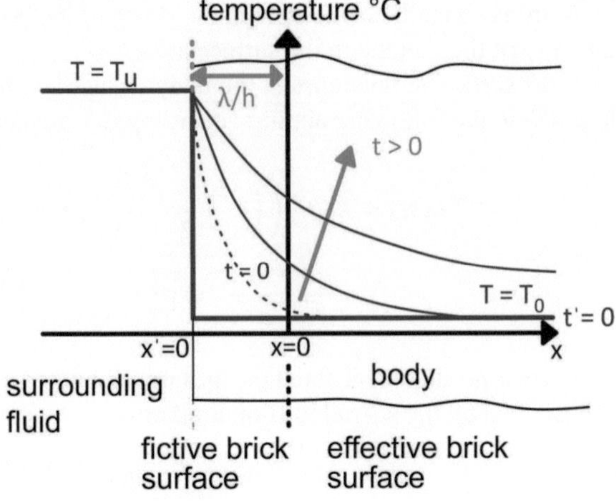

Fig. 2.20 Auxiliary construction for calculation of heating of a plate with remarkable influence of heat transition from the environment [18]

$$\left(\frac{d\theta}{dt}\right)_X = \frac{\text{Bi} \cdot \text{Fo}}{t} \cdot \exp(-\text{Bi} \cdot \text{Fo}) \tag{2.2.70}$$

and

$$\left(\frac{d\theta}{dt}\right)_X = \frac{\left(1 + \text{Bi}^{-1}\right) \cdot 4Fo}{2 \cdot t \cdot \left(4\text{Fo} + \text{Bi}^{-2}\right)^{3/2}} \cdot \exp\left(-\frac{1 + \text{Bi}^{-1}}{\sqrt{4\text{Fo} + \text{Bi}^{-2}}}\right)^2 . \tag{2.2.71}$$

The temporal heating and cooling can thus be calculated in a simple way and well approximated.

2.3 Introduction to Thermomechanics

2.3.1 Thermal Expansion

If a small force F [N] acts on a bar of length l and cross-sectional area A, it expands by the relative length

$$\frac{\Delta l}{l} = \frac{F}{A \cdot E} \equiv \varepsilon = \frac{\sigma}{E} \tag{2.3.1}$$

This law, first formulated by Hook, describes the relationship between stress σ [N/m^2], modulus of elasticity E [N/m^2] and strain ε if there is a linear elastic load.

As a result of the elongation, the diameter d of the bar decreases by

$$\frac{\Delta d}{d} = -v \cdot \frac{\Delta l}{l} \tag{2.3.2}$$

v is the transverse contraction or Poisson's ratio. The resulting relative volume change is:

$$\frac{\Delta V}{V} = \frac{\Delta l}{l} + 2 \cdot \frac{\Delta d}{d} = \varepsilon(1 - 2v) = \frac{\sigma}{E} \cdot (1 - 2v). \tag{2.3.3}$$

Since $\Delta V > 0$, v is < 0.5. For refractory ceramics $v \sim 0.2$.

If the same pressure $- \Delta P = \sigma$ acts on all sides of the body, its relative change in volume is

$$\frac{\Delta V}{V} = -3 \cdot \frac{\Delta P}{E} \cdot (1 - 2v). \tag{2.3.4}$$

With the compressibility

$$\kappa = \frac{1}{V} \cdot \frac{\Delta V}{\Delta P} \ \left[\text{m}^2/\text{N}\right], \tag{2.3.5}$$

one obtains

$$\kappa = \frac{3}{E} \cdot (1 - 2v) \ \left[\text{m}^2/\text{N}\right]. \tag{2.3.6}$$

As its temperature changes, the volume of a body changes:

$$\frac{\Delta V}{V} = 3 \cdot \alpha_T \cdot \Delta T \tag{2.3.7}$$

The linear change is

$$\frac{\Delta L}{L} = \alpha_T \cdot \Delta T. \tag{2.3.8}$$

$\alpha_T \sim 10^{-5} \ \text{K}^{-1}$ is the coefficient of linear thermal expansion.

If the expansion of the body is suppressed, the thermal stress is generated

$$\sigma = \frac{3 \cdot \alpha_T \cdot E \cdot \Delta T}{1 - 2v} \ [\text{N/m}^2 \equiv \text{Pa}]. \tag{2.3.9}$$

For one-dimensional stress, the denominator is $(1 - v)$ and the numerator omits the 3.

2.3.2   Fracture Mechanics

For practical reasons, thermal fracture mechanics concentrates on material stress by tensile forces and less on shear and torsional stress. A crack then propagates perpendicular to the maximum acting stress. We follow this simplification in the introductory to this chapter.

2.3.2.1   Linear Elastic Fracture Mechanics (LEFM)

Linear elastic fracture mechanics (LEFM) preferentially describes the fracture behavior of brittle materials, such as porcelain and glass. Prerequisites are:

- Reversible expansion, proportional to the occurring stress.
- Any irreversible deformation results in a fracture.
- The fracture energy is used exclusively to create new surfaces.

The characteristic parameter of linear elastic fracture mechanics is the critical fracture stress σ_f [MPa] at which a material fails. This concept is only approximately applicable to refractory materials, especially at elevated temperature because the above-mentioned conditions are not met. For example, a distinct process zone forms around the crack. The crack tip is not defined but microcracks already form before it as a result of deformations and after it interlocking of the grains with each other impedes crack propagation (refer to Fig. 2.21). These effects consume energy. A closer look at the individual processes therefore leads to nonlinear elastic fracture mechanics where the fracture energy G_f [J/m^2] is the focus of attention. For this reason, the separate presentation of linear elastic fracture mechanics is largely dispensed with and reference is made to the available relevant literature, e.g. [31].

2.3.2.2   Nonlinear Elastic Fracture Mechanics (NLEFM)

The reasons to extend linear elastic fracture mechanics were explained in the previous chapter.

A variety of measuring methods are used to test the mechanical properties of refractory materials [32]. Here, the wedge splitting test [33], developed by Tschegg, is used as the experimental basis because this method is particularly suitable for the very heterogeneously structured refractory material in that it provides a complete force/path diagram and, thus, enables the application of the fundamentals given in the following part in a relatively simple manner.

The fracture toughness σ_f of a cracked material is given by the so-called stress intensity factor K_{IC} described (LEFM):

$$K_{\text{IC}} = Y \cdot \sigma_f \cdot \sqrt{a} \left[\text{MPa m}^{-1/2}\right] \tag{2.3.10}$$

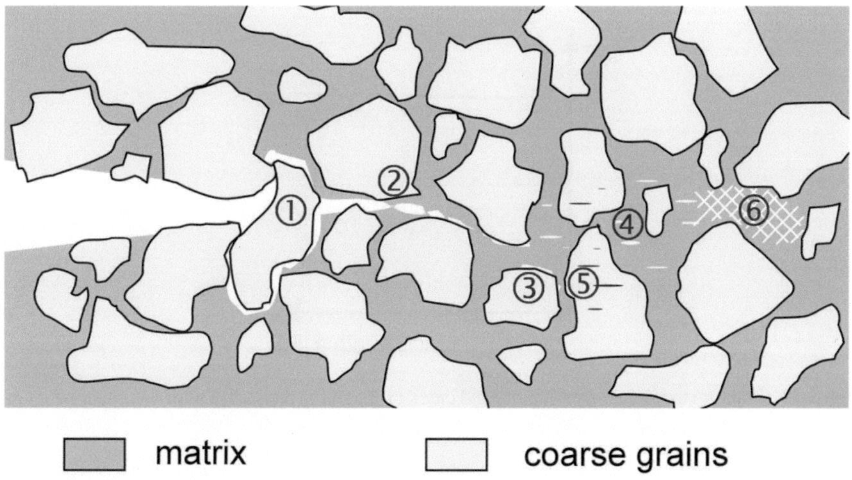

① formation of grain bridges
② ductile material bridge formation (specifically at high temperatures)
③ delamination
④ formation of microcracks in the matrix
⑤ formation of microcracks in the grain
⑥ plastic deformation

Fig. 2.21 Schematic illustration of the process zone at the crack front [18]

Here, the maximum tensile stress σ_f acts on the crack of length a [m]. Y stands for the parameter that takes into account the specimen shape and crack pattern. It can be calculated [34, 35].

If K_{IC} is exceeded, the crack starts to grow regardless of its length. The fracture is usually catastrophic because the stored elastic energy at the start of the crack is already greater than the fracture energy required to create the new surface.

Similarly, the so-called R-curve is calculated from a measured stress/strain curve in the case of nonlinear behavior of the material [35, 36]. It starts when the K_{IC} value is exceeded and reaches the limit value K_{IR}^{∞} with increasing stress which is likewise a characteristic material parameter (refer to Fig. 2.22).

A typical force/displacement progression, measured with the wedge-gap method, is shown in Fig. 2.23.

The fracture energy G_f and the adequate fracture work γ_{WOF} are calculated as the integral under the load/displacement measurement process:

$$G_f = \frac{1}{A} \int_0^{\delta} F(a) \cdot \mathrm{d}\delta \ [\mathrm{J/m^2}]. \tag{2.3.11}$$

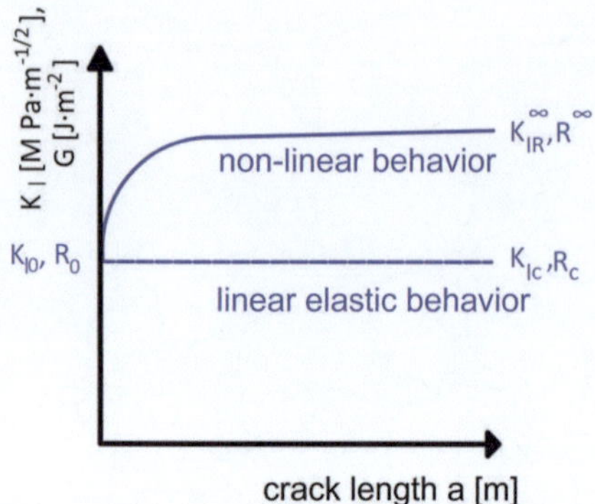

Fig. 2.22 Fracture toughness (resistance) and crack propagation resistance with nonlinear behavior [18]

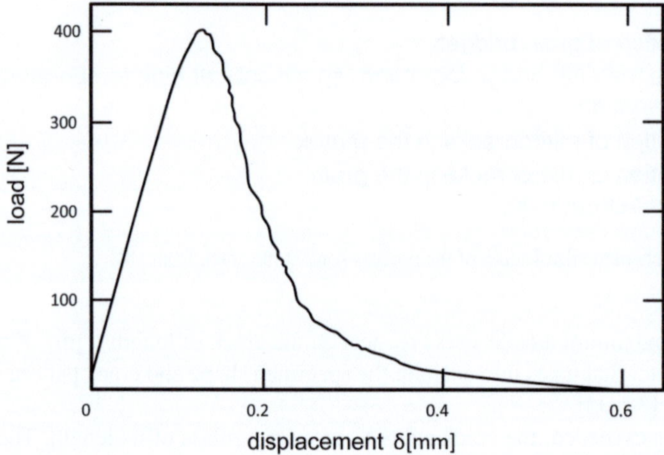

Fig. 2.23 Force/displacement curve

The work of fracture γ_{WOF} (correctly) refers to the entire newly created ligament area, i.e. to $2A$, which is why the following applies:

$$\gamma_{WOF} = G_f \big/ 2 \ [\text{J/m}^2].\qquad(2.3.12)$$

Looking at the progression in Fig. 2.23, in coarse materials with nonlinear elastic behavior, the crack does not yet appear visibly, but clearly before the maximum load (stress) is reached [37] and can be calculated from the measured force/path progression [38]. The energy required for the formation of the crack was calculated by Griffith [39, 40] to be

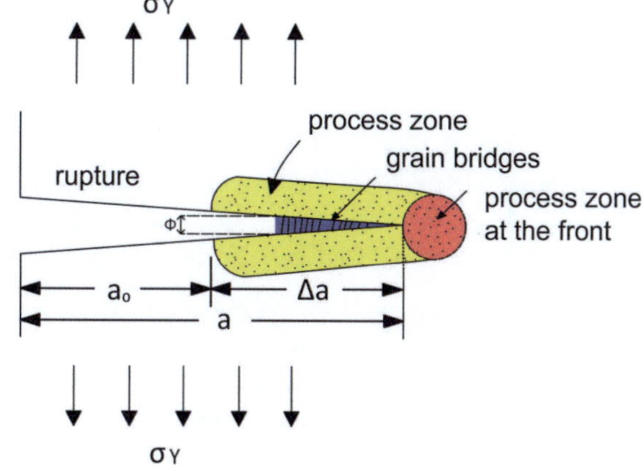

Fig. 2.24 Crack opening with process zone [19]

$$\gamma_{\text{NBT}} = \frac{K_{\text{IC}}^2}{2 \cdot E} = \frac{a_0 \cdot \sigma_f^2}{2 \cdot E} \cdot Y^2 \ [\text{J/m}^2]. \tag{2.3.13}$$

Here γ_{NBT} stands for the measurement technique used, namely the 3-point bending test with a notch of length a_0 [m] (**N**otch **B**ending **T**est). The parameter $Y^2/2$ will be considered as 1 in the future. It is a geometry-dependent parameter [34–36].

Figure 2.24 shows the schematic illustration of the crack opening Φ.

The process zone surrounding the crack is clearly visible in which, in addition to the creation of a new surface, further work is carried out, e.g. for the formation of microcracks, branching and separation of mechanical connections of the grains among one another [19].

Hasselmann [20, 41] calculated the crack propagation using the approach according to which, in the case of nonlinear elastic fracture behavior, the elastic energy stored in the material must be greater than the required fracture energy in the case of ideal elastic behavior in order to enable the additional work just described. It shows the relationship in Fig. 2.25, according to which a crack of length $\mathbf{a_0}$, after passing through the ΔT—minimum $\mathbf{a_m}$, only comes to a standstill on the subsequent branch of the curve, which is drawn in dotted lines and rises again. The critical temperature difference ΔT_C, which leads to crack growth Δa of already existing incipient cracks, is [20]:

$$\Delta T_C = \left(\frac{2 \cdot \gamma}{\pi \cdot \alpha_T^2 \cdot E_0 \cdot a_0} \right)^{1/2} \cdot \frac{E_0}{E} \ [\text{K}], \tag{2.3.14}$$

with

$$\frac{E_0}{E} = 1 + 2\pi \cdot N \cdot a_0^2. \tag{2.3.15}$$

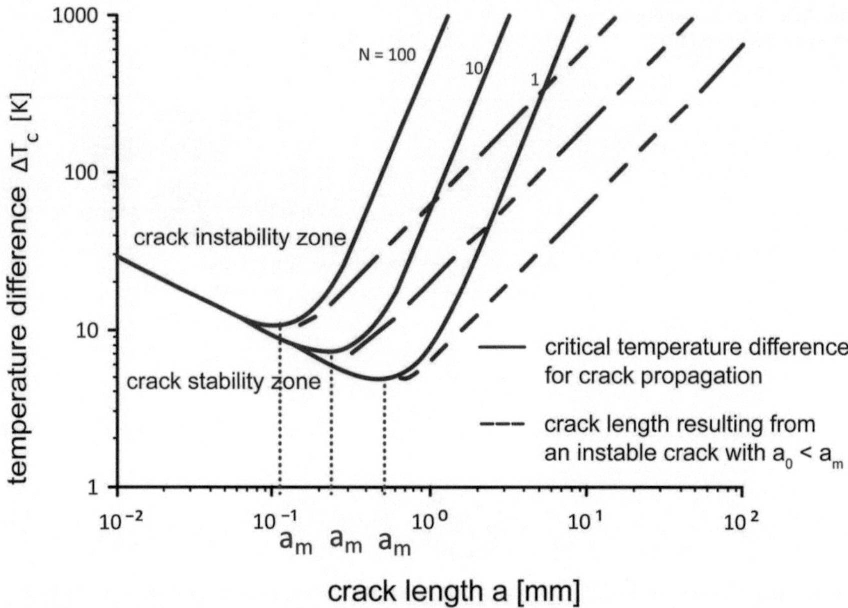

Fig. 2.25 Lengthening of crack length as a function of the critical temperature difference ΔT_C [18, 20]

γ [J/m^2] is the fracture surface energy or fracture toughness, α_T [1/K] is the coefficient of thermal expansion, E [GPa] is the current modulus of elasticity, E_0 [GPa] is the modulus of elasticity in the crack-free state, a_0 [m] is the incipient crack length at the start of the stress, and N [m^{-2}] is the initial crack density. Furthermore, the final crack length is inversely proportional to the initial crack length and crack density N: $a_f \sim 1/(a_0 \cdot N)$. A large number of long incipient cracks ($a \geq a_{0m}$) need not be detrimental because the larger ΔT_C is, the better the thermal shock resistance is despite pre-existing incipient cracks. Especially with rapid quenching, the stored elastic energy is often still so small that it is just enough to form incipient cracks, which relieve the stresses and do not grow further. Then several quenching cycles are required for fracture.

Whereas according to Griffith [40] the fracture must always occur as soon as the critical value K_{IC} is exceeded, an additional increase of the depicted driving temperature difference ΔT_C from the dotted line to the solid line above is now required. Then the crack continues to grow stepwise with the increase of ΔT_C. Consequently, if a crack comes to a halt before the dimension of the material is exceeded, it does not break.

The descending branch of the ΔT_C (a) progression describes the "kinetic" crack propagation at which, starting with the initial crack a_0, crack propagation is unstoppable and can terminate at the earliest when at least a_m is reached. The fracture work is calculated according to Griffith [40] to

$$\gamma_{\text{WOF}} = \frac{L_C \cdot \sigma_f^2}{E} \ [\text{J/m}^2]. \tag{2.3.16}$$

$L_C \geq a_0$ is the critical length of the body at which it will break if the crack does not come to a stop first. If comparing Eqs. (2.3.13) and (2.3.16) with each other, we obtain:

$$L_C = a_0 \cdot \frac{\gamma_{\text{WOF}}}{\gamma_{\text{NBT}}} \ [\text{m}], \tag{2.3.17}$$

where, according to experience, γ_{WOF} exerts the greater influence on the fracture (rupture) behavior. The greater this ratio, the better the so-called flexibility of the refractory material. Flexible behavior describes the ability of the material to react to external stress without failure. It should be as large as possible be [42, 43].

With ideal elastic behavior, the work of fracture $\gamma_{\text{WOF}} = \gamma$, i.e. equal to the surface energy. In the case of coarse ceramics, however, the dissipative effects in the process zone, such as the formation of microcracks, mechanical interlocking of individual grains, crack branching, etc. [39], add up.

Taking Eq. (2.3.12) into account, one obtains directly from Eq. (2.3.16) the damage tolerance parameter R'''' (thermal stress parameter) for the kinetic cracking process [20, 41]:

$$R'''' = \frac{G_f \cdot E}{2 \cdot \sigma_f^2} \ [m] \tag{2.3.18}$$

G_f [J/m^2] is the fracture energy, σ_f [MPa] is the initial strength and E [GPa] is the modulus of elasticity. The larger R'''' is, the more crack initiation is hindered.

The right ascending branch in Fig. 2.24 shows the steady-state crack propagation. The damage tolerance parameter (thermal shock resistance parameter) Eq. (2.3.19) applies which becomes clear by comparison with Eq. (2.3.14):

$$R_{\text{st}} = \sqrt{\frac{G_f}{2 \cdot \alpha_T^2 \cdot E_0}} \ [\text{K/m}^{1/2}] \tag{2.3.19}$$

With increasing R'''', the formation of incipient cracks becomes more difficult and with increasing R_{St} their growth. The possibilities, which result from this for the enhancement of the thermomechanical behavior of refractory materials, are discussed in Chap. 12 *"Thermal Shock Resistance."*

References

1. Frohberg, M.G.: Thermodynamics for Material Engineers and Metallurgists, 2nd edn. Deutscher Verlag für Grundstoffindustrie, Leipzig (1994)
2. Knacke, O., Kubaschewski, O., Hesselmann, K.: Thermochemical Properties of Inorganic Substances I/II, 2nd edn. Springer, Verlag Stahleisen, Berlin (1991)
3. Elliott, J.F., Gleiser, M., Ramakrishna, V.: Thermochemistry for Steelmaking I/II. Addison-Wesley Publishing Company (1960/63)
4. Kubaschewski, O., Alcock, C.B., Spencer, P.J.: Materials Thermo-Chemistry. Pergamon Press, New York (1993)
5. K. Hack: FACTSage. Thermfact & GTT Technologies (2007). www.factsage.com
6. Elliott, J.F., Gleiser, M., Ramakrishna, V.: Thermochemistry for Steelmaking I/II. Addison-Wesley Publication, New York (1960/63)
7. Schenk, H., Steinmetz, E.: Effective Parameters of Accompanying Iron Solutions and Their Interrelationships. Verlag Stahleisen, Düsseldorf (1966)
8. Volmer, M.: Kinetics of Phase Formation. Verlag Th. Steinkopf, Dresden (1939)
9. Cahn, R.W.: Physical Metallurgy. North Holland, Amsterdam (1965)
10. C. Gerthsen, D. Meschede: Physics. Springer, Berlin (2005)
11. Pawlek, F.: Metallurgy. Walter de Gruyter Verlag Berlin, New York (1983)
12. Mugele, F., Baret, J.C.: Electrowetting from the basis to application. J. Phys. Condensed Matter **17**, R705–R774 (2005)
13. Aneziris, C.G., Homola, F.: Electrically assisted high temperature wettability in oxide and carbon-based refractories. Refractories Manual Interceram 52–56 (2004)
14. Riaz, S., Mills, K.: Can the Application of Electric Potential Increase Refractory Life? Cfi/Reports DKG 79 (2003) No. 3 E 47–50
15. Darken, L.S., Gurry, R.W.: Physical Chemistry of Metals. McGraw-Hill Education, New York (1953)
16. Bird, R.B., Stewart, W.E., Lightfoot, E.N.: Transport Phenomena. Wiley, New York (1965)
17. Friedrichs, H.A., Knacke, O.: Process Engineering Basic Types of Isothermal Reactions in Blown Through and Overflown Fills. Research Reports of the State of North Rhein-Westphalia No. 2240, Verlag Opladen (1973)
18. Brochen, E.: Measurement and Modeling of Thermal Shock Resistance of Refractory Materials. Dr.-Ing. Dissertation. Bergakademie TH Freiberg in Freiberg, Germany (2011)
19. Sakai, M.: Fracture mechanics of refractory materials. Taikabutso Overseas **8**(2), 4–12 (1988)
20. Hasselman, D.P.H.: Role of fracture toughness connected to thermal shock resistance. Rep. Deutsche Keramische Gesellschaft **54**(6), 195–201 (1977)
21. Brüggmann, C.: A Contribution to slagging of MgO in secondary metallurgical slags. Dr.-Ing. Dissertation, TU Bergakademie Freiberg, Saxony, Germany (2011)
22. Newman, A.B.: The drying of porous solids: diffusion calculations. Trans. Am. Inst. hem. Eng. **27**, 310–333 (1931)
23. Carslaw, H.S., Jaeger, J.C.: Conduction of Heat in Solids, 2nd edn. Oxford University Press, Oxford (1959)
24. Crank, J.: The Mathematics of Diffusion., 2nd edn. Oxford University Press, Oxford
25. Oeters, F.: Metallurgy in Steelmaking. Verlag Stahleisen/Springer, Berlin (1989)
26. Levich, V.: Physiochemical Hydrodynamics. Longman Higher Education (1962)
27. Grassmann, P.: Basics of Chemical Technology. Salle + Sauerländer Verlag, Munich (1982)
28. Wurtz, M.: Theory and Practice of Vacuum Technology. Vieweg, Braunschweig (1965)
29. Kneer, R.: Heat and Material Transfer. Lecture of the Professorship for Heat and Material Transfer at the Rheinisch-Westfälische Technische Hochschule (RWTH) Aachen, Germany (2005)
30. Gröber, H., Erk, S.: Basics Laws of Heat Transfer. Springer, Berlin (1933)
31. Schott, G.: Material Fatigue, pp. 64–117. VEB Deutscher Verlag für Grundstoffindustrie, Leipzig (1985)

32. Routschka, G., Krause, O. (ed.): Refractory Materials and Engineering, DIN Standards, 2nd edn. Vulkan-Verlag (2010) specifically: DIN 51068 Testing of Ceramic Raw Materials and Materials—Determination of Resistance to Sudden Thermal Shock in Refractory Bricks

33. Tschegg, E.K.: Test equipment for determination of fracture mechanical characteristic values. Austrian Patent Specification AT 390328 (1986)

34. Irwin, G.R.: Fracture mechanics. In: Proceedings of 1st Symposium on Naval Structural Mechanics, Stanford University, Pergamon, New York, pp. 557–592 (1960)

35. Tan, D.M., Rieder, K., Harmuth, H., Tschegg, E.K.: Determination of R-curves of ordinary refractory ceramics from measurements of a new wedge splitting test method. Progress Rep. DKG **10**(3), 71–76 (1995)

36. Harmut, H.: Stability of crack propagation associated with fracture energy determined by wedge splitting specimen. Theory Appl. Fracture Mech. **23**, 103–108 (1995)

37. Hillerborg, A.: The theoretical basis of a method to determine the fracture energy G_F of concrete. RILEM Technical Committee 50, Fracture Mechanics of Concrete (1985)

38. Gera, S.: Development of a selection process for determining mechanical material parameters via fem simulation of a fracture mechanical testing. Diploma Thesis at the Institut für Gesteinshüttenkunde der Montanuniversität Leoben, Austria, June 2005

39. Griffith, A.A.: The phenomenon of rupture in liquids and solids. Philos. Trans. R. Soc. A **221**, 163–198 (1920)

40. Griffith, A.A.: Theory of rupture. In: 1st International Congress for Applied Mechanics, Delft, NL, pp. 55–63 (1924)

41. Hasselman, D.P.H.: Unified theory of thermal shock fracture initiation and crack propagation in brittle ceramics. J. Am. Ceramic Soc. **52**, 600–604 (1969)

42. Harmuth, H., Tschegg, E.K.: Fracture mechanical characterization of coarse ceramic refractory materials. Veitsch-Radex Rundschau **1–2**, 465–480 (1994)

43. Buchebener, G., Pirker, S.: New high-tech products for utilization in the converter. Veitsch-Radex Rundschau **2**, 3–14 (1996)

44. Oelsen, W., Schürmann, E.: Ironworks. Phys.-Chem. Basics 59–153. Verlag W. Ernst & Sohn, Berlin, Germany (1961)

Chapter 3
Steel Production

3.1 From Ore to Steel I

3.1.1 Production Route

Figure 3.1 shows schematically the production route of the liquid steel:

- Blast furnace (pig iron production by reduction of iron ore),
- Transport ladle (desulfurization),
- LD-converter (combustion of carbon, silicon, phosphorus), or
- Electric arc furnace (melting of scrap),
- Secondary metallurgy (setting the required final analysis).

  Continuous casting (solidification of the liquid steel).

3.1.1.1 Blast Furnace Process

The structure of the blast furnace is shown in Fig. 3.2. The burden and coke (approx. 500 kg/t pig iron) are charged via the throat of the blast furnace and the blast furnace gas is drawn off.

Two reactions are decisive for the description of the blast furnace process [1], which is complicated in detail:

- Reduction of iron ore:

$$(FeO) + \{CO\} = [Fe] + \{CO_2\} \tag{3.1}$$

- Conversion of coking coal:

$$\langle C \rangle + \{CO_2\} = 2\{CO\}. \tag{3.2}$$

© The Author(s), under exclusive license to Springer Nature Switzerland AG 2024
J. Pötschke, *Refractory Fundamentals in Metallurgical Practice*,
https://doi.org/10.1007/978-3-031-63709-4_3

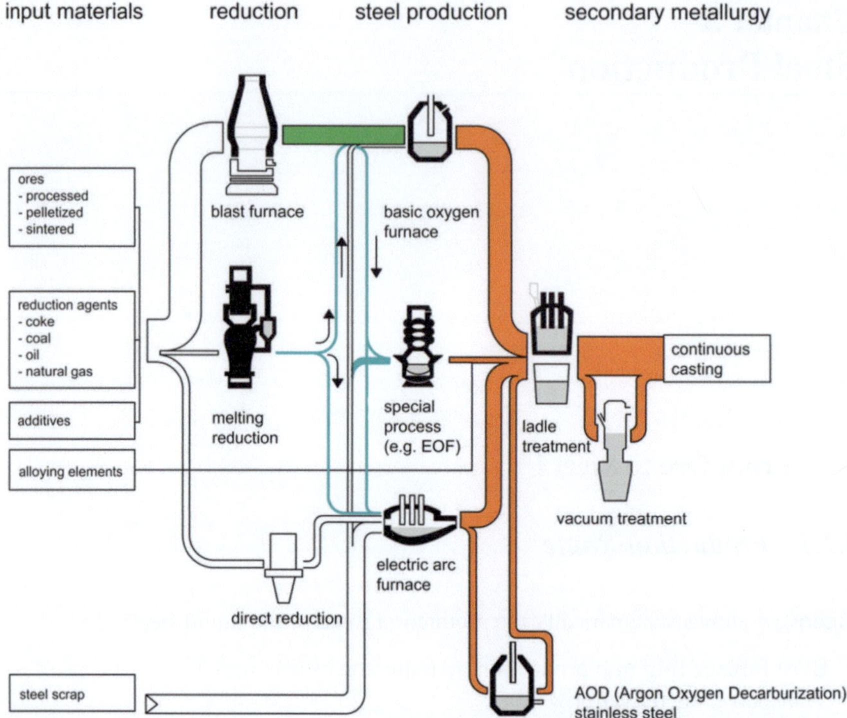

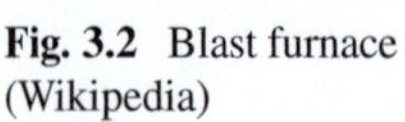

Fig. 3.1 Production line of liquid steel up to continuous casting (Stahl and Eisen 2007)

Fig. 3.2 Blast furnace (Wikipedia)

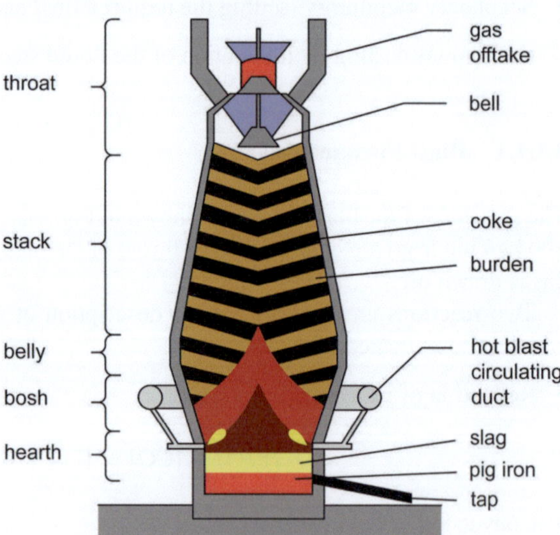

The second equation describes the so-called Boudouard reaction. As the temperature rises, its equilibrium shifts toward the CO gas. At low temperatures, e.g. 500 °C, CO decomposes with the excretion of graphite which can destroy the structure of the blast furnace lining in the shaft.

Another reaction sequence, the so-called alkali cycle, can influence the service life of the lining. The blast furnace burden contains alkalis. Their thermal stability is low, so that above 900 °C gaseous alkalis with a steam pressure of approx. 10^{-5} atm are formed. They reach the higher and colder areas with the gas flow, where they react to form compounds such as carbonate, cyanide, oxide and silicate, which in turn sink with the burden into hot areas. Although a part leaves the blast furnace with the slag and via the throat gas, a maximum enrichment exists in the range of about 1200 °C [2].

An example is:

$$\langle K_2O \rangle + \{CO\} = \{K\} + \{CO_2\} \tag{3.3}$$

If the gas penetrates into the structure of the lining, leucite ($KAlSi_2O_6$) (refer to Sect. 10.1.7) can be formed which destroys the structure by way of its high crystallization pressure.

If the mechanical abrasion caused by the burden sliding down the blast furnace shaft is also taken into account, the requirements for the lining of the shaft are as follows: High-fired, low-iron oxide fireclay bricks with the lowest possible porosity and a high degree of hardness are installed. The same applies to the lining of the shaft of a plant for the **direct reduction of** iron ore (refer to Sect. 3.6.6 Methane). Cooling boxes in the hot area of the blast furnace additionally reduce the wear caused by high temperatures.

Microporous graphite is used in the hearth because the pig iron is almost saturated in carbon (about 4.5% by weight) and the lining is hardly attacked.

The lining of the blast furnace channel (trough) with silicon carbide in SiC castable reduces corrosion by atmospheric oxygen.

The mechanisms addressed, which can lead to the destruction of the lining, are briefly explained.

3.1.1.2 Carbon Monoxide-Bursting

The thermodynamic stability of a phase is described by its free enthalpy of formation (refer to Sect. 2.1.3.3).

$$\Delta G = \Delta G^0 + RT \ln K. \tag{3.4}$$

At equilibrium, $\Delta G = 0$. ΔG^0 is the free standard enthalpy and K is the equilibrium constant. It follows from the chemical reaction, e.g. $2CO = C + CO_2$ ($T = $ const.) to

$$K = \frac{a_C \cdot p_{CO_2}}{p_{CO}^2}.$$ (3.5)

If the pure, solid-phase graphite is excreted, its activity is in equilibrium $a_C = 1$. However, if excretion is impaired, e.g. by delayed nucleation or an environment, which mechanically impedes growth, the activity of the impaired phase may considerably exceed the value 1: $a_C \gg 1$ (refer to Sect. 2.1.3.3).

The free enthalpy (Eq. 3.4) represents the external work $\Delta G = (\pi - 1) \cdot \Delta V$, which results from the volume change $\Delta V = (V_C + V_{CO_2} - 2V_{CO})$ as a consequence of the excretion of the new phase graphite. $(\pi - 1)$ describes the so-called crystallization pressure. Thus, Eq. (3.4) now reads

$$\Delta G^\circ + (\pi - 1)\left(V_C + V_{CO_2} - 2V_{CO}\right) \cdot R_1/R_2 + R_1 T \ln K_C = 0 \ [\text{J/FU}]$$ (3.6)

$R_1 = 8.31$ J/mol K and $R_2 = 84.8$ atm cm^3/mol K are the ideal gas constants. The gas exerts no mechanical force on its surroundings. Consequently, the crystallization pressure π is (refer to Eq. 2.1.22, Sect. 2.1.3.2):

$$\pi - 1 = \frac{-\Delta G^\circ - RT \ln K_C}{V_C \cdot R_1/R_2} \ [\text{atm/FU}].$$ (3.7)

The calculation is made for 500 °C because this corresponds to the temperature of the standardized test for CO resistance of refractory materials (ISO 12676, ASTM C 288). The following is taken from literature [3] for the Boudouard reaction

$$2CO = C + CO_2, \ \ \Delta G^\circ (773 \, K) = -35{,}522 \, \text{J/mol}.$$ (3.8)

As catalysts of CO decomposition, iron oxide, but above all metallic iron, which is always found next to the graphite excretion, have been detected [4]. Furthermore, small gas portions of hydrogen also accelerate the CO decay [5, 6].

The Baur–Glaessner diagram (Fig. 3.3) [1] shows that at 500 °C magnetite (Fe$_3$O$_4$) can be reduced directly to metallic iron only if the volume shares of CO and CO$_2$ are just equal. Then the equilibrium constant of the reaction (3.8) at the total pressure is 1 atm:

$$K_c = \frac{P_{CO_2}}{P_{CO}^2} \cdot a_C = 2 \cdot a_C$$ (3.9)

The required activity of the supersaturated carbon is calculated with $\Delta G^\circ = -R_1 T \cdot \ln(2 - a_C)$ to $-35{,}522 = -8.31 \times 773 \cdot \ln(2 - a_C)$, i.e. $a_C = 126$.

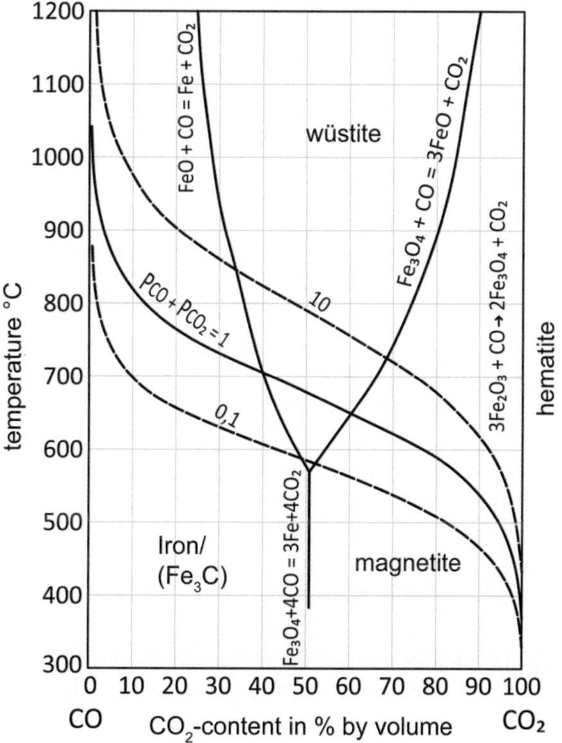

Fig. 3.3 Influence of temperature, gas pressure and CO/CO$_2$ ratio on reduction capacity of carbon [1]

The molar volume of the graphite is $V = 5.3$ cm^3/mol. With $R_1/R_2 = 0.1$ [J/atm cm^3] and $a_C = 1$, it follows from Eq. (3.7) that the crystallization pressure is

$$\pi - 1 = \frac{+35{,}522 - 8.31 \times 773 \times \ln 2}{5.3 \times 0.1} = 58 \times 10^3 \text{ atm.} \qquad (3.10)$$

It destroys the structure of the refractory material, whereby it depends on the form of growth of the crystallizing graphite. If it is equiaxed (onion-shaped), the structure is only slightly damaged, but if single-crystalline needles (whiskers) are formed, their large space requirement and extraordinary mechanical strength drive the structure apart [5] (also refer to Sect. 10.1.1).

3.1.1.3 Alkali-Bursting

As an example of alkali decay, the bursting of fireclay as a result of the formation of leucite (K$_2$O·Al$_2$O$_3$·4SiO$_2$) at 1200 °C is calculated [7] because the content of potassium in the blast furnace (throat) gas is significantly higher than that of sodium [2]:

$$K_2O + Al_2O_3 + 4SiO_2 = 2KAlSi_2O_6;$$

$$\Delta G_L^\circ(1473) = -44.7 \times 10^3 \, \text{J/mol}$$

$$\Delta V = \left(2V_L - V_{K_2O} - V_{Al_2O_3} - 4V_{SiO_2}\right)$$
$$= (177.2 - 40.5 - 25.7 - 4 \times 23.1) = 17.9 \, \text{cm}^3/\text{mol}$$

Both phases, mullite with 72% Al_2O_3 and leucite, are stoichiometric.

$$\pi - 1 = \frac{-\Delta G_L^\circ}{\Delta V \cdot R_1/R_2} = \frac{44,700}{17.9 \times 0.1} = 25 \times 10^3 \, \text{atm.} \qquad (3.11)$$

The activity of the supersaturated leucite is $a_L = \exp(44,700/(8.31 \times 1473)) = 38$.

For the decomposition by excretion of nepheline ($NaAlSi_2O_6$) comparable conditions apply (refer to Sect. 10.1.6).

3.1.1.4 Hearth Wear

The premature wear of the refractory lining in the hearth of a blast furnace can be counteracted by the use of titanium ore [8]. The addition takes place in the burden or advantageously together with the combustion air in the tuyere level. Ilmenite ($FeTiO_3$) or, better, synthetically produced RUTILIT (rutile, TiO_2, approx. 35% by weight Ti) with fine grains are used.

Titanium is known to react with carbon dissolved in pig iron (RE) to form titanium carbide $\langle TiC \rangle$ and with dissolved nitrogen to form titanium nitride $\langle TiN \rangle$. With the participation of the iron, compounds of the type $\langle Fe_x Ti_y C_z N_{1-z} \rangle$ can be formed. According to what can be observed, they accumulate in the zones of the highest wear, strengthen there the partially worn wall and, consequently, prolong the operating time of the blast furnace.

Approx. 500 kg of coke and 2300 kg of wind are required to produce one ton of pig iron. The resulting blast furnace gas consists of approx. 45–60% by volume N_2, 20–30% by volume CO and 20–25% by volume CO_2 at the throat. Reduction of the iron ore by coke produces about 300 kg of slag/tRE (pig iron). Both liquid phases collect in the hearth and are separated from each other during tapping. Carbon bricks are used to line the hearth. The pig iron is almost saturated in carbon and the wear is, therefore, less. During tapping, the temperature is approx. 1400–1550 °C [9, 10].

The following examination [11] attempts to explain the protection process quantitatively on the basis of thermochemical calculations and a simple kinetic model, for which the highly enriched RUTILIT with 80–85% by weight TiO_2 and max. 15% by weight FeO is used. In principle, the result also applies for other titanium ores, e.g. ilmenite.

An initial thermodynamic information provides the stoichiometric reaction equation, $P = 1$ atm:

$$2TiO_2 + 5C + 0.5N_2 = TiC + TiN + 4CO$$
$$(\Delta G^\circ\ 1470\,°C = -\,130\,kJ/FU).$$

Considering 100 kg (TiO_2) in the slag, its activity at 1470 °C is $a_{TiO_2} = 0.16$ [7]. At a total pressure of 4 atm at the tuyere level of the blast furnace, the compounds TiC and TiN are formed only if the activity of the carbon $a_C > 0.92$. Consequently, the formation of the titanium compounds is not self-evident.

The calculations, which are more precisely adapted to practical conditions, are carried out with "FactSage" 6.2/Equilib. [7]. 10 or 100 kg of titanium ore per ton of pig iron are used. The gas composition just above the hearth, i.e. in the melting zone of the blast furnace, is estimated to be 50–70% by volume N_2 and 30–50% by volume CO. Since the gas pressure is about 4 atm, the gas equilibrium shifts to counts of 2.5–3.5 atm CO and correspondingly 1.5–0.5 atm N_2. The temperature can rise to about 2000 °C in the melting zone. 54.3 kg carbon/t pig iron corresponds to C-saturation at 1470 °C. Depending on the sulfur and silicon content, pig iron contains 4.2–4.8% by weight carbon dissolved [10] (refer to Sect. 2.1.3.3, Fig. 2.4). The average flow speed of the pig iron melt in the hearth is approx. 0.2 cm/s [9].

If we include the slag in the consideration, which is 300 kg/1000 kg pig iron and add 100 kg of titanium oxide, its approximate composition in kg is: 135 CaO, 15 MgO, 120 SiO_2, 30 Al_2O_3, 100 TiO_2, which adds up to a total of 400 kg. In addition, 6 kg of nitrogen is considered, which is introduced with the blast. Upon carbon saturation (54.3 kg C), 1470 °C and 4 atm gas pressure, the calculation results in no excretion of TiC, TiN and titanium carbonitride [7].

Due to its higher specific gravity, the pig iron freshly reduced in the shaft falls in the form of individual droplets through the liquid slag into the pig iron bath. In the process, both phases react with each other. No information is available on the size, i.e. the weight of the droplets. Depending on the weight of the iron, there will be varying concentration ratios in both phases, slag and metal. For example, if only 100 kg pig iron reacts with the above slag, which contains 100 kg/t TiO_2, which is 25% TiO_2, the carbonitride $\langle Fe_{0.2}Ti_{0.8}C_{0.4}N_{0.6}\rangle$ is formed even before carbon saturation is achieved. If only 10 kg of pig iron reacts with this slag and following the mass balance, the entire mass (material) can react to form titanium carbonitride, leaving no liquid pig iron left.

If only 10 kg/t TiO_2 is used, which is 3.2%, a "droplet" with 10 kg mass will not be completely transformed, but titanium carbonitride $\langle Fe_{0.2}Ti_{0.8}C_{0.4}N_{0.6}\rangle$ is formed even before carbon saturation is reached. If, however, 100 kg pig iron reacts with 10 kg TiO_2/t pig iron, no titanium carbonitride is formed even upon carbon saturation.

In summary, the calculation states that if sufficient TiO_2 is used, the formation of titanium carbonitride in pig iron droplets can be expected because the concentration of titanium dissolved in the pig iron reaches sufficiently high values. In principle, the smaller the droplets are, the more completely they can react to form titanium carbonitride.

For reasons of reaction kinetics, however, it must be assumed that the titanium carbonitrides grow into the metal in phase contact with the slag, i.e. starting from the droplet surface. Carbide represents a strong impediment to diffusion, i.e. to the exchange of substances between the slag and the metal. A droplet, which is completely coated with carbide, will, therefore, only change in the long term. It remains liquid inside for a very long time.

The falling drops that reach the slag/metal phase boundary are surrounded by slag in addition to the carbide. Both phases delay the redissolution of the carbide in the pig iron so that an accumulation of droplets occurs in a layer which forms a heterogeneously structured transition zone between the two phases $\rightarrow$ slag/metal.

So far, the observation that titanium carbonitride accumulates preferentially in the worn areas has not been explained. As a first step, the temperature dependence of the excretion of titanium carbonitride is considered. It is found that with decreasing temperature, more and more titanium carbonitride is formed. However, the temperature dependence of the excreted carbide mass is small, about 15×10^{-3} at %/K, and comparable to the excretion of carbon along the $C'–D'$ line in the iron-carbon diagram (about 9×10^{-3} at %/K). The massive carbide excretion observed in practice can only be expected if the inner wall temperature deviates very strongly from the pig iron temperature. This is only to be expected to a limited extent because the heat transfer (334 W/m^2 K) is good and the thermal conductivities of pig iron and carbon bricks are comparable at approx. 20–30 W/mK.

It is much more important to look for a mechanism which causes a strong, permanent and purposeful circulation of both melts toward the wear zone, whereby the carbide agglomerates present as droplets can get there. Not experimentally proven, but conceivable, is the so-called Marangoni convection (refer to Chap. 7).

Interfacial convection, named after C. Marangoni (1871), is the result of a changing interfacial tension along the phase surface of two liquids in contact with one another, in this case steel/slag. Possible causes, for example, include gradients in temperature and oxygen content. Oxygen is enormously capillary-active in liquid iron.

The inside surface of a wear zone is about 100 °C colder than its surroundings, which are not affected by corrosion, because the wear zones are additionally cooled on the outside. In addition, the molten pig iron can become fully saturated in carbon if in contact with the graphite lining, which further reduces the activity of dissolved oxygen by about 2.7×10^{-6} between 1500 and 1400 °C. Both effects result in higher interfacial energy in the wear zone than at some distance. Therefore, with respect to energy minimization, a flow of slag and metal occurs from the center toward the wall, i.e. toward the wear zone. It can be assumed that this flow transports the already existing excretions of iron–titanium carbonitride toward the wear zone. Since the carbonitride is not wetted by either melt, it remains firmly attached to the wall after contact and is sintered there.

The bath movement shall be roughly estimated [11]. It is calculated according to Levich [12] (refer to Chap. 7, Eq. 7.8) as follows:

$$u_{\mathrm{M}} = \frac{d}{4\eta}\left(\frac{\mathrm{d}\sigma}{\mathrm{d}T}\,\mathrm{grad}\,T + \frac{\mathrm{d}\sigma}{\mathrm{d}C}\,\mathrm{grad}\,C\right)\;(\mathrm{cm}\,\mathrm{s}^{-1}) \qquad (7.8)$$

There is no information about the characteristic length d which is the thickness of the flowing layer. It is calculated with $d = 0.02$ cm. This is the thickness of the flow edge layer upon the dissolution of graphite in the pig iron [13]. The viscosity of the pig iron is $\eta_{\mathrm{SL}} = 4$ cPoise [4×10^{-2} g/cm s] and that of the slag is $\eta_{\mathrm{SL}} = 4$ Poise [g/cm s]. The temperature gradient is estimated to be grad $T = -30$ K/cm. The change in interfacial tension is dσ/d$T = -1.55$ mN/m K [14–16]. At equilibrium with pig iron nearly saturated with carbon, the dissolved oxygen activity at 1500 °C is $C_{1500} = 12.5 \times 10^{-6}$ and at 1400 °C it is $C_{1400} = 9.85 \times 10^{-6}$ [7]. The mean value is $C_0 = 11.2 \times 10^{-6}$. The activity gradient of dissolved oxygen in pig iron is approximately grad $C = -2.65 \times 10^{-6}$ cm^{-1}. The change of the pig iron/slag interfacial energy with the activity of the oxygen dissolved in the pig iron follows the equation dσ/d$C = -314/C_0$ mN/m [17], and thus it is approximately dσ/d$C = -2.81 \times 10^7$ mN/m.

If the calculation is based on the experience gained in numerous laboratory tests looking at wear involving Marangoni convection (refer to Chap. 7), according to which the convection is determined by the components dissolved in the metal bath, a flow velocity of approx. 12 cm/s follows from Eq. (7.8). If the excretion of carbonitride completely covers the graphitic lining, only the influence of the temperature gradient is taken into account and the flow speed $u_{\mathrm{M}} = 5$ cm/s is obtained. It should be remembered that the slag has the same velocity in contact with the pig iron. However, because of its much greater toughness, the thickness of its flow boundary layer is approx. 1 cm.

In summary, carbide excretion in conjunction with Marangoni convection could provide an explanation for the observed phenomenon of extended blast furnace operation time due to the use of material containing TiO_2. At the same time, the Marangoni convection could explain the premature wear affecting the wear zone at preferred locations, since the mass (material) transfer between the wall and the melts was already enhanced there before, i.e. before the application of TiO_2 due to the increased convection.

3.1.1.5 Pig Iron Transport

The torpedo ladle serves to transport the pig iron from the blast furnace to the steel mill. In addition to approx. 4.8% by weight carbon and 1% by weight silicon and 0.3% by weight phosphorus, the pig iron also contains approx. 0.03% by weight dissolved sulfur. In addition to magnesia chromite bricks, the lining consists largely of alumina-silicon carbide bricks (ASC-bricks), i.e. the entire environment is reducing. Therefore, it is obvious to carry out the pre-desulfurization here. The remaining accompanying elements are slagged in the converter.

Desulfurization

For this purpose, in many cases lime dust is injected into the melt with a carrier gas, e.g. methane. Punctual equilibrium is [18]:

$$K_{CH_4/S} = \frac{[a_S] \cdot P_{CH_4} \cdot (a_{CaO})}{P_{CO} \cdot P_{H_2}^2 \cdot (a_{CaS})} = 5 \times 10^{-5} \quad \text{at } 1500\,°C. \tag{3.12}$$

With $a_{CaO} = a_{CaS} = 1$ and $P_{H_2}/P_{CO} = 2$, one obtains at the partial pressure of methane $P_{CH_4} = 0.8$ atm and the total pressure $P_{total} = 1$ atm, the sulfur activity $a_{[S]} = 4 \times 10^{-8}$. The activity coefficient of sulfur in pig iron is $f = a_{[S]}/[S] = 5$, which is why $[S] = 4 \times 10^{-8}/5 = 8 \times 10^{-9}\%$ by weight. Experience shows that such low values are not achieved in the entire melt.

CaC_2 is also widely used for desulfurization because of its good efficiency.

The attack on the lining is low in all cases. Along the slag line, premature wear occurs which would probably be explained primarily by the prevailing interfacial convection. It increases the mass (material) transfer considerably (refer to Chap. 7).

From the torpedo ladle, the desulfurized pig iron is emptied into a transport ladle to be transported therein and discharged into the converter. The main wear zone is the impact surface of the melt. So-called SMAC bricks are used there (SiC, MA spinel with pitch bond). They are particularly resistant to abrasion.

Pig Iron-Emulsion in the Slag

When the pig iron is purged with gas for better mixing, the individual gas bubbles burst during phase passage into the slag and entrain a large number of small iron droplets. This not only means a loss of iron, but also, depending on the composition of the slag, changes its FeO_n content and, thus, its aggressiveness toward the refractory lining. Practical experience shows that less iron is emulsified by using fluorspar (CaF_2) or cryolite (Na_3AlF_6). Laboratory tests show, in agreement with theoretical considerations, that the increase in interfacial tension between pig iron and slag is partly responsible for this. Empirically, the mass of iron transported into the slag can be described by the numerical value equation in the cgs–system [19]

$$m_{Fe} = 1.8 \cdot \rho_{Fe} \cdot g^{0.25} \cdot d_e^{3.25} \cdot \sqrt{\frac{\eta_{Fe}}{\sigma_{Fe/Sl}}} \; [\text{g/bubble}] \tag{3.13}$$

Nothing is stated about the distribution of the droplet size and its change over time. The average droplet size is 1 mm and the average residence time is 1 s [19]. The larger the droplets, the shorter their dwelling time ($t \sim 1/d_{Fe}$). The frequency of the individual bubbles has no influence on the dwelling time [19].

The density of the pig iron is calculated on the basis of its content of carbon and the temperature ϑ [°C] to [14]

$$\rho_{Fe} = 8.75 - 6.96 \times 10^{-2}\,[wt\%C] - 1.15 \times 10^{-3} \cdot \vartheta\,\left[g/cm^3\right]. \tag{3.14}$$

The equivalent bubble diameter d_e of argon blown into the melt through a pipe of clear width d_n [mm] is, to a good approximation, [20]

$$d_e = 7.5 + 1.5 \cdot d_n\,[mm]. \tag{3.15}$$

The viscosity of the molten pig iron is [14]

$$\eta_{Fe} = 0.01 \cdot \left(1.32 \times 10^{\frac{1418}{T}} - 0.63\,\left[\%\,by\,weight\,C\right]\right)\,\left[g/cm\,s\right]. \tag{3.16}$$

Using [C] = 4.5% by weight and $\vartheta = 1400$ °C ($T = 1673$ K), we get $\rho_{Fe} = 6.83$ g/cm^3 and $\eta_{Fe} = 0.065$ g/cm s.

Taking a slag of composition (in % by weight): 41% CaO, 9% MgO, 15% Al$_2$O$_3$, 35% SiO$_2$, the interfacial tension $\sigma_{Fe/Sl}$ against pig iron with 4% carbon was determined at 1300–1500 °C in the laboratory (DIFK). The result is shown in Fig. 3.4. It can be described for $0.1 \leq (CaF_2\%\,by\,weight) \leq 10$ by the equation

$$\sigma_{Fe/Sl} = -2.1\vartheta + 4072 + 63 \cdot \log(10 \cdot (wt\%CaF_2))\,[mN/m] \tag{3.17}$$

At $\vartheta = 1400$ °C and for (CaF$_2$% by weight) we get $\sigma_{Fe/Sl} = 1195$ mN/m. Since dissolved carbon does not influence the surface tension of iron [14] and its solubility in the slag is often negligible, carbon does not noticeably change the interfacial tension.

Using Eq. (3.15), the bubble size $d_e = 15$ mm follows from the clear width of the pipe $d_n = 5$ mm. From Eq. (3.13), the amount of iron transferred into the slag from each bubble of this size is calculated to be

$$m_{Fe} = 1.8 \times 6.83 \times 981^{0.25} \times 1.5^{3.25} \cdot \sqrt{\frac{0.065}{1195}} = 1.9\,\left[gFe/Blase\right]. \tag{3.18}$$

For 1000 kg of iron, this adds up to 530 drops with an average diameter of 15 mm.

At a gas flow rate of 0.01 Nm^3/min-t$_{pig\,iron}$

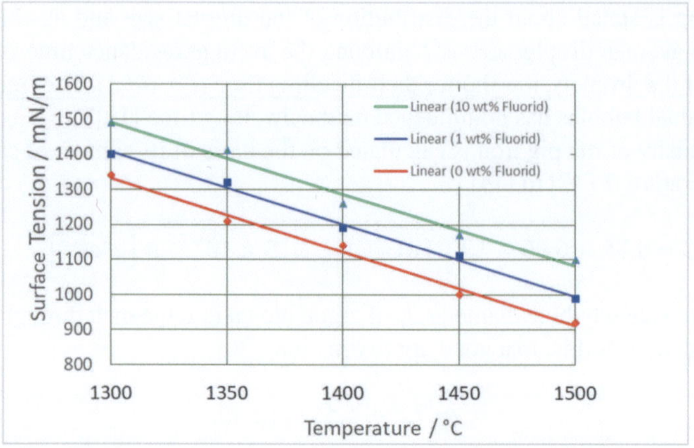

Fig. 3.4 Influence of the fluoride content on the interfacial tension slag/pig iron (laboratory measurement/DIFK)

$$\dot{Z} = \frac{\dot{V}_{Gas}}{\frac{4}{3}\pi \cdot r_e^3} = \frac{0.01 \times 10^6 \, \text{cm}^3/\text{min} \cdot t}{\frac{4}{3}\pi \cdot (0.75 \, \text{cm})^3/\text{Blase}}$$

$$= 5.7 \times 10^3 \, [\text{Blasen/min} \cdot t_{RE}]$$

$$= 11 \, \text{kgFe/min} \cdot t_{\text{pig iron}}$$

are carried into the slag and react there. This explains the good kinetic effect of purging on desulfurization. A large proportion of the iron droplets "rains" again after a short time (seconds) and do not remain in the slag so that the loss is kept within narrow limits. This process can only be described very imperfectly by applying Stoke's equation. Through collision, the droplets coagulate, sink rapidly and are absorbed by the molten iron. However, the quantitative description of this process would have no significance for the corrosion of the refractory material.

This emulsion process takes place wherever gas is used for purging, e.g. also in the OBM and AOD processes (refer to Sect. 3.1.1.3.2). However, higher gas flow speeds are used there. Thus, the results are not directly comparable.

3.1.1.6 LD Process (Linz–Donawitz)

By injecting oxygen intensely into the molten iron by means of a lance (LD) and/or through bottom tuyeres (OBM—Oxygen Bottom Maxhütte), iron and, above all, its constituents with stronger oxygen affinity, such as carbon, silicon, manganese and phosphorus, are oxidized at temperatures above 1800 °C in a very turbulent flow and dissolved into the foamed slag. The environment is strongly oxidizing. The reactions

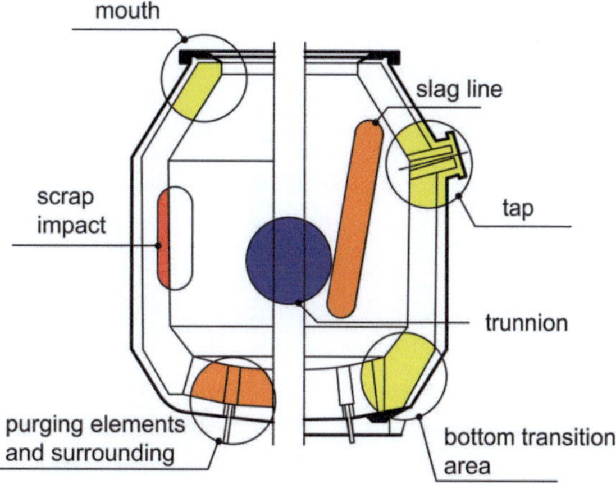

Fig. 3.5 Zones of the strongest wear in a LD converter (A. Eschner, RHI)

take place in a short time because the mass (material) transfer rates in the metal/slag emulsion are very high. The rising CO bubbles cause the formation of a metal/slag emulsion.

After slagging, the oxygen content is adjusted or set by deoxidation with ferrosilicon and/or aluminum and subsequent feeding of a lime-aluminate slag. The lining consists of unshaped MgO and MgO–C bricks. The latter are preferably used in the main wear zones. The zones subjected to the most intensive wear are shown in Fig. 3.5.

Among other things, scrap is charged into the emptied converter, which has a temperature of 1200–1400 °C, in order to reduce the high process temperature. This leads to enormous mechanical stress on the lining. Just like for the chemical wear, it is a rough calculation.

Impact Wear

During charging, a piece of scrap of mass m [kg] falls from height h [m] to the bottom of the lining. The transferred energy is $W = m \cdot g \cdot h$ [N m]. It is divided into three components, elastic deformation $W_E = 0.5 \cdot C \cdot x^2$ [N m], plastic deformation W_P and the formation of n cracks of the total area A [m^2]. Let the linear deformation be $x = 1$ mm. According to measurements by Hampel and Anezieris [21], the elastic and plastic deformation are approximately equal (Fig. 3.6). $C = F/x = E \cdot x$ [N/m] is the stiffness of the microstructure, E [N/m^2] is the modulus of elasticity, F [N] is the acting force, and x [m] is the deformation. The energy balance is:

$$m \cdot g \cdot h = E \cdot x^3 + A \cdot n \cdot G_f \ [\text{N m}]. \tag{3.19}$$

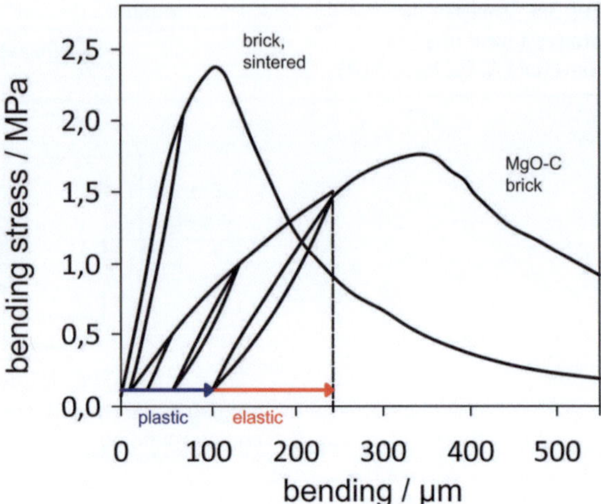

Fig. 3.6 Comparison of bending stresses of refractory materials (9% by weight C) [21]

The height of fall is approx. $h = 2$ m, and the mass of the scrap piece is $m = 100$ kg. The fracture strength is $G_f = 200$ J/m^2 [22], so that with

$$A \cdot n = \frac{m \cdot g \cdot h - E \cdot x^3}{G_f}$$

$$= \frac{100 \times 9.81 \times 2 - 4 \times 10^{+9} \times \left(10^{-3}\right)^3}{200} = 10 \left[\text{m}^2\right] \qquad (3.20)$$

a fracture surface of $10 \, \text{m}^2$ is created which is distributed over n cracks. This is usually not a disaster because in the following blowing period there is a high probability that the cracks will "heal" as a result of the high temperature.

Despite all inaccuracies, the calculation shows that crack formation predominates if the deformation resistance of the refractory material is low. Cracking tends not to occur if ductility is high. In the present example, the ductility should be at least 8 mm to avoid cracking. Then $A \cdot n = 0$.

The progression of the stress/expansion curve in Fig. 3.6 suggests replacing Young's modulus by the V-modulus (modulus of deformation), which takes into account the nonlinear behavior of refractory material. The deformation modulus results from the linear connection of the coordinate origin with the stress maximum.

Metal/Slag-Emulsion in the Converter (OBM, AOD, etc.)

If reaction or purging gases, e.g. oxygen/argon, are injected into a converter through bottom or side tuyeres, this is done with a high gas flow speed in order to achieve rapid and thorough mixing of the two phases slag and metal. In front of each tuyere, a pulsating gas flow is created which can consist of large individual bubbles or is coherent. The pulsation is the result of the collapse of the bubbles emerging from the tuyere in rapid succession. In this process, the high-velocity, abrupt backflow of the liquid metal transfers its kinetic energy to the refractory material (back attack). This process shatters the refractory material over time. As a result, the destruction of the lining is comparable to that caused by cavitation. The mechanisms, however, differ from each other.

The majority of the bubbles rise, the ferrostatic pressure decreases, the gas heats up, the bubbles rapidly increase in size, lose their stability and break down into individual bubbles, which in turn behave in a similar way. This process stirs the melt and an emulsion of steel droplets forms in the slag.

The emulsion represents a strong enlargement of the slag/metal phase boundary and, therefore, accelerates the reactions between the two. It cannot yet be calculated in detail, but reaches a steady state when its formation speed and segregation proceed at the same rate. Since, in addition to the "back attack," it is, above all, the composition of the slag which has a decisive influence on refractory material corrosion, an estimate of the emulsion in the case of strong flow is given in simplified form.

Based on practical experience, laboratory experiments and process engineering considerations, a detailed description of these processes can be found in Deo and Boom [23], among others. A numerical model is used by Odenthal et al. [24]. Here, a simplified calculation is done, which describes the process in principle.

An argon/oxygen mixture is injected through five tuyeres into a converter with 80 t steel and 8 t slag. 80 Nm^3/min are used, which is 1 Nm^3/(t min) or $\dot{V}_G = 2.7 \times 10^5$ cm^3/s per tuyere. The outlet opening of the tuyere has a clear width of $d = 1.5$ cm. A sequence of bubbles is formed which is referred to as "bubble gases." The primary bubbles have an average size [23].

$$d_B = 0.54 \cdot \left(\dot{V}_G \cdot \sqrt{d} \right)^{0.289}$$

$$= 0.54 \times \left(2.7 \times 10^5 \times \sqrt{1.5} \right)^{0.289} = 21 \, \text{cm}. \tag{3.21}$$

They grow rapidly and decay into a bubble column. The average ascent speed of a bubble is [24].

$$u_B = \left(\frac{2 \cdot \sigma}{\rho_{Fe} \cdot d_B} + 0.5 \cdot g \cdot d_B \right)^{1/2}$$

$$= \left(\frac{2 \times 1500}{7_{Fe} \times 21} + 0.5 \times 981 \times 21 \right)^{1/2} = 102 \, \text{cm/s} \tag{3.22}$$

The surface tension of the steel is $\sigma = 1500$ mN/m and its density is $\rho = 7$ g/cm^3. The swarm of bubbles generates a flow. Its mean velocity is [25]

$$\bar{u}_{St} = 3.37 \cdot \dot{V}_G^{1/4} \cdot z^{-0.12}$$

$$= 3.37 \times \left(2.7 \times 10^5 \right)^{1/4} \times 100^{-0.12} = 44 \, \text{cm/s}. \tag{3.23}$$

The calculation is performed at the point $z = 1$ m above the tuyere exit, i.e. just below the slag. If the flow meets the slag, the metal emulsifies in it, if the condition [18]

$$u_{\text{St,crit}} \geq \left(\frac{8}{\rho_{\text{St}}}\right)^{1/2} \cdot \left(\frac{2}{3} \cdot \sigma \cdot g \cdot (\rho_{\text{St}} - \rho_{\text{Sl}}) \cdot \cos\theta\right)^{1/4} = 44\,\text{cm/s} \qquad (3.24)$$

is fulfilled. The interfacial tension slag/steel is $\sigma = 1200$ g/s^2, the density difference $\Delta\rho = 4$ g/cm^3 and the wetting angle between both phases $\theta = 30°$. The critical, i.e. maximum size of the metal droplets [18], not their mean value, is

$$\begin{aligned}
d_{\text{crit}} &= \left(\frac{6 \cdot \sigma}{g \cdot \Delta\rho \cdot \cos\theta}\right)^{1/2} \\
&= \left(\frac{6 \times 1200}{981 \times 4 \times \cos 30}\right)^{1/2} = 1 \times 46\,\text{cm}.
\end{aligned} \qquad (3.25)$$

Smaller droplets form more easily so that the mean value of the droplet size distribution is about 1–2 mm [18, 19, 26].

With a size of 1 mm, each bubble passage from the metal to the slag produces a maximum of $Z_{\text{Fe}} = 3000$ droplets of mass 3.7 mg, i.e. of total mass $m_{\text{Fe}} = 11$ g/bubble. The number of bubbles formed with critical volume is at least [19].

$$\begin{aligned}
\dot{Z} &= \frac{\dot{V}_G}{\frac{4}{3}\pi \cdot r_B^3} = \frac{1\,\text{Nm}^3/\text{min t}}{\frac{4}{3}\pi \cdot \left(\frac{0.21\,\text{m}}{2}\right)^3/\text{Blase}} \\
&= 206\,\text{Blasen/min t} = 2.3\,\text{kg/min t}.
\end{aligned} \qquad (3.26)$$

There are about 100 kg of slag per ton of steel. Consequently, about 23 kg of steel droplets are emulsified in one ton of slag. This corresponds to the data from practice [18, 19, 26].

Back Attack-Mechanism

The large gas bubbles collapsing in the case of pulsating bubbles cause the melt to hit the lining; just like a shock wave at up to its speed of sound. To estimate this, the pressure, at which the melt hits the wall, is compared with the hot crushing strength $\sigma_{\text{HCS}} \approx 10^7$ N/m^2 [27] of the refractory material. The speed of sound in condensed phases is calculated from the density ρ [kg/m^3] and the compressibility κ [Pa^{-1}] to give

$$c_{\text{Fe}} = \sqrt{\frac{1}{\rho \cdot \kappa}}. \qquad (3.27)$$

The result for the melt is $c_{Fe} = 1/\sqrt{7000 \times 5 \times 10^{-14}} \simeq 5 \times 10^4$ m/s and for the refractory material $c_{FF} = 1/\sqrt{3500 \times 4 \times 10^{-12}} \simeq 8450$ m/s.

The highest impact pressure of the melt at the speed of sound c_{Fe} is, therefore, [28].

$$P = \frac{1}{2} \cdot \rho_{Fe} \cdot c_{Fe}^2 = 0.5 \times 7000 \times (50{,}000)^2$$
$$= 9 \times 10^{12} \, \text{N/m}^2. \tag{3.28}$$

However, the material may already fail when the impact pressure exceeds the hot crushing strength $\sigma_{HCS} \sim 10^7$ Pa. The bond phase loosens and the wear grains fall out. This is the case if the impact speed u is exceeded:

$$u \geq \sqrt{\frac{2 \cdot \sigma_{HDF}}{\rho}} = \sqrt{\frac{2 \times 10^7}{7000}} = 53 \, \text{m/s}. \tag{3.29}$$

According to measurements with the water model, the frequency of the shocks is 7–18 Hz [29]. However, only one out of 30,000 shocks is supposed to be successful. The mean bubble diameter is assumed to be $2r = 1$ cm. The melt volume V_{Fe} [m^3] corresponds to the bubble volume transferring its kinetic energy to the volume V_{Ref} of the refractory material. The ratio of both volumes is

$$\frac{V_{FF}}{V_{Fe}} = \frac{\rho_{Fe}}{\rho_{FF}} \cdot \left(\frac{u}{c_{FF}}\right)^2 = \frac{7000}{3500} \times \left(\frac{53}{8450}\right)^2 = 7.9 \times 10^{-5} \tag{3.30}$$

The depth of the wear zone corresponds to twice the radius of the impinged refractory material

$$2r_{FF} = \left(\frac{V_{FF}}{V_{Fe}} \cdot (2r_{Fe})^3\right)^{\frac{1}{3}} = \left(7.9 \times 10^{-5} \times \left(10^{-2}\right)^3\right)^{\frac{1}{3}} = 4.3 \times 10^{-4} \, [\text{m}] \tag{3.31}$$

Taking into account 10 impacts/s and a success rate of 1/30,000, one gets the wear rate $\dot{x} = 1.4 \times 10^{-7}$ m/s which is 0.5 mm/h. The single process takes about 20 min. If part of the gas is simultaneously blown onto the melt with a lance, this reduces the corrosion caused by the back attack [29].

MgO-Saturation

In Asian steel mills in particular, the converter is rinsed with a liquid slag saturated with MgO after each charge to reduce wear so that a coating is formed. According to [30], the MgO saturation in contact with air at 1600 °C is calculated from the relationship

$$(\%\text{MgO})_{\text{sat}} = 0.00816 \cdot x^2 - 1.404 \cdot x + 62.31, \tag{3.32}$$

with $x = (\%\text{CaO}) + 0.45 \cdot (\%\text{Fe}_2\text{O}_3 + \text{FeO}) + 0.55 \cdot (\%\text{MnO})$.

The range of validity is given as $1 \leq (\%\,\text{CaO})/(\%\,\text{SiO}_2) \leq 5$, $10 \leq (\%\,\text{FeO}_n) \leq 35$, $1 \leq (\%\,\text{MnO}) \leq 13.2 \leq (\%\,\text{Al}_2\text{O}_3) \leq 12$.

One of the factors responsible for the wear of the lining is the iron content of the slag, in particular the ratio $(\text{Fe}/\text{Fe}^{2+3+})$. It is calculated at 1600 °C to [30]

$$\log\left(\text{Fe}^{2+}/\text{Fe}^{3+}\right) = -0.00107x^2 + 0.0721x - 1.982 \tag{3.33}$$

where $x = (\%\,\text{CaO}) + 0.38(\text{FeO}_n) + 3.2\,(\%\,\text{MnO})$.

FeO and MgO form the gapless solid solution magnesio-wüstite. As the (FeO) content increases, its melting temperature falls. At the same time, the melting temperatures of the slag and its viscosity decrease; i.e. it attacks the lining more intensively [31].

Example The slag has the following composition: 42% (CaO), 28% (SiO$_2$), 20%

(FeO$_n$), 5% (% MnO), 5% (Al$_2$O$_3$), $B \equiv (\%\,\text{CaO})/(\%\,\text{SiO}_2) = 1.5$, resulting in $x = 53.75$ (MgO)$_{\text{sat}} = 10.4\%$.

Chemical Wear of the Converter Lining

The estimation of the wear of the converter lining is imprecise in that in practice there is an emulsion of steel droplets in the slag whose physical and chemical properties are not precisely known. Here it is assumed that the MgO-based lining reacts chemically with the slag. Its viscosity η and density ρ determine the kinematic toughness

$$\nu = \eta/\rho.$$

The LD-slag consists of 49% by weight CaO, 31% by weight FeO$_n$, 13% by weight SiO$_2$, 2% by weight MnO and 2% by weight MgO, 1% by weight Al$_2$O$_3$ and 2% by weight P$_2$O$_5$. Their saturation with MgO is according to Eq. (3.32) $(\%\text{MgO})_{\text{sat}} = 0.00816 \cdot x^2 - 1.404 \cdot x + 62.31 = 6\,\text{wt}\%$ by weight with $x = (\%\text{CaO}) + 0.45 \cdot \%\text{FeO}_n + 0.55 \cdot (\%\text{MnO}) = 64.05$.

The viscosity of the slag is calculated for 1950 °C from the model of Urbain [31, 32] modified by Mudersbach to $\eta = 0.03$ Pa s. With density $\rho = 3000$ kg/m^3, this gives us $\nu = 10^{-5}$ m^2/s $= 0.1$ cm^2/s. The mean diffusion coefficient of the MgO in the slag is calculated according to Stokes–Einstein [28] based on the viscosity:

$$D = \frac{k \cdot T}{6\pi \cdot r \cdot \eta} = \frac{1.38 \times 10^{-23} \times 2223}{6\pi \times 10^{-10} \times 0.03}$$
$$= 5 \times 10^{-10}\left[\text{m}^2/\text{s}\right] = 5 \times 10^{-6}\left[\text{cm}^2/\text{s}\right]. \tag{3.34}$$

$k = 1.38 \times 10^{-23}$ J/K is the Boltzmann constant and $r = 10^{-10}$ is the effective ion radius [31].

In process engineering, one works with dimensionless ratios (refer to Sect. 2.2.1) in order to be able to establish the analogy between operational practice and model tests. The mass (material) transfer coefficient β [cm/s] is given by the Sherwood number

$$\mathrm{Sh} \equiv \frac{\beta \cdot y}{D} = \frac{2}{\sqrt{\pi}} \cdot \mathrm{Re}^{\frac{1}{2}} \cdot \mathrm{Sc}^{\frac{1}{3}} \tag{3.35}$$

The Reynolds number $\mathrm{Re} \equiv u \cdot y/v$ and the Schmidt number $\mathrm{Sc} \equiv v/D$ can be calculated from the above values if the mean speed (velocity) u [cm/s] of the circulating flow in the converter and the height of the slag emulsion are known. Practical observations give the estimated values [33] $u \approx 500$ cm/s and $y \approx 20$ cm. One obtains $\mathrm{Re} = 100{,}000$, i.e. a highly turbulent flow and $\mathrm{Sc} = 20{,}000$. From this follows with

$$\mathrm{Sh} = \frac{2}{\sqrt{\pi}} \times \left(1 \times 10^5\right)^{\frac{1}{2}} \times \left(2 \times 10^4\right)^{\frac{1}{3}} = 10^4 \tag{3.36}$$

the mass transfer coefficient with wall friction $\beta \equiv \mathrm{Sh} \cdot D/y = 2.5 \times 10^{-3}$ cm/s.

According to Nernst (refer to Sect. 2.2.4), the corrosion speed (rate) is

$$V_{\mathrm{CORR}} = 360 \cdot \frac{\rho_{\mathrm{Sl}}}{\rho_{\mathrm{FF}}} \cdot \beta \cdot (C_{\mathrm{S}} - C_0) \, [\mathrm{mm/h}] \tag{3.37}$$

$$\frac{\mathrm{d}x}{\mathrm{d}t} = 360 \times \frac{2.9}{3.5} \times 2.5 \times 10^{-3} \times (6 - 2) = 3 \, \mathrm{mm/h}$$

The result agrees quite well with the experience. The process takes about 20 min.

3.1.1.7 Electric Arc Furnace Process

With the help of electrical energy, steel scrap is melted with the addition of oxygen (Fig. 3.7). The process takes about one hour.

The arc between the graphite electrodes and the metal bath reaches over 3000 °C. To protect the lining, a foaming slag rich in lime and iron oxide (CO formation, starting from the graphite electrode) is, therefore, generated. It reduces the direct radiation on the lining. Nevertheless, the temperature is the most important parameter for the wear of the wall (slag zone) and lid. Bricks made of MgO–C, MgO–Cr_2O_3 (picrochromite) and MgO–Al_2O_3 mixtures are used as refractory material. In addition, the walls and lid are water-cooled.

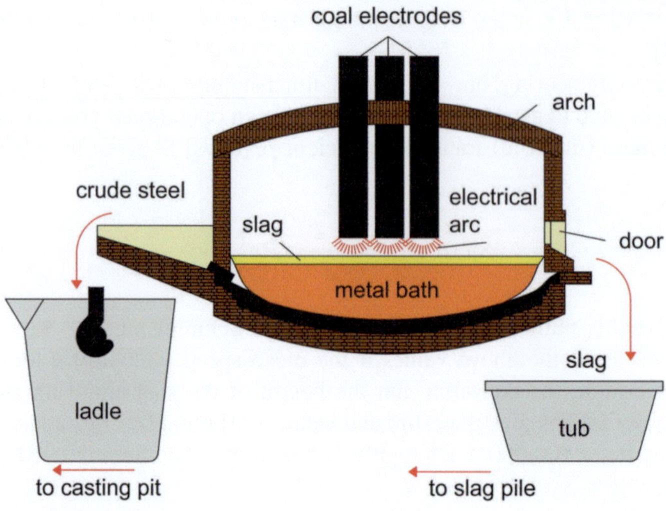

Fig. 3.7 Electric arc furnace, schematic illustration (Wikipedia)

Chemical Wear in the Electric Arc Furnace

The chemical attack on the lining based on MgO is a part of the total wear picture.
It can be estimated using the equations (Sects. 2.2.3 and 4.1):

$$\frac{\Delta x}{\Delta t} = 360 \cdot \beta \cdot \frac{\rho}{\rho_R} \cdot (C_s - C_0) \, [\text{mm/h}], \tag{3.37}$$

with $\beta \equiv D/\delta$ [cm/s] the defined mass (material) transfer coefficient. The Sherwood
number describes the mass (material) transfer with wall friction:

$$\text{Sh} = \frac{2}{\sqrt{\pi}} \cdot \text{Re}^{\frac{1}{2}} \cdot \text{Sc}^{\frac{1}{3}}, \tag{3.36}$$

$$\beta \equiv \text{Sh} \cdot D/y. \tag{3.38}$$

The required values are taken from the literature [33]:

The slag consists of 40% by weight each of CaO and FeO_n, 16% by weight SiO_2,
1% by weight MnO and 3% by weight MgO. Its saturation in MgO is according to
Eq. (3.32):

$(\%\text{MgO})_{\text{sat}} = 0.00816 \cdot x^2 - 1.404 \cdot x + 62.31 = 8 \, \text{wt\%}$, whereby $x = (\%\text{CaO}) + 0.45 \cdot \%\text{FeO}_n + 0.55 \cdot (\%\text{MnO}) = 58.55$.

The kinematic viscosity of the very hot E-furnace slag is $v = 0.15 \, \text{cm}^2/\text{s}$ [32,
34] and the mean diffusion coefficient $D_{\text{MgO}} = 4 \times 10^{-6} \, \text{cm}^2/\text{s}$ calculated from this
according to Stokes–Einstein [28], Eq. (3.34), so that the Schmidt number results in
$\text{Sc} = v/D = 3.75 \times 10^4$. Let the mean flow speed (velocity) of the circulating slag be

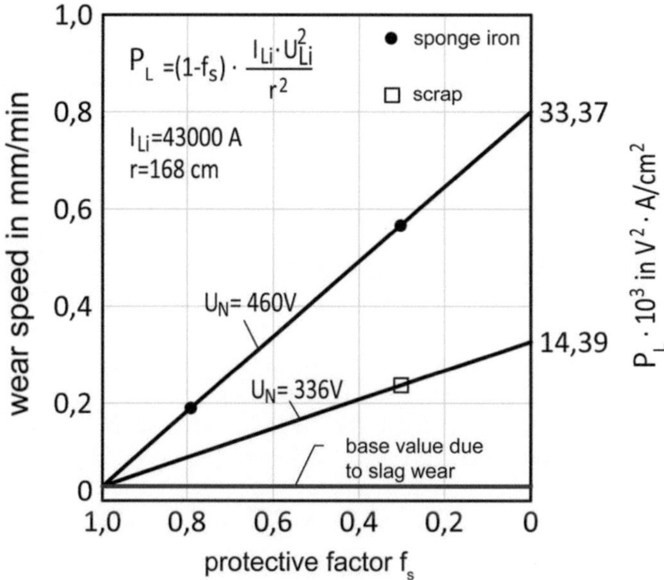

Fig. 3.8 Measured wall wear above protection factor f_S [18]

$u = 40$ cm/s and the thickness of the slag layer $y = 15$ cm, giving Reynolds number $\mathrm{Re} = u \cdot y/v = 4 \times 10^3$. The mass (material) transfer coefficient is calculated from the Sherwood number (Eq. 3.36)

$$\mathrm{Sh} = \frac{2}{\sqrt{\pi}} \times \left(4 \times 10^3\right)^{\frac{1}{2}} \times \left(3.75 \times 10^4\right)^{\frac{1}{3}} = 2.4 \times 10^3$$

to $\beta \equiv \mathrm{Sh} - D/y = 6.4 \times 10^{-4}$ cm/s. The corrosion speed (rate) is

$$\frac{\Delta x}{\Delta t} = 360 \times 6.4 \times 10^{-4} \times \frac{2.9}{3.2} \times (8 - 3) = 1 \; [\mathrm{mm/h}]$$

and agrees quite well with practice (compare Fig. 3.8).

Radiation Wear

The wear of the wall of an arc furnace is primarily proportional to the radiation load. This corresponds to the square of the applied electrical voltage U_L and the resulting current intensity I_L. Empirically, the proportional radius R_L of the furnace is a measure of the radiation wear [33]:

$$P_L = \frac{U_L^2 \cdot I_L}{R_L} \; \left[\mathrm{V^2 \, A/cm^2}\right]. \tag{3.39}$$

The current density of the radiation is $D = K \cdot P_L \cdot K \approx 10^{-3}$ [1/V] is a proportionality factor dependent on the voltage and to be determined empirically. The radiation striking the wall acts in two ways. One part melts the refractory material and the adhering slag, the other part is dissipated through the wall to the outside [35]:

$$K \cdot (1 - f_S) \cdot P_L = \Delta H \cdot \frac{dx}{dt} + \lambda \cdot \frac{T_W - T_A}{d} \; \left[W/cm^2\right] \qquad (3.40)$$

The protection factor $0 < f_S < 1$ takes into account the shielding of the wall by way of the foaming slag. ΔH [J/cm^3] is the required melting enthalpy and dx/dt [cm/s] is the melting speed (rate). λ [W/cm K] is the thermal conductivity of the wall. T [K] is the wall temperature inside (W) and outside (A), and d [cm] is the wall thickness. The slag foams as a result of intensive CO formation.

Without the foaming slag present, the 3000 °C hot graphite electrodes would radiate the power density

$$q_{12} = \varepsilon_{12} \cdot \sigma_{SB} \cdot T^4 = 1 \times 5.67 \times 10^{-8} \times (3273)^4$$
$$= 6.5 \times 10^6 \; \left[W/m^2\right] \qquad (3.41)$$

$\sigma_{SB} = 5.67 \times 10^{-8}$ [W/m^2 K] is the radiation constant (Stefan–Boltzmann). The emission coefficient of graphite is $\varepsilon_1 \approx 1$, that of refractory material $\varepsilon_2 < 1$, e.g. 0.3. Thus, only a fraction of the radiation constant acts in the radiation exchange [36], viz.

$$\varepsilon_{12} = \frac{1}{\frac{1}{\varepsilon_1} + \frac{1}{\varepsilon_2} - 1} = \frac{1}{\frac{1}{1} + \frac{1}{0.3} - 1} = 0.3. \qquad (3.42)$$

The effective power density of the radiation is therefore $2.1 \times 10^6 \left[W/m^2\right]$, i.e. about 30–35% of the supply.

In order to heat refractory material (MgO) from room temperature to 3000 °C and melt it, approximately $\Delta H_{Ref} = 5.8 \times 10^6$ J/kg is required [7]. Consequently, per second and m^2-radiated area, $m_{Ref} \approx 2.1 \times 10^6/5.8 \times 10^6 = 0.36$ kg/s m^2 refractory material could be lost. Consequently, in the first place, it is the foaming slag and cooling equipment that make continuous operation possible.

The emission coefficient of the molten foam slag is expressed by the so-called radiation conductivity λ_{rad} [31]:

$$\lambda_{rad} = \frac{16}{3} \cdot \frac{n^2 \cdot \sigma_{SB} \cdot T_{rad}^3}{\alpha^*} = \frac{16}{3} \times \frac{1.6^2 \times 5.67 \times 10^{-8} \times (3273)^3}{1000}$$
$$\approx 27 \; [W/m\,K] \qquad (3.43)$$

$\alpha^* \approx 1000$ m^{-1} is the optical absorption coefficient of slags at high temperature [31] and $n \approx 1.6$ is their refractive index. Thus, the average power density is

$$q_{12} = \lambda_{\text{rad}} \cdot \Delta T / d = 27 \times 175/2.5 \approx 1890 \ \left[\text{W/m}^2\right]. \qquad (3.44)$$

$\Delta T = 175$ °C is the temperature difference between the radiator and the melting point of the MgO and $d = 2.5$ m is the mean distance to the wall. Consequently, $m_{\text{FF}} = \frac{1890 \times 3600}{5.8 \times 10^6} = 1.2$ kg refractory material/m^2 radiant surface can be lost in one hour. The foam slag reduces the wear by a factor of 10^3.

This calculation does not take into account the resistance heating which occurs as soon as the foamed slag transports the current between the electrode and the steel bath as a result of its electrical conductivity. The electrical efficiency increases with the formation of the foamed slag, i.e. the radiation losses decrease.

In addition, water cooling is provided by the use of cooling boxes. Wear progresses until the lowered surface temperature of the cooled wall allows a protective layer of solidified slag to form. Then, above all, the heat dissipated to the outside destroys the radiant energy.

Figure 3.8 shows the wear of a wall measured in practice above the protection factor f_S. It can be seen that the wear speed (rate) is mainly determined by melting and heat dissipation but only slightly by chemical dissolution (slag wear).

3.1.2   From Pig Iron to Steel II

3.1.2.1   Secondary Metallurgy

For the cost-effective production of the highest qualities, the converter and the electric arc furnace are nowadays used as primary melting units and the metallurgical fine work is shifted to downstream units. Figure 3.9 shows a selection of secondary metallurgical processes. The use of vacuum allows the lowest contents of carbon, oxygen, sulfur, phosphorus, hydrogen and nitrogen to be selectively adjusted. The injection of gases, e.g. argon, mixes the melt and standardizes its temperature and composition.

With regard to the behavior of refractory lining, some processes are discussed here. Vacuum metallurgy is dealt with specifically when considering the induction furnace. It should be noted that most of the statements are more or less applicable to all processes and are, therefore, only presented once.

Slags

Chemical Basicity

The metallurgical efficiency of slags is traditionally expressed by the empirical term slag basicity. There are a variety of definitions, which are adapted to the particular application under consideration [31]. The simplest basicity is the concentration ratio in % by weight of basic lime to silica

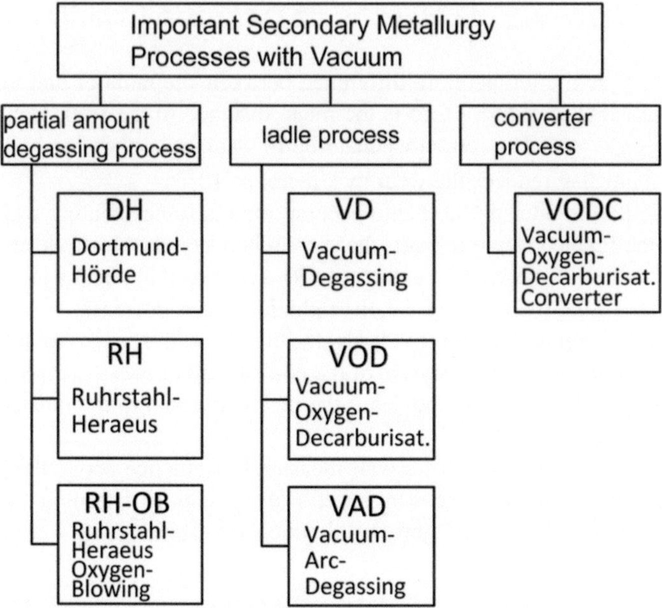

Fig. 3.9 Selection of secondary metallurgical processes [37]

$$B = \frac{(\% \, \text{CaO})}{(\% \, \text{SiO}_2)}. \tag{3.45.1}$$

The main components of the calcium aluminate slags preferred in secondary metallurgy are lime, alumina, and silica, which is why the basicity

$$B = \frac{(\% \, \text{CaO})}{(\% \, \text{SiO}_2) + (\% \, \text{Al}_2\text{O}_3)} \tag{3.45.2}$$

frequently tends to be favorable. Figure 3.10 shows the position of these slags in the relevant four-component system together with MgO.

The concentration quadrangle outlined in blue is in the area of lowest MgO saturation. This reduces the corrosion of the MgO-based ladle lining. A typical composition is % by weight is: 50% CaO, 29% Al_2O_3, 5% SiO_2, 8% MgO, 5% MnO, 3% FeO, i.e. $B = 1.5$.

The melting point of such slags follows its basicity (Fig. 3.11).

Based on many measurements and calculations, there are a large number of equations for the saturation concentration of magnesium oxide, which refer to basicities. Figure 3.12 shows measurement results for the solubility of MgO in CA slags. There is a close correlation to the basicity $B \equiv \text{CaO}/(\text{Al}_2\text{O}_3 + \text{SiO}_2)$ [39]. For the above slag ($B = 1.5$), $(\% \, \text{MgO})_{\text{sat}} = 7.4\%$ by weight.

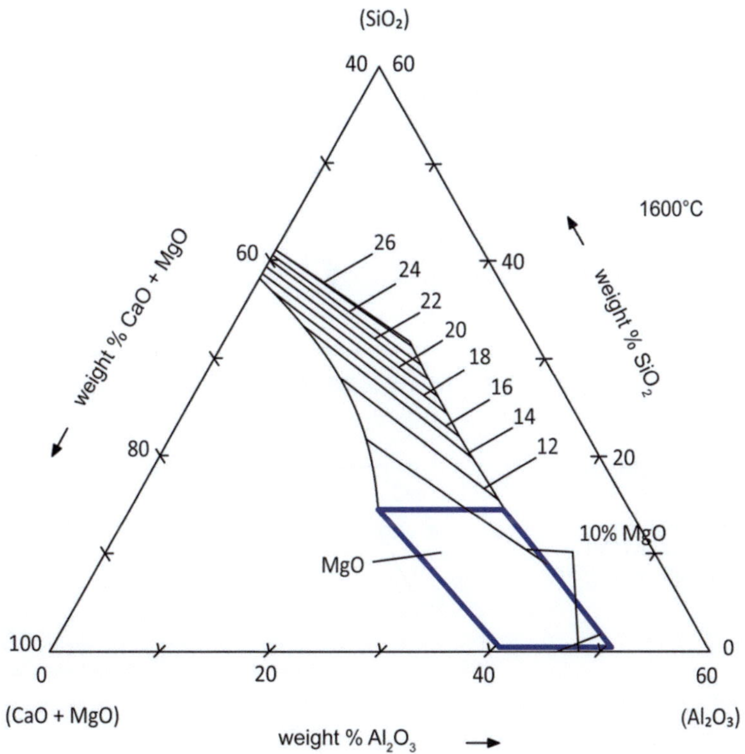

Fig. 3.10 Calcium aluminate slags in the system CaO + MgO, Al$_2$O$_3$, SiO$_2$ at 1600 °C, working range framed in blue [31]

Fig. 3.11 Melting temperatures of calcium aluminate slags [38]

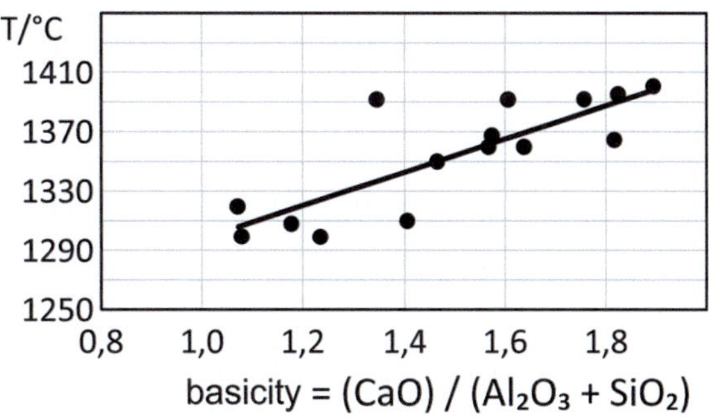

The equations of both progressions are:

$$\text{For } B > 0.76 : (\% \, MgO)_{sat} = 10.572 \cdot B^{-0.9867} \rightarrow \langle \text{Periclase} \rangle,$$
$$\text{for } B < 0.76 : (\% \, MgO)_{sat} = 45.262 \cdot B^{+4.2879} \rightarrow \langle \text{MA spinel} \rangle. \qquad (3.46)$$

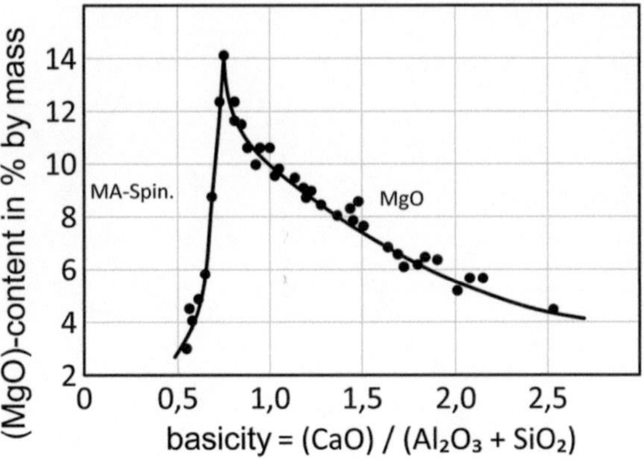

Fig. 3.12 Saturation concentration of MgO in lime-aluminate slags at steel mill temperatures [39]

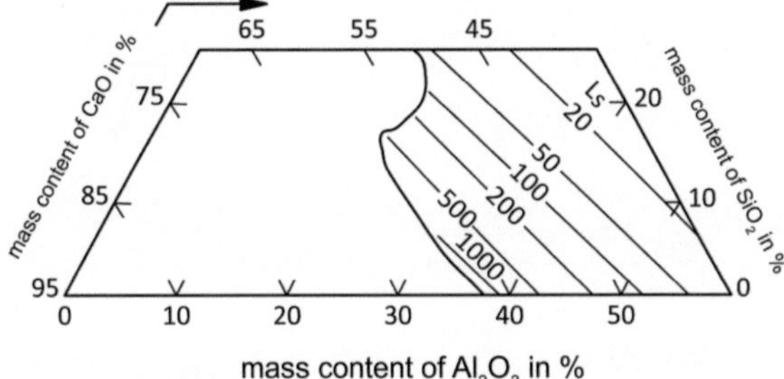

Fig. 3.13 Sulfur distribution L_S between slag (5% by weight MgO) and Al-deoxidized steel ([Al] = 0.07% by weight) at 1625 °C [18]

The following two relationships were calculated with "**Fact-Sage**":
In the limits $0 < SiO_2° < 10\%$ by weight and $0.7 < B < 1.4$ this applies:

$$(\% \, MgO)_{sat} = 20.7 - 11.5 \cdot B + 0.02 \cdot \left(T\left[°C\right] - 1600\right) \tag{3.47.1}$$

and for $10 < SiO_2° < 20\%$ by weight approximately

$$(\% \, MgO)_{sat} = 9.5/B + 0.02 \cdot \left(T\left[°C\right] - 1600\right). \tag{3.47.2}$$

$B = CaO°/(Al_2O_3° + SiO_2°)$ is the defined basicity of the slag. It is calculated by converting the components of basicity to 100%, neglecting minority components such as FeO, MnO and CaF₂. For $B < 0.76$, MA spinel excretes, above which periclase excretes. The temperature range is 1600–1700 °C [40]. For 1650 °C and $B = 1.5$, $(\% \, MgO)_{sat} = 7.4\%$ by weight is calculated.

The distribution of sulfur dissolved in slag and steel bath also follows basicity (Fig. 3.13) [18, 31]. With basicity, the sulfur distribution $L_S \equiv (S)/[S]$ increases. Lime-based slags favor the desulfurization of the steel.

Phosphorus is bound in the slag as calcium phosphate ($Ca_3(PO_4)_2$), which is why dephosphorization is improved during the production of crude steel as the basicity of the lime-rich slag increases [18].

The slagging of iron and manganese decreases with increasing basicity [18].

These few examples already show that the empirical term "basicity" has proved its worth for the quantitative description of metallurgical reactions, which also include chemical corrosion. For the sake of completeness, two physically justifiable basicities should also be mentioned: The concept of the pO value derived from the ion theory of slags [41] and the optical basicity B_{OP} [42].

pO Value as Basicity Measure

The measure of basicity

$$pO \equiv -\log a_{O^{2-}} \tag{3.48}$$

was defined following the basis of the pH value of aqueous solutions by Frohberg and Kapoor [41]. The pO value is based on the ion theory of slags and describes the stability of an oxygen compound in the slag. The pO value is calculated from activity measurements of the oxides of a multicomponent system. Depending on the activity and the position of the ionic oxygen on the left or right side of a given reaction equation, changing the pO value shifts the equilibrium more to the left or to the right. For example, sulfur dissolved in a steel bath reacts with a slag according to

$$\left(O^{2-}\right) + [S] = [O] + \left(S^{2-}\right). \tag{3.49}$$

The equilibrium constant is for $a_{S^{2-}} \equiv (\%S)$

$$K = \frac{(\%S)}{[S]} \cdot \frac{[O]}{(a_{O^{2-}})}. \tag{3.50}$$

The sulfur capacity of the slag is defined as

$$C_S \equiv [O] \cdot \frac{(\%S)}{[S]}, \tag{3.51}$$

from which it follows immediately by Eq. (3.50):

$$\log C_S = \log K - pO. \tag{3.52}$$

The equilibrium constant K depends only on the temperature. In calcium aluminate slags at 1600 °C, $K = 0.06$ [18]. At lime saturation, this applies: $a_{(CaO)} = a_{Ca^{2+}} \cdot a_{O^{2-}} = 1$, i.e. $a_{Ca^{2+}} = a_{O^{2-}} = 1$ and, consequently, pO = 0. As the pO value increases, C_S becomes smaller, i.e. for a given oxygen content of the melt, the desulfurization effect of the slag decreases (Eq. 3.51). It is calculated taking into account the oxygen content previously set by the deoxidation of the melt. For example, after aluminum deoxidation $[O] = 6$ ppm ($6 \times 10^{-4}\%$ by weight) is reached. Then, because

Table 3.1 Calculated optical basicities [31]

Oxides	Electronegativity	Λ
K_2O	0.8	1.40
Na_2O	0.9	1.15
BaO	0.9	1.15
Li_2O	1.0	1.00
CaO	1.0	1.00
MgO	1.2	0.78
Al_2O_3	1.5	0.61
SiO_2	1.8	0.48
B_2O_3	2.0	0.42
P_2O_5	2.1	0.40
CO_2	2.5	0.33
SO_3	2.5	0.33

of $K = C_S$, the sulfur distribution at lime saturation is $L_S \equiv (\% \text{ S})/[\text{S}] = C_S/[\text{O}] = 0.06/0.0006 = 100$ (refer to Fig. 3.13).

Optical Basicity

Duffy and Ingram [42] found out that the increasing activity of oxygen ions in slags with chemical basicity is directly linked to the tendency to discharge electrons. Therefore, if the slag is irradiated with UV light, the change in absorption frequency can be measured in comparison with a standard. By definition, the standard is CaO. If the Pauling electronegativities are denoted by χ, then the optical basicity of a slag component is

$$\Lambda_i = \frac{0.75}{\chi - 0.25}. \tag{3.53}$$

The attached Table 3.1 allows the simple calculation of each component.
For the total slag:

$$\Lambda_{\text{gas}} = \frac{\sum_1^m x_i \cdot n_i \cdot \Lambda_i}{\sum_1^n x_i \cdot n_i}. \tag{3.54}$$

An example is the slag 55% CaO, 30% Al_2O_3, 15% SiO_2. Converted to mole fractions, we get $x_{CaO} = 0.64$, $x_{Al_2O_3} = 0.19$, $x_{SiO_2} = 0.17$. The number of oxygen atoms in the molecule is $n_{CaO} = 1$, $n_{Al_2O_3} = 3$, $n_{SiO_2} = 2$. It follows

$$\Lambda = \frac{0,64 \cdot 1.1 + 0,19 \cdot 3 \cdot 0,61 + 0,17 \cdot 2 \cdot 0,48}{0,64 \cdot 1 + 0,19 \cdot 3 + 0,17 \cdot 2} = 0,74 \tag{3.55}$$

In the **Slag Atlas** [43] there are some examples of the calculation of slag capacities for sulfur and phosphorus with the assistance of optical basicity.

Both physically based (explained) basicities are hardly used in practice because, on the one hand, too few measured values are available and, on the other hand, people in metallurgy have become accustomed to the empirically defined chemical basicity.

AOD-Process

Decarburization

After the scrap is melted in the electric arc furnace, the steel is decarburized in the AOD converter from initially approx. 1.5% by weight to 0.02% by weight [C]. In addition, deviating from the LD process, a mixture of argon and oxygen is blown into the melt. The argon reduces the partial pressure of the carbon monoxide produced and, following the Vacher–Hamilton product [1], enhances decarburization with sparing use of oxygen. In this way, the oxidation of the chromium is limited, which must finally be reduced by deoxidation with silicon from the slag back into the melt. The following reactions are decisive for this:

$$
\begin{aligned}
3[C] + 3[O] &= 3\{CO\} & \Delta G^0(1600\,°C) &= -2.80 \times 10^5 \,[J/FU] \\
(Cr_2O_3) &= 2[Cr] + 3[O] & \Delta G^0(1600\,°C) &= +1.38 \times 10^5 \,[J/FU] \\
\hline
(Cr_2O_3) + 3[C] &= 2[Cr] + 3\{CO\} & \Delta G^0(1600\,°C) &= -1.41 \times 10^5 \,[J/FU]
\end{aligned}
$$

$$\tag{3.56}$$

The equilibrium constant at 1600 °C is

$$
K = \frac{a_{[Cr]}^2 \cdot p_{\{CO\}}^3}{a_{(Cr_2O_3)} \cdot a_{[C]}^3} = 8.96 \times 10^3. \tag{3.57}
$$

In the course of the process, the proportion of argon is increased, causing the CO partial pressure to drop continuously. The carbon content decreases and the chromium oxide content increases only slightly. Finally, pure argon is blown and thereby deoxidized, thus recovering the initial chromium content of the molten steel from out of the slag.

The lining consists of magnesia chromite and pitch-bonded sintered magnesia in the slag and impact area of the melt, and dolomite bricks in the melt area. Due to the very high temperature during the initial refining (oxidation phase), wear is high because the solubility of the refractory material in the slag increases with rising temperature. In addition, as a result of CO formation, enormous convection occurs, which facilitates erosion (refer to Sect. 3.1.1.5.3.2.1). Argon injection alone drives convection only via the thermal expansion of the gas, i.e. much less. Tar-bonded MgO–C material is less durable than magnesia chromite because at high temperatures

MgO and C clearly react with each other and no protective MgO layer is formed on the brick surface during intensive deoxidation [44].

In some cases, fluorspar is used to accelerate slag formation. The slag becomes more liquid. As a result, the mass (material) transfer coefficient increases and refractory wear increases likewise. Calculated with "**FactSage**" and experimentally confirmed, the saturation equation at steel mill temperatures is [45]:

$$(\% \, \text{MgO})_{\text{sat}} = 8.2/B + 0.06 \cdot (\% \, \text{FeO}) + 0.2 \cdot (\% \, \text{CaF}_2) + 0.019 \cdot (T - 1550\,^{\circ}\text{C}) \tag{3.58}$$

with basicity $B \equiv \text{CaO}/(\text{Al}_2\text{O}_3 + \text{SiO}_2)$.

Wear Due to Bubble Blowback (Back Attack)

Wear is particularly high in the nozzle area. There, the argon/oxygen mixture is fed into the melt at high pressure through bottom or side nozzles under a nitrogen fog. The result is a swarm of bubbles pulsating at high frequency. The large bubbles in particular are recirculated by the flow and burst abruptly once they hit the wall. During this process, the rebounding molten steel transmits a strong impulse to the lining. This has an effect comparable to cavitation because the refractory material is shattered and subjected to intense wear (refer to Sect. 3.1.1.3.3).

This effect is major (approx. 1 mm/h) and must be considered in addition to the chemical wear caused by the molten steel. The chemical wear is also particularly intense due to the heavy flow in the area of the nozzles. The mass (material) transfer coefficient is known to increase with the root of the flow velocity: $\beta \sim \sqrt{u}$ [36].

Ladle Furnace-Process

The steel is treated in the ladle furnace in a ladle which is closed with a lid. Graphite electrodes penetrate through the lid all the way close to the molten steel. When the electrical voltage is switched on, an electric arc is generated. In this way, in addition to the metallurgical fine-tuning of the melt, the temperature can be precisely adjusted. The use of plasma torches has been successfully investigated, but has not become established in practice. Relatively abrasion-resistant magnesia chromite bricks have been installed.

VOD-Process

For stainless steel production, the final setting of the alloy composition is often carried out in the Vacuum-Oxygen-Decarburization (VOD) process. For this purpose, the ladle is placed in a vacuum vessel or sealed vacuum-tight with a lid. More consistently than in the AOD process, the steel is treated with oxygen and argon at reduced pressure in order, on the one hand, to bring the carbon and phosphorus contents to the lowest possible counts and, on the other hand, to greatly reduce the gas contents (hydrogen and nitrogen).

The solubility of gases follows the law of Sievert [18]

Hydrogen:

$$\frac{1}{2}\{H_2\} = [H], \quad K_{[H]} = \frac{[H]}{\sqrt{P_{H_2}}}, \quad \log K_{[H]} = -\frac{1823}{T} - 1.63. \tag{3.59}$$

Nitrogen:

$$\frac{1}{2}\{N_2\} = [N], \quad K_{[N]} = \frac{[N]}{\sqrt{P_{N_2}}}, \quad \log K_{[H]} = -\frac{285}{T} - 1.21. \tag{3.60}$$

The contents of the dissolved gases decrease with the square root of their partial pressure. The solubility of argon is vanishingly small.

The same applies for the lining and its wear as is the case in AOD process. The wear pattern (picture) suggests erosion [44] (refer to Sect. 3.2.1.3).

RH-Process

In the Ruhrstahl–Heraeus process (Fig. 3.14), a vacuum circulation process, a partial quantity of the melt is conveyed in a continuous process, driven by a bubble column (argon), into an evacuated space above the ladle and degassed there (mammoth pump). Next the partial quantity flows back into the ladle where it mixes with the melt. The efficiency is high and extremely low gas contents are achieved. By blowing in oxygen, decarburization takes place at ultra-low values (RH-OB process).

Magnesia chromite has proven to be the best for the refractory lining in the snorkel area. All other materials wear much faster, i.e. they erode (see above). Wear occurs mainly by erosion (refer to Sect. 3.2.1.3). However, due to the high flow velocity of the melt and its high gas load, cavitation is also conceivable as a wear mechanism (refer to Sect. 3.2.1.4).

Erosion A simple calculation can help to estimate the abrasion:

It is assumed that the turbulent flow exerts shear forces on the grains of the refractory lining and breaks them out of the binding matrix. A spherical grain of diameter D [m] is looked at. The portion h_1/D protrudes into the melt, the portion h_2/D is anchored in the structure (refer to Fig. 3.15).

The critical flow velocity u above where the grain breaks out results from the force balance

$$C_W \cdot \pi \cdot D \cdot h_1 \cdot \frac{\rho}{2}u^2 = \pi \cdot D \cdot h_2 \cdot \sigma_B \ [N] \tag{3.61}$$

and is [46]

Fig. 3.14 Schematic illustration of the Ruhrstahl–Heraeus process (Wikipedia)

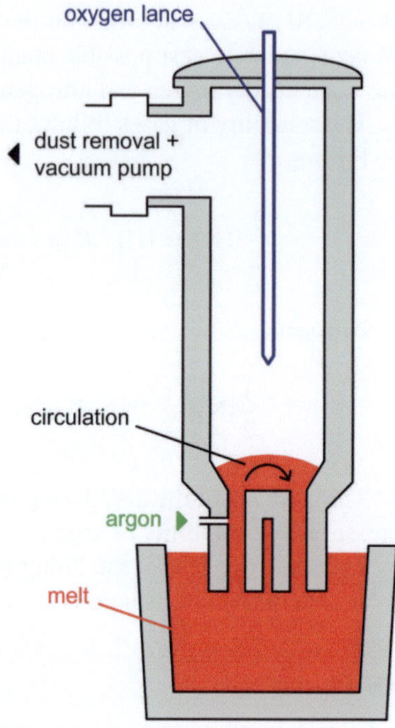

Fig. 3.15 Model of the grain subjected to flow [46]

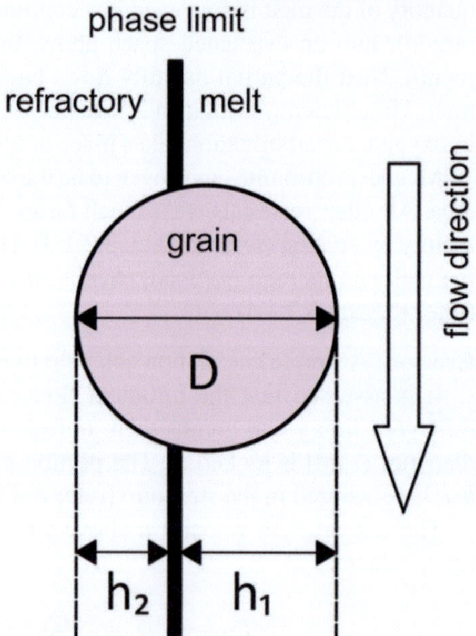

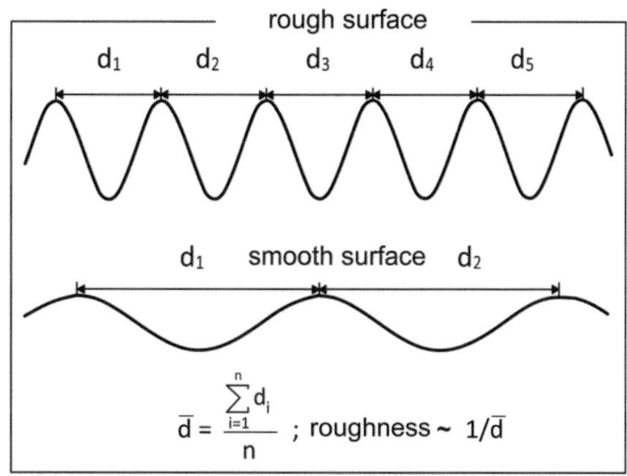

Fig. 3.16 Definition of surface roughness [47]

$$\bar{d} = \frac{\sum_{i=1}^{n} d_i}{n} \quad ; \text{ roughness} \sim 1/\bar{d}$$

$$u \geq \left(4 \cdot \frac{h_2}{h_1} \cdot \frac{\sigma_B}{\rho} \right)^{\frac{1}{2}} [\text{m/s}] \tag{3.62}$$

With the values $h_2/h_1 = 0.1$, $\rho = 7 \times 10^3$ kg/m^3 steel, the flow coefficient $C_W = 0.5$ and $\sigma_B = 0.5$ N/mm$^2 = 0.5 \times 10^6$ kg m/(s^2 m^2), $u \geq 5$ m/s is obtained. However, especially if the bond between the matrix and the grain is already loosened by chemical attack, it breaks out more easily [47].

Cavitation The high flow velocity of the melt, its loading with gases and the surface roughness of the refractory lining can lead to cavitation, which destroys the microstructure [46]. Gas bubbles are formed which remain stable only for a short time. Their collapse occurs at an approximate sonic velocity with such a high momentum being transferred to the refractory lining that the microstructure shatters and crumbles.

The equation of Bernoulli [28]

$$\rho \cdot g \cdot h + P_a = P_S + \frac{\rho}{2} \cdot u^2 \left[\text{N/m}^2 \right] \tag{3.63}$$

explains cavitation with the requirement that the remaining static pressure P_S must be smaller than the steam pressure P_D of the liquid at a given total pressure $\rho \cdot g \cdot h + P_a$ and locally prevailing flow pressure (dynamic pressure) $\frac{\rho}{2} \cdot u^2 \left[\text{N/m}^2 \right]$. The pressure of dissolved gases, e.g. hydrogen and nitrogen, must be added to the steam pressure.

The defined roughness of the surfaces is illustrated in Fig. 3.16 and follows the empirical proportionality $R \sim n / \sum_{i-1}^{n} d_i$. d_i is the distance between two adjacent grain peaks (spikes).

This surface roughness facilitates the formation of voids because the wetting angle $\theta \approx 120°$, so that the capillary pressure $P_\sigma = 4 \cdot \sigma \cos \theta / d$ is added. Next comes equation [47] follows:

$$\rho \cdot g \cdot h + P_\mathrm{a} = P_\mathrm{S} + \frac{\rho}{2} \cdot u_\mathrm{C}^2 - \frac{4 \cdot \sigma \cdot \cos\theta}{d} \; \left[\mathrm{N/m^2}\right]. \qquad (3.64)$$

The critical flow velocity is therefore:

$$u_\mathrm{krit} = \sqrt{\frac{2 \cdot \left[\rho \cdot g \cdot h + P_\mathrm{a} - P_\mathrm{S} + \frac{4 \cdot \sigma \cdot \cos\theta}{d}\right]}{\rho}} \; \left[\mathrm{m/s}\right] \qquad (3.65)$$

The steam pressure of the steel and the partial pressures of dissolved gases are generally very low ($< 10^{-4}$ atm) which is why the even lower static pressure $P_\mathrm{S} = 0$ is set for an estimate. With the known numerical values, at the pressure $P_\mathrm{a} = 0.01$ atm $= 1013 \; \frac{\mathrm{kg\,m}}{\mathrm{m^2\,s^2}}$ in the recipient and the roughness $d = 10^{-4}$ m in the vicinity of the bath surface, i.e. at $h = 0.5$ m, the critical velocity is calculated to be $u_\mathrm{crit} = 1.24$ m/s:

$$u_\mathrm{krit} = \sqrt{\frac{2 \times \left[7 \times 10^3 \times 9.81 \times 0.5 + 1013 - 0 + \frac{4 \times 1.5 \times \cos 120°}{10^{-4}}\right]}{7 \times 10^3}} = 1.24\,\mathrm{m/s}$$

$$(3.66)$$

Since the critical flow velocity is smaller than the average bath velocity (Eq. 3.85: 3 m/s), cavitation cannot be excluded. If roughness is not taken into account, $u_\mathrm{crit} = 3.2$ m/s. The surface roughness is, therefore, an important influencing variable. Experience shows that it increases with time [48].

Degree of Purity

The degree of purity, a term not precisely defined, describes the non-metallic inclusions present in the solidified material. The smaller their number and size during casting and the more favorably they behave during deformation, the higher the degree of purity can be classified. A distinction is made between the oxidic and the sulfidic degree of purity. Often, excretions of both types are present side by side and associated with each other. Dissolved gases or gases present as compounds also have a negative influence on the degree of purity, e.g. hydrogen and nitrogen (nitrides).

The degree of purity is subject to many influences. Examples are:

- The used raw materials.
- The metallurgical process sequence.
- The used deoxidizing and desulfurizing agents.
- The deposition behavior of the excretions.
- The used slags.
- The atmosphere and its pressure.
- The refractory material.

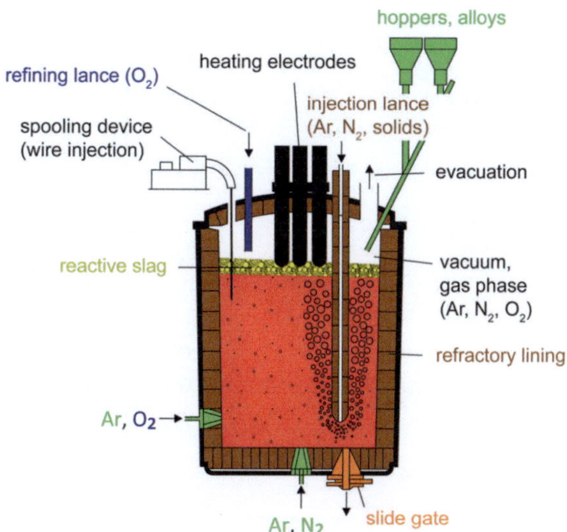

Fig. 3.17 Steel ladle as metallurgical reactor [37]

In the present case, the refractory material is the focus of interest.

For the production of steels of the highest quality, for about 25 years the secondary metallurgy is becoming increasingly important. The steel ladle is used as a metallurgical reactor, also under vacuum. Figure 3.17 illustrates the multiple possibilities for influencing the metallurgical process: Heating, alloying, refining, slag changing, purging, evacuating and casting are some examples [37].

Deoxidation

During refining, some of the oxygen used remains dissolved in the steel.

A distinction is made between the dissolved oxygen content and the total present, the so-called total content. The difference is the excreted oxides. Even with a low, electrochemically determined dissolved oxygen content, the degree of purity can be unsatisfactory if, in terms of quality, the total oxygen content is too high. This is counteracted by purging with inert gas. These problems occur particularly during deoxidation with aluminum. Since nucleation of the alumina in the steel is inhibited, equilibrium is not achieved during deoxidation and new inclusions are continuously formed during cooling and solidification.

The refractory material reacts with each process step. Especially in a vacuum, the lining of the ladle furnace is subjected to additional stress. An obvious example is the deoxidation of steel melts with carbon. With decreasing pressure, the so-called Vacher–Hamilton equilibrium [1] between the carbon and oxygen components dissolved in the iron increasingly shifts toward low contents. At 1600 °C, the following applies [1] (refer to Sect. 3.2.5)

$$[C] \cdot [O] = 2.4 \times 10^{-3} \cdot P_{CO} \tag{3.67}$$

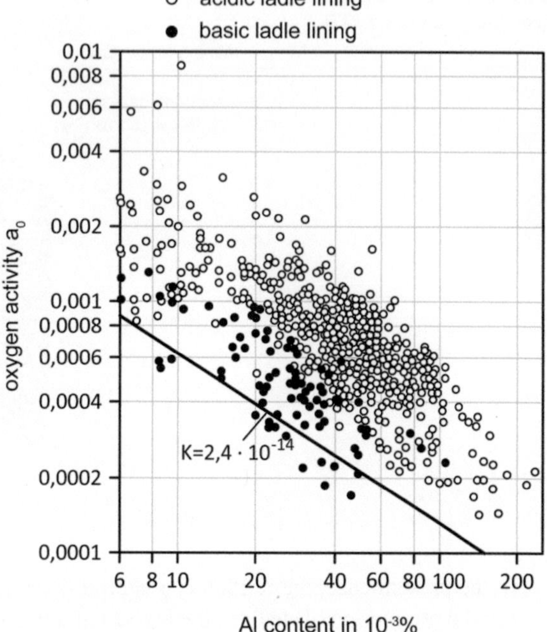

Fig. 3.18 Influence of the ladle lining on the achievable equilibrium during deoxidation with aluminum under normal pressure [37]

As a result, the tendency of the refractory material to go into solution increases; the more so the less negative its free enthalpy of formation ΔG^0 is (refer to Sect. 2. 1.3.3 Fundamentals: Richardson–Ellingham Diagram Fig. 2.5). Therefore, the use of silica in vacuum furnaces is declining. Preference is given to more stable oxides such as chrome magnesia, alumina, magnesia and MA spinel [49, 50]. Figure 3.18 [37] illustrates this on the basis of the result of a large number of operational deoxidations in acidic and basic ladle linings under normal pressure and Fig. 3.19 on the basis of the deoxidation equilibria calculated with "**FactSage**"

The acidic lining does not allow to set oxygen contents as low as the basic lining, because it is thermodynamically less stable. The plotted points are measured electrochemically; i.e. they represent the dissolved oxygen content. The reason for the deviation from the drawn thermodynamic equilibrium shown can be explained partly by kinetic impairments during oxide nucleation, but mainly by the re-oxidation of the melt due to the dissolution of the refractory material.

For comparison, the dissolution and deoxidation equilibria of the most important oxides in iron at 1600 °C are compared [1, 7, 51–53]: The basis is the reaction

$$m[\text{Me}] + n[\text{O}] = \langle \text{Me}_n\text{O}_n \rangle;$$
$$K = [\text{Me}]^m \cdot [\text{O}]^n;$$
$$[\text{O}] = \left(\frac{K}{[\text{Me}]^m} \right)^{1/n}.$$

(3.68)

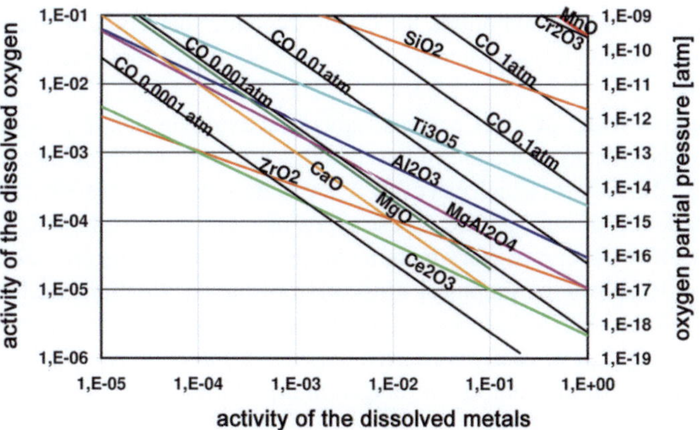

Fig. 3.19 Deoxidation equilibria in iron at 1600 °C

$$K_{\text{Fe,MnO}} = [\text{Mn}] \cdot [\text{O}] = 5 \times 10^{-2}, \ (1600\,^{\circ}\text{C})$$

$$K_{\text{Cr}_2\text{O}_3} = [\text{Cr}]^2 \cdot [\text{O}]^3 = 1.4 \times 10^{-4}, \ (1600\,^{\circ}\text{C})$$

$$K_{\text{P}} = [\text{C}] \cdot [\text{O}] = 2.4 \times 10^{-3} \cdot P_{\text{CO}}, \ (1600\,^{\circ}\text{C})$$

$$K_{\text{SiO}_2} = [\text{Si}] \cdot [\text{O}]^2 = 2 \times 10^{-5}, \ (1600\,^{\circ}\text{C})$$

$$K_{\text{Ti}_3\text{O}_5} = [\text{Ti}]^3 \cdot [\text{O}]^5 = 1.6 \times 10^{-19}, \ (1600\,^{\circ}\text{C})$$

$$K_{\text{Al}_2\text{O}_3} = [\text{Al}]^2 \cdot [\text{O}]^3 = 8 \times 10^{-14}, \ (1600\,^{\circ}\text{C})$$

$$K_{\text{Spin}} = \left[\text{Mg}\right] \cdot [\text{Al}]^2 \cdot [\text{O}]^4 = 5 \times 10^{-20}, \ (1600\,^{\circ}\text{C})$$

$$K_{\text{ZrO}_2} = [\text{Zr}] \cdot [\text{O}]^2 = 1.1 \times 10^{-10}, \ (1600\,^{\circ}\text{C})$$

$$K_{\text{Ce}_2\text{O}_3} = [\text{Ce}]^2 \cdot [\text{O}]^3 = 1 \times 10^{-17}, \ (1600\,^{\circ}\text{C})$$

$$K_{\text{MgO}} = \left[\text{Mg}\right] \cdot [\text{O}] = 2 \times 10^{-10}, \ (1600\,^{\circ}\text{C})$$

$$K_{\text{CaO}} = [\text{Ca}] \cdot [\text{O}] = 1.4 \times 10^{-12}, \ (1600\,^{\circ}\text{C})$$

In the order from top to bottom, the thermodynamic stability of oxides increases. Of the most common lining materials, MgO and CaO are the most stable for resisting steel melts.

Deoxidation with CaC_2 or Ce attacks any refractory material. If deoxidation with aluminum takes place in a magnesia-based lining, not deformable inclusions of spinel ($MgO \cdot Al_2O_3$) can form, which can cause a devastating decline in quality, specifically in the production of wire grades. These types of inclusions are feared. The use of zirconium silicate as a refractory material helps to minimize this problem [54].

Figure 3.19 shows the graphical illustration of the deoxidation equilibria calculated above for 1600 °C. At low contents, the activity and concentration of oxygen are the same and can be easily converted (calculated) into oxygen partial pressures. With the reaction of the solution of oxygen in liquid iron [53]

$$\frac{1}{2}\{O_2\} = [O], \quad \Delta G^0 = -137{,}118 + 7.79 \cdot T \; [\text{J/FU}] \tag{3.69}$$

is obtained at 1600 °C with $\ln K = -\Delta G^0 / R \cdot T$

$$K_{[O]} = 2.62 \times 10^{+3} \text{ and } P_{O_2} = 1.54 \times 10^{-7} \cdot [O]^2 \; [\text{atm}]. \tag{3.70}$$

Under reduced CO pressure, the oxygen content can be decreased further, thereby reducing increasingly stable oxides.

Desulfurization

Pig iron contains approx. 0.03% by weight sulfur. A large part of the desulfurization already takes place in the torpedo ladle. The influence of the refractory material on the deterioration of the cleanliness level triggered by an excessive number of sulfides can be derived from the desulfurization reaction:

$$\begin{aligned} [S] + (CaO) &= (CaS) + [O], \\ [O] + x[Me] &= (Me_xO) \\ \hline [S] + (CaO) + x[Me] &= (CaS) + (Me_xO) \end{aligned} \tag{3.71}$$

The reaction scheme shows firstly that the sulfur in the slag is primarily bound as calcium sulfide and, secondly, that the oxygen content in the steel must be as low as possible in order to desulfurize successfully. Consequently, fierce deoxidation can lower the dissolved sulfur content sufficiently before solidification to minimize sulfide excretions. It is obvious that an acidic lining of the ladle is unsuitable because its oxygen potential is high. Rather, dolomite is recommended as refractory material and the use of a lime-based top slag.

Example Deoxidation with metallic aluminum enables the setting $[O] = 2$ ppm dissolved oxygen in the steel melt. The top slag has the following composition in %

in weight: 58% CaO, 32% Al$_2$O$_3$, 8% SiO$_2$ and 2% MgO. This slag is saturated in lime; i.e. its activity is $a_{CaO} = 1$. Consequently, [52]

$$[S] + (CaO) = (CaS) + [O],$$
$$\Delta G^0 \,(1600\,°C) = +45.7 \times 10^3 \,J/mol$$
$$[O] + 2/3[Al] = 1/3(Al_2O_3),$$
$$\Delta G^0 \,(1600\,°C) = -162.5 \times 10^3 \,J/mol$$

$$\overline{[S] + (CaO) + 2/3[Al] = (CaS) + 1/3(Al_2O_3),}$$
$$\Delta G^0 \,(1600\,°C) = -116.8 \times 10^3 \,J/mol \qquad (3.72)$$

The equilibrium index is:

$$\frac{a_{(CaS)}}{[S]} \cdot \frac{a_{(Al_2O_3)}^{1/3}}{a_{[Al]}^{2/3} \cdot a_{(CaO)}} = 1.82 \times 10^3 \qquad (3.73)$$

From Eq. (3.71) one takes the close correlation between the contents of sulfur [S] and oxygen [O] in the melt which is why the sulfur concentration can generally be determined by electrochemical measurement of the dissolved oxygen after appropriate calibration [55].

Example In equilibrium with the oxygen dissolved in the steel $[O] = 2 \times 10^{-4}\%$ by weight there is $[Al] = 0.07\%$ by weight aluminum. The activity of the lime is $a_{(CaO)} = 1$; that of the aluminum oxide in the slag is $a_{(Al_2O_3)} = 0.03$ [43]. The maximum solubility of sulfur in the lime-saturated slag at 1600 °C is $(CaS)_{max} = 4.8\%$ by weight, and the activity is then $a_{(CaS)} = 1$ [43]. With this data, from (Eq. 3.73) the sulfur content in the steel is calculated to be $[S] = 1 \times 10^{-3}\%$ by weight which corresponds to operational practice.

The use of calcium carbide for desulfurization is also very effective:

$$\langle CaC_2 \rangle + [S] = \langle CaS \rangle + 2[C], \ \ \Delta G^0 \,(1600\,°C) = -1.51 \times 10^5. \qquad (3.74)$$

If volatile metals, such as calcium and magnesium, are used the application of positive pressure increases their effect, as Fig. 3.20 shows.

Desulfurization using a vacuum can be achieved indirectly via the formation of carbon monoxide and the associated decarburization or deoxidization of the melt. The injection of argon is also a common method of lowering the partial pressure of the dissolved gases, specifically carbon monoxide.

In addition, the melt is stirred by the bubbles. The effect of the utilized gas quantity $\dot{Q}$ [Nl/ min] on the desulfurization follows a first-order reaction:

$$[S] = [S_0] \cdot \exp(-kt) \,[wt\%] \qquad (3.75)$$

Fig. 3.20 Steel desulfurization a_S by dissolved metals a_x in % by weight (1600 °C) [52]

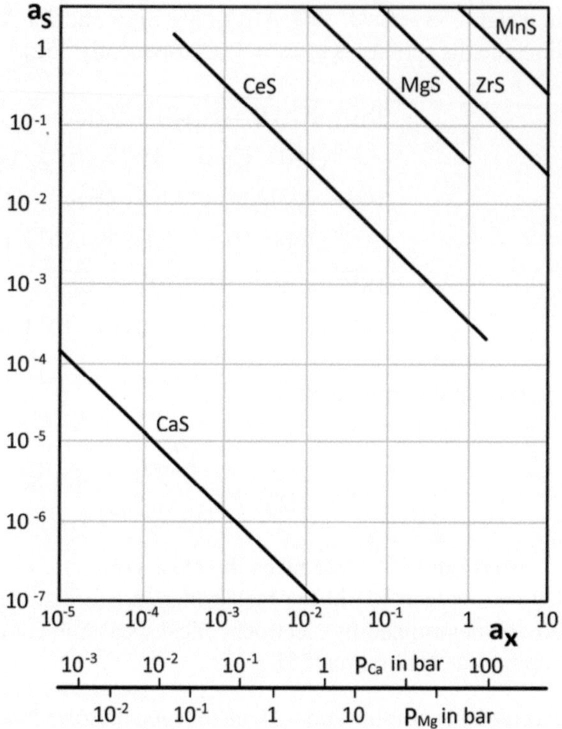

The rate constant k [1/s] for light stirring is [56]

$$\dot{Q} \le 140\,\mathrm{Nl/\,min} : k_< = 1.9 \times 10^{-4} \cdot \dot{Q}^{0.29}\,\left[\mathrm{s}^{-1}\right] \qquad (3.76.1)$$

and for strong stirring [56]

$$\dot{Q} > 140\,\mathrm{Nl/\,min} : k_> = 1.2 \times 10^{-8} \cdot \dot{Q}^{2.26}\,\left[\mathrm{s}^{-1}\right] \qquad (3.76.2)$$

With strong bath agitation, slag and steel bath emulsify, increasing the mass (material) transfer. If the initial content [S] is$_0$ = 0.005% by weight and the rinsing time t = 15 min with $\dot{Q}$ = 70 Nl/ min, then according to Eq. (3.76) $k_<$ = 6.5 × 10^{-4} s^{-1} and the sulfur content declines by about half to [S] = 2.7 × 10^{-3}% by weight.

Premature Wear at the Submerged Nozzle

Experiments

The wear of submerged nozzle material was investigated in the laboratory [57]. It consists (in % by weight) of 60% Al_2O_3, 25% graphite, 20% SiO_2 and 5% SiC. Round bars with a diameter of 12 mm are immersed under inert gas at 1600 °C to a depth of approx. 40 mm in iron melts and are rotated around their axis at circumferential speeds of up to 360 cm/min. The melts contain different carbon contents up to saturation. On the melt one sees CAS slags of varying composition whose basicity is about 1, up to saturation with Al_2O_3. After 15 min, the tests are stopped because the depth of the corrosion channel (trough) formed at the three-phase line is already more than 1 mm (Fig. 7.2c).

Results

The rotational speed of the specimen has no influence on the corrosion. The depth of the corrosion channel increases linearly in time and increases with falling viscosity of the slag. On average, in low-carbon melt, 6.5 mm of submerged nozzle material wears in one hour over an area of 4 cm^2. This is 2.6 cm^3/h. Wear decreases as the amount of carbon dissolved in the iron increases. If the melt is saturated with carbon, the premature wear ceases completely. Slag saturated with alumina, on the other hand, does not prevent wear. On the surface of the part of the rod immersed in the iron, a sealing layer of α-corundum, or calcium hexaaluminate (CaO–6 Al_2O_3) is formed, depending on the composition of the slag. Typical are 50% Al_2O_3, 40% CaO, 10% SiO_2, T_m ~ 1350 °C.

Discussion

The arrangement and shape of the corrosion channel, as well as the speed of its formation, allow the conclusion that it is a corrosion process caused by wetting effects, namely Marangoni convection [57, 58]. This mechanism is described in detail in Sect. 2.2.5.2 and calculated using the example of corrosion of AMC material in ladles (Sect. 7.6.3). The detailed quantitative treatment is, therefore, not done here.

For the wear of AMC material, the dissolution of the oxide is the speed-determining process. Submerged entry nozzle material contains more than twice as much graphite which is why the dissolution of the carbon into the slag is the decisive partial step [57] (refer to Sect. 13.1.5.2). As soon as the slag in the gap between the submerged entry nozzle and the melt is saturated with carbon; i.e. its activity is one everywhere, the mass (material) transfer driving force disappears completely. This is the case if the iron is saturated with carbon.

The test results allow us to directly calculate the change in the mass transfer coefficient: In one hour, about 2.6 cm^3 submerged nozzle material wears out. It has a density $\rho_{TA} = 2.8$ g/cm^3. Consequently, 7.3 g of submerged nozzle material wears out in one hour. Since it is 25% graphite, 1.83 g of graphite enters the slag every hour which is 5.1×10^{-4} g/s. Therefore, in relation to the area of the corrosion zone $F = 4$ cm^2, the mass (material) flux density of carbon is $j = 1.25 \times 10^{-4}$ g/cm^2 s. The solubility of carbon in the slag is 0.1% by weight [59] (refer to Sect. 13.1.5.2). This value represents the driving concentration difference ΔC of carbon. The density of the slag is $\rho_{Sl} = 2.8$ g/cm^3, i.e. $\Delta C = 2.8 \times 10^{-3}$ g/cm^3. Since $j = \beta \cdot \Delta C$ (Nernst), the mass transfer coefficient is calculated to be $\beta = 0.045$ cm/s. For natural convection,

Schubert and Schwerdtfeger [59] determined the mass (material) transfer coefficient during the dissolution of graphite in slags to be $\beta = 0.004$ cm/s. As a result of the Marangoni convection, the mass (material) transfer increases by a factor of 11 in the present case and causes the premature wear.

The formation of the oxidic top layer at the interface between the submerged nozzle and the iron is the result of the reduction of the alumina or CaO in the structure to the volatile suboxide Al_2O or rather to the gases Ca and CO by graphite at 1600 °C (refer to Sect. 13.1.3). The gases diffuse to the surface of the material and react there from out of the iron at the interface with the submerged nozzle to form the observed oxide (refer to Sects. 13.1.2 and 13.2.3).

3.1.3 Opening Rate of the Slide Gate

Normally ceramic slide gates are used during continuous casting to close ladles (refer to Fig. 12.6). Before pouring the steel into the ladle and with a shut slide gate the sleeve is filled with sand in order to prevent that steel flows prematurely, solidifies and closes off the opening. If the slide gate nonetheless does not happen to open the nozzle is "burned free". The opening rate should be above 95 % in order to ensure sufficient efficiency. How can this requirement be achieved reliably?

Five partial steps must be coordinated:

– Filling in the slide gate sand
– Moving the ladle to the melting furnace (waiting time)
– Emptying of the melt
– Moving the ladle into casting position (waiting time)
– Opening of the slide gate (casting)

The sand must fill the nozzle completely and normally it forms to a little mound at the bottom of the ladle. During movement of the ladle to the melting furnace the sand starts to sinter. The incoming melt normally pushes the sand sideways so that the passage to the nozzle is levelled. While the filled ladle is moving into casting position the sand continues to sinter and forms a lid. This lid breaks upon the sand running out as soon as the slide gate opens. The steel flows into the tundish and from there through the submerged nozzle into the ingot mold.

The sintering behavior of the sand and the design of the sleeve are the determining factors for a successful opening of the slide gate. Two characteristic dependencies must be observed for its enhancement:

$$\dot{\rho}\; prop \cdot \rho \cdot d^{-n} \tag{3.77.1}$$

$$\sigma_{B\max} \leq 0,3 \cdot \rho_{Fe} \cdot g \cdot h_{Fe}\left(\frac{D}{H}\right)^2 \tag{3.77.2}$$

Equation (3.77.1) [81] describes the sintering process of the sand. Its bulk density P increases the faster the smaller the grain diameter d is. The exponent n depends on the transportation mechanism of the sintering process. For grain boundary diffusion (compaction) and surface diffusion (grain growth) n is equal 4. For lattice diffusion (compaction) n is equal 3 and for transportation over the gas phase (grain growth) n is equal 2 [81].

However, in this case reaction sintering and liquid phase sintering are major factors. The iron oxide in the sand reacts with the silica enabling the formation of a deep-melting phase ($2FeO \cdot SiO_2$, 1205 °C) which is surrounded by two eutectics (1177 °C) [43]. A standard slide gate sand consists of approximately 70 % chrome ore with portions of iron oxide, some alumina, magnesia and quartz which react with one another and form new phases. The coarsening of the microstructure occurs by way of Ostwald ripening (refer to Section 6.3.2.2.2). In addition, there is a small portion of soot. It influences the sintering atmosphere and acts as a separating agent. This prevents too intensive microstructural bonding.

In addition, it cannot be excluded that some steel penetrates the boundary layer (refer to Sect. 3.2.2). Finally, it must be considered that chrome oxide and alloying elements, such as manganese, (refer to Fig. 3.29). Generate enormous steam pressure and penetrate the packed bed (fill) together with the vaporized iron

Equation (3.77.2) [82ρ] describes the fracture strength of the sintering layer. It must be overcome by the ferrostatic pressure of the melt once the slide gate has been opened and the sand has flowed out. The following numerical counts (values) are deemed typical:

$Ó B$ max = maximum fracture stress identical with hot modulus of rupture of the sintering layer.
$\rho \cdot g \cdot h = 7$ g/cm3 $\cdot$ 981 cm/s2 $\cdot$ 400 cm (ferrostatic pressure of the melt)
$D = 8$ cm (largest diameter of the nozzle channel)
$H = 1.5$ cm (height of the sintering layer)

Taking this example, the hot modulus of rupture at a casting temperature of approximately 1550 °C must be smaller then $2 \cdot 10^7$ dyn/cm2 = 2MPa. This count (value) corresponds approximately to that of a high-fired magnesia-chrome brick [18]. Subsequently, the sintering layer should be significantly weaker.

How does one achieve the intended high opening rate of 95 % reliably if dealing with a steadily changing steel quality (grade)? The sintering behavior of the sand must be to a large extent independent of the steel quality (grade) and reproducible. Consequently, the utilized raw materials must always be from the same mine, the specific portions mixed very well together and the grain size distribution may not change. The portion of silica in relation to chrome ore should be as small as necessary. This is also a financial consideration. Alloying elements with high steam pressure could in low amounts and in the form of powder be mixed into the sand in order to consider their influence on the sintering process from the start. The down (resting) times of the ladles should not change in order to ensure the consistent quality of the sintering layer. The diameter of the sleeve D (Eq. 3.77.2) must not be changed because it flows quadratically into the calculation of the fracture stress just as is the case for the thickness of the sintering layer [82].

3.2   Vacuum Metallurgy

3.2.1   Crucible Induction Furnace

A particular advantage of this method is that very precise analysis specifications can be reliably set at the lowest gas contents.

At the beginning, the function of the induction furnace shall be described [46]. Figure 3.21 shows the schematic illustration of a vacuum induction furnace for precision casting.

3.2.1.1   Bath Movement

The winding of an induction furnace, which is used to melt metals, encloses the crucible, which consists of refractory material. Alternating magnetic fields induce eddy currents in the metal, which cause it to melt and keep it liquid [61].

The output active power of the generator is

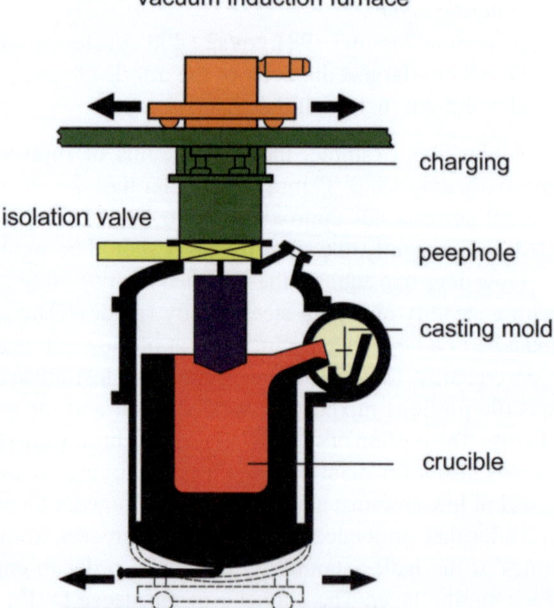

Fig. 3.21 Vacuum induction melting and casting (VIMP) [60]

$$P_W = U \cdot I_W \, [\text{W}]. \tag{3.78}$$

It is read on the generator and heats the metal.
The inductor current is calculated as

$$I_{BI} = U \cdot \omega \cdot C \, [\text{A}]. \tag{3.79}$$

Consequently, the resonant circuit can be matched to the material and, thus, the inductance **L** of the coil by changing the capacitances **C** (older design) or/and the angular frequency **ω** (modern design):

$$\omega = \left(\frac{1}{L \cdot C} \right)^{1/2} [1/\text{s}] \tag{3.80}$$

Figure 3.22 shows the dependence of the induced volume power density on the ratio d/δ. Here d is the workpiece diameter and δ the penetration depth of the magnetic field.

If $d/\delta \leq 0.1$, the power consumption is negligible because the material does not couple sufficiently to the electromagnetic field. This effect is known when metal threads penetrate cracks in the refractory material: They solidify and are coarser the lower the frequency (see below).

The penetration depth of the magnetic field into the melt (equivalent conductive layer thickness) is inversely proportional to the frequency **f** [61]:

$$\delta = \left(\frac{\rho_{el}}{\pi \cdot f \cdot \mu_r \cdot \mu_0} \right)^{1/2} [\text{m}]. \tag{3.81}$$

With

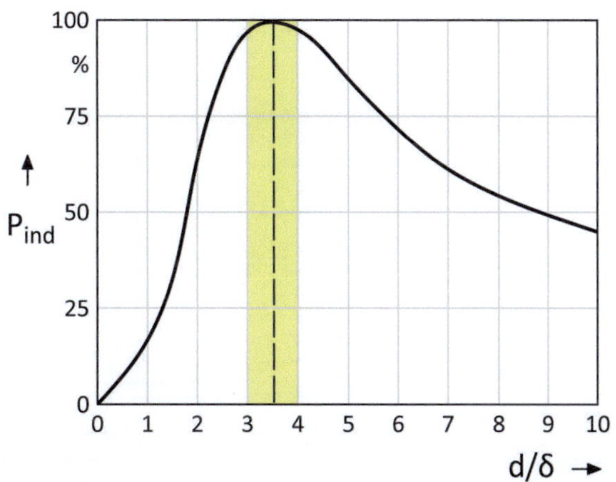

Fig. 3.22 Induced volume power density as a function of the ratio d/δ at constant frequency and variation of the workpiece diameter. *Source* RWE-Information Prozesstechnik

ρ_{el} $= 1.006 \times 10^{-6} \frac{V\,m}{A}$, resistivity of liquid iron
μ_0 $= 1.26 \times 10^{-6} \frac{V\,s}{A\,m}$, magnetic field constant
μ_r $= 1$, relative permeability of the melt
f $= 250\ s^{-1}$

follows

$$\delta = 502.6\left(\frac{1.006 \times 10^{-6}}{1 \times 250}\right)^{1/2} = 3.2 \times 10^{-2}\ m. \tag{3.82}$$

The penetration depth into liquid iron in a production plant is only approx. 3 cm. Using a laboratory furnace with graphite recipients ($\rho_{el} = 10 \times 10^{-6}$ V m/A) and f = 10 kHz, $\delta = 1.6$ cm [61].

The structure of a crucible induction furnace is shown in Fig. 3.23.

Characteristic of the crucible induction furnace is the convection rollers directed against each other. By moving the coil downward, the lower ones can be minimized.

The force of the magnetic field acts radially inward on the melt at the circumference of the crucible wall, resulting in a dome-shaped bath elevation. Its height is calculated from equation [61]

$$h_{\ddot{u}} = \frac{P_i}{2F \cdot \rho \cdot g} \cdot \sqrt{\frac{\mu_r \cdot \mu_0}{\pi \cdot \rho_{el} \cdot f}}. \tag{3.83}$$

For a production plant with a high energy input, the following data apply, among others

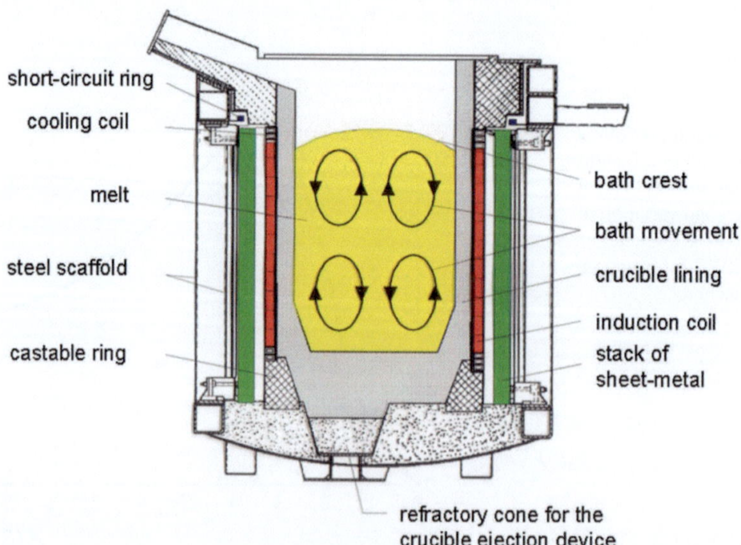

Fig. 3.23 Crucible induction furnace (E. Baake, personal communication, 2012)

Specifications (**ABP-Induction**)

m_{Fe} = 24.5 t, operating weight
P_i = 16 MW = 16 × 10^6 V A = 16 × 10^6 kg m²/s³, induced power
F = 9.93 m², perimeter area
f = 250 Hz, frequency
ρ = 7000 kg/m³ density of the melt
g = 9.81 m/s².

From this, Eq. (3.83) calculates the bath superelevation

$$h_{\ddot{u}} = \frac{16 \times 10^6}{2 \times 9.93 \times 7 \times 10^3 \times 9.81} \left(\frac{1 \times 1.26 \times 10^{-6}}{\pi \times 1.006 \times 10^{-6} \times 250} \right)^{1/2} = 0.47 \, \text{m}$$

(3.84)

The bath superelevation is 47 cm.

As an approximation, the mean flow speed of the melt can be calculated from the bath superelevation by following Torricelli's law:

$$u = \sqrt{2gh_{\ddot{u}}} = \sqrt{2 \times 9.81 \times 0.47} = 3 \, \text{m/s}.$$

(3.85)

This value also confirms quite well with observations made in practice.

3.2.1.2 Infiltration

The bath superelevation is the result of the electromagnetic pressure acting uniformly on the entire shell surface of the melt:

$$P_m = \rho \cdot g \cdot h_{\ddot{u}} \, \text{N/m}^2.$$

(3.86)

In principle, this pressure counteracts the infiltration of the refractory material. However, the effect on the infiltrate can only be expected if it docks to the field. As Fig. 3.22 shows, metal tongues with dimensions $(d/\delta) < 0.1$ practically dock no longer. Inflowing metal tongues narrower than 3 mm, therefore, solidify as soon as their melting point is reached and do not penetrate further into the structure of the lining. For the total metallostatic height, the bath superelevation $h_{\ddot{u}}$ must also be taken into account.

On the other hand, a wetting angle $\theta > 90°$ of the melt against the refractory, which is practically always the case with metals, has a strong effect, specifically in the range of small dimensions. The capillary equilibrium (refer to Sect. 5.3, Eq. 5.13) is as follows:

$$P_{\mathrm{a}} + \rho \cdot g \cdot (h + h_{\ddot{u}}) = -\frac{4\sigma \cdot \cos\theta}{d} \left[\mathrm{N/m^2}\right] \tag{3.87}$$

With

P_{a} $\quad= 1\,\mathrm{atm} = 101{,}330\,\mathrm{N/m^2}$, external pressure (gas pressure in the recipient)

σ $\quad= 1.5\,\mathrm{N/m}$, surface tension of iron

θ $\quad= 120°$, wetting angle iron/refractory material

$h_{\ddot{u}}$ $\quad= 0.47\,\mathrm{m}$, bath superelevation

h $\quad= 2.7\,\mathrm{m}$, bath height

P_{Fe} $\quad= \rho \cdot g \cdot (h + h_{\ddot{u}}) = 7 \times 10^3 \times 9.81 \times 3.17 = 2.18 \times 10^5\,\mathrm{N/m^2}$, ferrostatic pressure

P_{m} $\quad= \rho \cdot g \cdot h_{\ddot{u}} = 7 \times 10^3 \times 9.81 \times 0.47 = 3.23 \times 10^4\,\mathrm{N/m^2}$, magnetic pressure

and Eq. (3.87), one gets the pore diameter d, upon it being exceeded the infiltration starts

$$d = -\frac{4\sigma \cdot \cos\theta}{P_{\mathrm{a}} + \rho \cdot g(h + h_{\ddot{u}})}\,[\mathrm{m}] \tag{3.88}$$

$$d = -\frac{4 \times 1.5 \cdot \cos 120°}{101{,}330 + 2.18 \times 10^5} = 10^{-5}\,\mathrm{m} = 0.01\,\mathrm{mm}$$

In vacuum, the critical pore diameter would be 0.014 mm because $P_{\mathrm{a}} = 0$.

If the wetting angle $\theta < 90°$, infiltration could not be avoided at all.

The calculated infiltration slope applies at the bottom of the crucible. The further up the melt is considered, the shorter the bath height h. The relative influence of the magnetic pressure P_{m} on the melt bath increases. In some places and momentarily, the melt detaches completely from the crucible shell (pinch effect). This creates a locally confined cavity into which the melt falls back. This "cavitation" possibly leads to erosion and destruction of the refractory material [47].

The length of the infiltrated metal tongues can be calculated from the integration of the equation of Hagen–Poiseuille [28] (refer to Sect. 5.3). Under the existing conditions, [47] is valid:

$$l = \sqrt{\left(\frac{d \cdot \sigma \cdot \cos\theta}{4 \cdot \eta} + \frac{d^2 \cdot (P_{\mathrm{a}} + \rho \cdot g \cdot (h + h_{\ddot{u}}))}{16 \cdot \eta}\right) \cdot t}\,[\mathrm{m}]. \tag{3.89}$$

With the values already used and the dynamic viscosity of the melt $\eta = 3 \times 10^{-3}\,\mathrm{Pa\,s}$, the inflow length of 4.2 cm in one minute is obtained from Eq. (3.89) at the bottom of the crucible.

3.2.1.3 Erosion

Shear forces caused by the bath flow act on the rough surface of the lining (refer to Sect. 3.1.1.5.3.2.1 and Fig. 3.15). The portion h_1/D of a grain protrudes into the melt, whereas h_2/D is anchored in the microstructure. The critical flow velocity of the melt above which the grain breaks out is [46]

$$u \geq \left(4 \cdot \frac{h_2}{h_1} \cdot \frac{\sigma_B}{\rho}\right)^{\frac{1}{2}} \text{[m/s]}. \tag{3.90}$$

Using the values $h_2/h_1 = 0.1$, $\rho = 7 \times 10^3$ kg/m^3 steel and $\sigma_B = 0.5$ N/mm^2 $= 0.5 \times 10^6$ kg m/(s^2 m^2) [27] gives $u \geq 5$ m/s. Especially if the bond between matrix and grain is already loosened as result of chemical attack, it can break out even at lower flow speed [47]. This is the case even with careful lining of the rammed channel (trough) of the induction channel furnace (refer to Sect. 3.1.1.5.3.2.1).

3.2.1.4 Cavitation

At sufficiently high flow speeds, e.g. in turbulence, cavitation can occur and, as a consequence, premature wear of the refractory material (refer to Sect. 3.1.1.5.3.2.2). Edges, roughness of the lining and the electromagnetic pressure in induction melting plants support this process [46]. (The numerical values used below apply to molten steel and refractories.)

Cavitation is explained by Bernoulli's equation, according to which the total pressure P_0 of a flow is the sum of the static pressure P_S and the dynamic pressure $P_u = 0.5 \cdot \rho_1 \cdot u^2$ (refer to Fig. 3.24). ρ_1 [kg/m^3] is the density of the fluid, e.g. liquid steel, and u [m/s] its local flow speed. The following applies (19):

$$P_0 = P_S + \frac{\rho_1}{2} \cdot u^2 \text{ [Pa]}. \tag{3.91}$$

The total pressure is generally composed of the external pressure, e.g. $P_a = 1 \times 10^5$ Pa (1 atm), and the gravity pressure $P_g = \rho \cdot g \cdot h$ [Pa]. At 73 cm depth, the pressure in the steel bath is $P_g = 7000$ kg/m^3 $\cdot$ 9.81 m/s^2 $\cdot$ 0.73 m $= 0.5 \times 10^5$ Pa. If the flow increases to the point where the static pressure becomes lower than the steam pressure of the steel (P_D (1650 °C) ≈ 15 Pa), the liquid may rupture and steam bubbles will form [62]. This critical velocity is calculated with $P_D = P_S = 15$ Pa to be

$$u = \sqrt{\frac{2(P_a + \rho_1 \cdot g \cdot h - P_S)}{\rho_1}}$$

$$= \sqrt{\frac{2(10^5 + 0.5 \times 10^5 - 15)}{7000}} = 6.5 \text{ [m/s]}. \tag{3.92}$$

The steam bubble is stable if the following equilibrium of pressures acting on it is given:

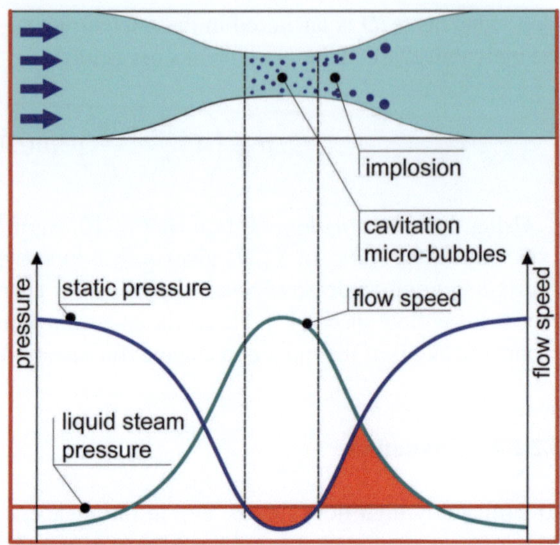

Fig. 3.24 Cavitation, flow speed and static pressure (Google, Caleffi Armaturen GmbH)

$$P_D = P_a + \rho_l \cdot g \cdot h + \frac{2\sigma}{r} \ [\text{Pa}]. \tag{3.93}$$

$P_\sigma = 2 \cdot \sigma/r$ is the capillary pressure, $\sigma = 1.5$ [N/m] is the surface tension of the steel, and r [m] is the radius of the spherical steam bubble. The radius r is calculated by setting Eqs. (3.91) and (3.93) equal. This is possible for the limiting case where the steam pressure P_D and the static pressure P_S are equal:

$$r = \frac{4 \cdot \sigma}{\rho \cdot u^2} = \frac{4 \times 1.5}{7000 \times 6.5^2} = 2 \times 10^{-5} \ [\text{m}]. \tag{3.94}$$

As soon as the bubble enters a region of low flow, so that the static pressure P_S becomes greater than the steam pressure P_D, the steam condenses and the bubble spontaneously collapses, thus imploding. This produces a jet of liquid which hits the wall, on which the bubble possibly sits, at high velocity (Fig. 3.25). The pressures of the right-hand side of Eq. (3.93) drive the liquid jet, while the vanishing steam pressure P_D seeks to stop it. The velocity of the jet is therefore:

$$u_j = \sqrt{\frac{2(P_a + \rho_l \cdot g \cdot h + 2 \cdot \sigma/r - P_D)}{\rho_l}}$$

$$= \sqrt{\frac{2\big((1 + 0.5 + 1.5) \times 10^5 - 15\big)}{7000}} = 9.3 \ [\text{m/s}]. \tag{3.95}$$

The pressure of the jet on the wall is proportional to the speed of sound c_l in the liquid [63].

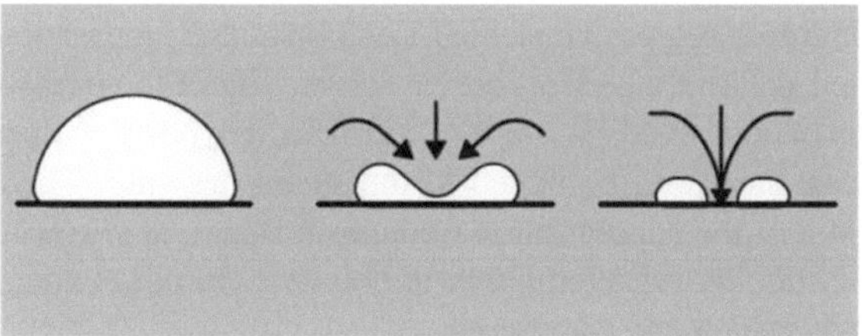

Fig. 3.25 Development of the liquid jet during implosion of a bubble sitting on the wall (jet) (Google/Samson: Techn. Info. 3: Cavitation in Control Valves)

$$P_j = \rho_l \cdot c_l \cdot u_j = 7000 \times 50{,}000 \times 9.3 = 3.2 \times 10^9 \, [\text{Pa}]. \qquad (3.96)$$

For condensed phases, the speed of sound is calculated as [28]

$$c = \sqrt{\frac{1}{\rho \cdot \kappa}} \, [\text{m/s}]. \qquad (3.97)$$

For refractories $\rho_{\text{Ref}} \simeq 3500 \, \text{kg/m}^3$ and $\kappa_{\text{Ref}} \simeq 4 \times 10^{-12} \, \text{m}^2/\text{N}$ and for liquid iron $\rho_{\text{Fe}} \simeq 7000 \, \text{kg/m}^3$ and $\kappa_{\text{Fe}} \simeq 5 \times 10^{-14} \, \text{m}^2/\text{N}$, the compressibility of liquids is much lower than that of solid bodies because of the denser packing of their building blocks [28]. Consequently, $c_{\text{Ref}} = 8450 \, \text{m/s}$ and $c_{\text{Fe}} = 50{,}000 \, \text{m/s}$ for the calculation.

The jet transfers its momentum to the wall and the energy is absorbed there. The energy balance is $P_j \cdot \Delta V_j = \Delta m \cdot c_{\text{Ref}}/2$ with $\Delta V_j = 1.33\pi \cdot r^3 = 3.3 \times 10^{-14} \, [\text{m}^3]$. Since the pressure of the jet ($\sim 10^9 \, [\text{Pa}]$) is much higher than the hot crushing strength of the refractory material ($\sigma_{\text{HCS}} \sim 10^7 \, \text{Pa}$), the material could be shattered and destroyed in a limited area:

$$\Delta m_{\text{FF}} \simeq \frac{2(P_j \cdot \Delta V_j)}{c_{\text{FF}}^2} = \frac{2(3.2 \times 10^{+9} \times 3.3 \times 10^{-14})}{8450^2} = 3 \times 10^{-12} \, [\text{kg}]. \quad (3.98)$$

The wear speed $\dot{x}$ is with consideration of the impact area $F \sim \pi \cdot r^2 = 1.25 \times 10^{-9} \, [\text{m}^2]$:

$$\dot{x} \simeq \frac{\Delta m}{\rho_{\text{FF}} \cdot F_j} \cdot f_j = \frac{3 \times 10^{-12}}{7000 \times 1.25 \times 10^{-9}} \times 10 = 3.4 \times 10^{-6} \, [\text{m/s}]. \qquad (3.99)$$

From experiments on the "back attack" mechanism, the frequency of bubble formation is given as $f_j \leq 20/\text{s}$, whereby not every impact is "successful" [64]. From this follows with $f_j \sim 10/\text{s}$ the wear rate to $\dot{x} \simeq 3.4 \times 10^{-6} \, \text{m/s}$, corresponding to 1.2 cm/h.

With the "**back attack**" mechanism, some of the gas bubbles, created by injection of gases into the melt, strike back against the wall, implode there and damage the lining to a considerable degree. Thus, the formation and disintegration of the bubbles differ but not their devastating effect on the service life of the lining.

The pressure of dissolved gases, e.g. nitrogen, supports the formation of the bubbles and must, therefore, be added to the steam pressure. The collapse of the bubbles, as a result of the condensation of the iron steam, can, however, be counteracted for a short time by excreted gases in that they do not condense but dissolve back into the melt, which can take longer.

The surface roughness $1/d$ (refer to Fig. 3.14) of the refractory material facilitates the formation of voids because the wetting angle $\theta > 90°$. If the bubble forms directly on the wall, add the capillary pressure as $P_\sigma = 4 \cdot \sigma \cdot \cos\theta/d$ (refer to (3.64)). The surface roughness $1/d$ increases with the operating time [65]. The critical flow speed leading to the formation of the bubble is reduced as result [47].

If the pressure in the furnace vessel is lowered, e.g. from 1 to 0.01 bar (10^3 N/m^2), the critical flow speed of the melt is reduced from 5.5 to 1.24 m/s according to Eq. (3.65) and the wear speed increases because bubble formation is facilitated.

3.2.2 Induction Channel Furnace

An induction channel furnace is mainly used as a holding tank for liquid metal in foundries. Figure 3.26 shows the setup.

The channel (trough) filled with molten metal is in contact with the bath and serves as the secondary winding of a transformer with an iron core.

The bath movement in the channel is divided into two flow directions:

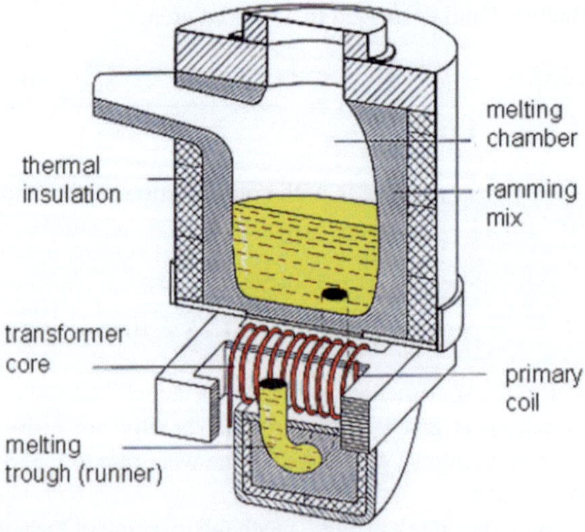

Fig. 3.26 Channel-type induction furnace (E. Baake, personal communication, 2012)

- The transit speed for mass transfer between the channel and the bath is regulated solely by density convection and occurs only slowly at approx. 5–10 cm/s.
- In contrast, the flow of the transverse vortices in the channel itself, as a result of the electromagnetic Lorenz forces, is highly turbulent and occurs at approx. 0.5–1 m/s. In addition to chemical corrosion, erosion and cavitation can, consequently, also play a role as wear mechanisms.

In Fig. 3.27, the lower left partial figure at the lower left shows the coating formation. The two upper partial figures show on the left the calculated very low exchange speed between channel and bath, as well as on the right the simultaneously acting high kinetic energy of the turbulent swirls in the channel itself. The case shown is one in which oxide deposits are very likely to form in the transition area between the channel and the bath, which would subsequently lead to blockage. This is most likely the case if the overall temperature is low and mass (material) transfer with the bath can only take place very slowly because the temperatures at the inlet and outlet differ only slightly from each other. If at the same time the turbulence in the channel is high, the impact frequency of the oxides suspended in the melt with each other and with the lining is increased. At these conditions, oxide solutions and excretions accumulate in the transition region, dwell there and lump together since they are not wetted by the melt. They sinter firmly together. These conditions correspond to those for clogging in the submerged nozzle in continuous casting of steel (refer to Sect. 9.3).

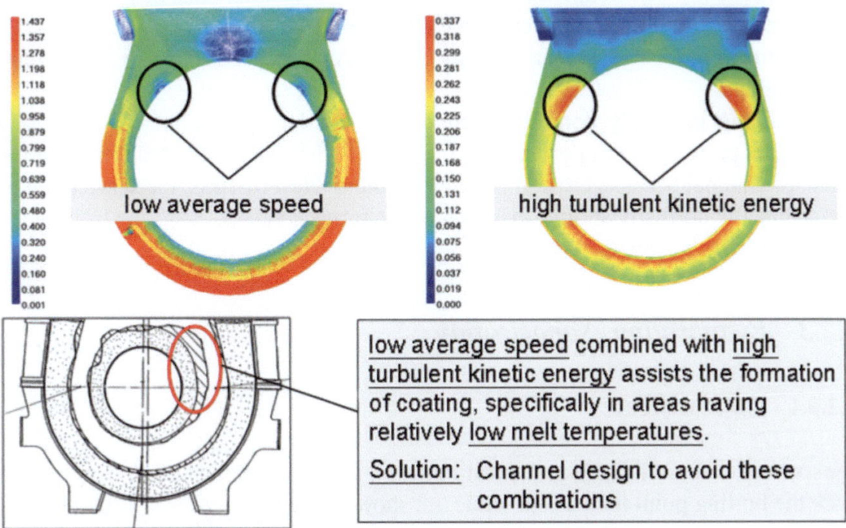

Fig. 3.27 Coating formation in the channel-type induction furnace [66]

Although coating formation is the most common cause of channel failure, chemical corrosion, erosion, and cracks resulting from thermal stresses also play a role. The liquid metal infiltrates into the cracks, and the skin effect and the temperature gradient in the lining limit the tongue thickness and depth—just as in the case for the induction furnace. The excess temperature in the channel ΔT, that is the difference between the bath temperature at the end parts of the channel and the highest temperature in the channel itself, is calculated from the inductor power P_i as follows:

$$\Delta T = \alpha \cdot P_i^{0.67} \qquad (3.100)$$

with α as empirical proportionality factor [67].

Finally, wear occurs due to cavitation, triggered by the pinch effect (refer also to Sect. 1.1.3.3). This is understood to mean the localized detachment of the melt from the wall of the channel, possibly up to the complete constriction of the melt itself, as a result of the high electromagnetic pressure. The metal vaporizes into the cavities. The steam cools and condenses abruptly. The cavity collapses at the speed of sound, and the melt hits the wall and shatters the structure. This effect is also similar to that in the induction crucible furnace and is calculated there (refer to Sect. 2.1.4). If the pinch effect occurs, which can be noticed by vibrations of the plant, the lining can be permanently destroyed. To avoid this, the electrical power must be limited.

Minimization of wear is based on practical experience and numerical turbulence models [67]. Closed-loop solutions are not known.

The high symmetry of the energy distribution, as shown in Fig. 3.27, can be avoided by an asymmetrical design of the channel, which improves the mass (material) transfer [67]: The temperature distribution in the channel must be asymmetrical to ensure mass (material) transfer with the bath [68].

Movement in the bath is slow. Mainly infiltration there leads to premature failure of the lining.

The lining is mainly done with ramming mixes based on alumina, especially spinel. It is particularly important to achieve the highest possible compaction in the inductor channel because erosion is intensive there. Nevertheless, the service life of the channel is normally shorter than that of the furnace vessel.

3.2.3 Evaporation, Vaporization

3.2.3.1 Molten Metal

We speak of evaporation and condensation below the boiling point and of vaporization once the boiling point is reached. Table 3.2 shows the temperatures at which metals boil at a pressure of 1 bar. Figure 3.28 shows the steam pressure of metals between 700 and 1700 °C.

Table 3.2 Boiling temperature of pure metals at 1 atm vapor pressure

Metal	°C	Metal	°C
Zn	908	Co	2925
Mg	1093	Si	3210
Ca	1500	Ti	3358
Mn	2060	V	3417
Al	2514	Zr	4432
Cu	2564	Mo	4678
Sn	2585	Hf	4690
Cr	2669	Nb	4858
Fe	2859	Ta	5504
Ni	2911	W	5884

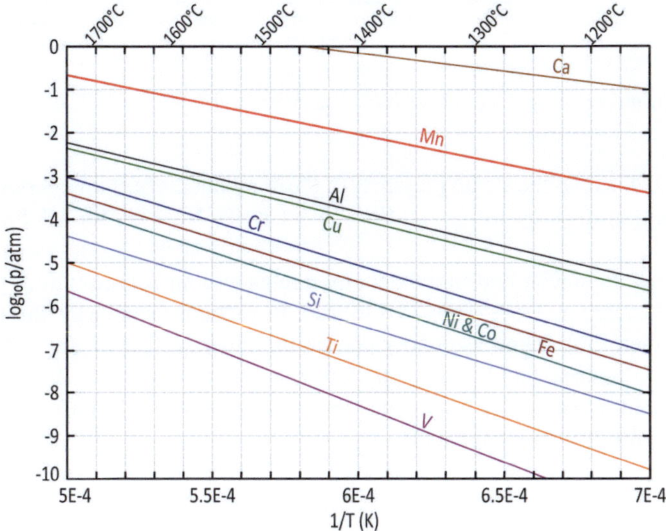

Fig. 3.28 Steam pressure of pure metals from 1200 to 1700 °C

The calculation was made with "*FactSage*" [7] on the basis of the equation of Clausius–Clapeyron [51]:

$$\ln P_{\mathrm{Me}} = -\frac{A}{T_{\mathrm{K}}} + B\,[\mathrm{bar}]. \tag{3.101}$$

$A\,[\mathrm{K}] = \Delta H_{\mathrm{V}}/R$ is the quotient of the enthalpy of vaporization ΔH_{V} [J/mol] and the ideal gas constant R [Joule/mol K]. B [bar] is an integration constant.

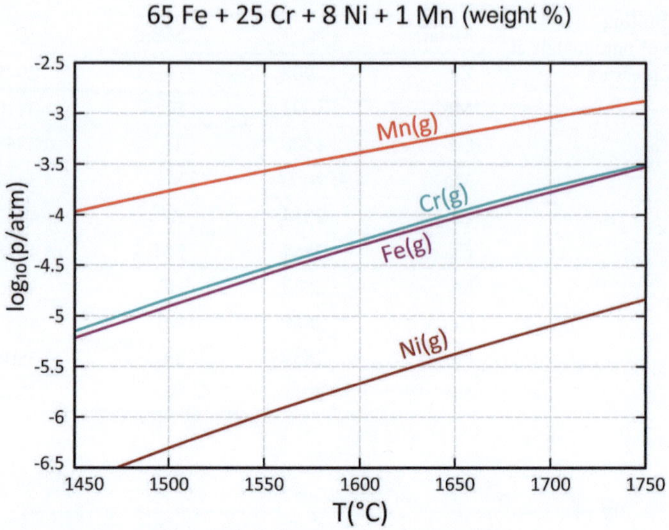

Fig. 3.29 Steam pressures of the components of stainless steel 1.4501

Figure 3.29 shows the steam pressures of the components of stainless steel **1.4501**. Its composition in % by weight is: Fe—25% Cr, 8% Ni, 3% Mo, 1% Si, 1% Mn, 0.01% C.

In contrast to silicon and molybdenum, manganese and chromium in particular have a high steam pressure compared to iron. In vacuum metallurgy, these components are, therefore, found as condensate on the inner surface of the water-cooled recipient. On the one hand, the condensate can lead to electric flashovers and, on the other hand, it can easily ignite once the vessel is opened. It must, therefore, be removed (carefully) after each melt.

In a first approximation, the partial pressure of the components can be calculated using Raoult's solution law [51]. This is especially valid for the elements Mn, Cr, Ni, Co because their molecular weight is not very different from that of iron. If one approximates the mole fraction X_{Me} by the concentration in % by weight, i.e. $X_{Me} \approx$ % by weight/100 follows $P_{Me} = P_0 \cdot X_{Me}$. P_0 [bar] is the steam pressure of the pure alloying element at the temperature in question.

As a first example, the partial pressure of manganese dissolved in iron at 1650 °C is calculated: Its steam pressure is about $P_0 = 0.11$ bar (Fig. 3.28) and the concentration is 1% by weight. Then the partial pressure in the iron is $P_{Mn} \approx 0.08 \times 1/100 = 8 \times 10^{-4}$ bar, which agrees well with Fig. 3.29. Recall the definition of activity, where $a \equiv P_i/P_0$. The activity of manganese in our example is $a_{Mn} = 8 \times 10^{-4}/0.08 = 10^{-2}$. Taking a second example, consider chromium. At 1650 °C, $P_0 = 4 \times 10^{-4}$ bar (Fig. 3.28) and the concentration in 1.4501 is 25% by weight Cr, i.e. $X_{Cr} \approx 0.25$. It follows that $P_{Cr} \approx 10^{-4}$ bar (refer to Fig. 3.29).

At reduced pressure, the evaporation rate can be calculated with the equation of Hertz–Knudsen [69]:

$$ j = \alpha \cdot (P_i - P_R) \cdot \sqrt{\frac{M_i}{2\pi \cdot R \cdot T_K}} \ \left[\text{kg}/(\text{m}^2\,\text{s})\right] \tag{3.102} $$

P_i [Pa] is the steam pressure of the observed component at the surface of the melt and P_R is the pressure in the recipient. M_i [kg/mol] is the molecular weight, $R = 8.31$ [J/mol K] is the ideal gas constant, T [K] is the absolute temperature, and $\alpha \leq 1$ is the evaporation or condensation coefficient.

In equilibrium between melt and atmosphere, the speeds of evaporation and condensation on the melt bath are equal. However, if the evaporating component, e.g. manganese, condenses on the cold recipient, $P_R \ll P_i$ and the equation [69] applies in practice:

$$ j = 44 \cdot \alpha \cdot P_i \cdot \sqrt{\frac{M_i}{T_K}} \ \left[\text{g}/(\text{cm}^2\,\text{s})\right] \tag{3.103} $$

Here, the pressure P_i is inserted in [bar] and the molecular weight M in [g/mol].

The condensation coefficient is $\alpha \leq 1$ and is often between 0.2 and 0.9 [69]. It depends essentially on the amount of dissolved capillary-active elements, such as oxygen and sulfur. For example, $\alpha = 0.5$ is obtained for the evaporation of manganese in a vacuum induction furnace with the above numerical values $j_{Mn} = 44 \times 0.5 \times 10^{-3} \times (55/1923)^{1/2} = 3.7 \times 10^{-3}$ g/cm² s. This is a considerable manganese loss of 2.3 kg/h for a bath surface of the considered 20 kg laboratory furnace with 170 cm² free surface.

If we observe that the evaporation rate is independent of the pressure in the recipient, the transport of the evaporating component from out the interior of the melt to the bath surface determines the speed. Then this applies

$$j = \beta \cdot \rho_{\mathrm{M}} \cdot \frac{C_0 - C_{\mathrm{s}}}{100} \left[\mathrm{g/cm^2\,s}\right] \tag{3.104}$$

$\rho_{\mathrm{M}} = 7$ g/cm^3 is the density of the melt, β [cm/s] is the mass (material) transfer coefficient, C_0 is the concentration in the bath and $C_{\mathrm{s}} \to 0\%$ by weight the concentration at the free surface to the atmosphere. The mass (material) transfer coefficient for flow without friction is [70].

$$\beta = \left(\frac{4 \cdot D \cdot u}{\pi \cdot h}\right)^{1/2} \mathrm{[cm/s]}. \tag{3.105}$$

The diffusion coefficient in liquid steel of dissolved elements is $10^{-5} < D < 10^{-4}$ cm^2/s. We choose $h = 1$ cm as the characteristic length.

Let the mean flow speed of the melt be $u = 35$ cm/s [58], resulting in

$$\beta = \left(\frac{4 \times 5 \times 10^{-5} \times 35}{\pi \times 1}\right)^{1/2} = 0.05 \,\mathrm{[cm/s]}. \tag{3.106}$$

The evaporation rate of manganese at 1650 °C follows to

$$j = \frac{1}{100} \times 0.05 \times 7 \times (1 - 0) = 3.5 \times 10^{-3} \left[\mathrm{g/cm^2\,s}\right]. \tag{3.107}$$

With a free surface of 170 cm^2 this is 2 kg of vaporized manganese in one hour. Both mechanisms result in approximately the same evaporation rate in this example.

This evaporation rate appears very high and is rarely observed in practice. The reason is that capillary-active elements accumulate (enrich) on the surface of the melt and form an extremely thin "slag film" that impedes mass (material) transfer.

Another reason for delayed evaporation, which has not been considered so far, may be the transport of the required evaporation heat. A typical example is the drying of refractory ramming mixes [71] (refer to Sect. 1132). In strongly stirred melts, e.g. in induction furnaces, this effect has no influence, however.

Again, the evaporation of manganese from a molten steel at 1650 °C serves as an **example**. To evaporate $j = 3.7 \times 10^{-2}$ kg manganese/m^2 s, the heat flux density $\dot{q} = j \cdot \Delta H_{\mathrm{V}} \left[\mathrm{W/m^2}\right]$ is required. The enthalpy of vaporization of manganese is $\Delta H_{\mathrm{V}} = 4.2 \times 10^6$ J/kg manganese. Consequently, $\dot{q} = 1.6 \times 10^5 \left[\mathrm{W/m^2}\right]$. To maintain this heat flux density in strongly stirred molten steel, the temperature difference $\Delta T = \dot{q} \cdot \delta / \lambda$ [K] is required. Let the thickness of the flow boundary layer be $\delta = 10^{-5}$ m and the thermal conductivity $\lambda = 35$ W/m K. It follows that $\Delta T = 0.04$ K.

3.2.3.2 Oxides

The "evaporation" of solids is called sublimation. This process is often linked to a chemical reaction, e.g. the decomposition of molecules. An example is the reaction $2\langle SiO_2\rangle = 2\{SiO\} + \{O_2\}$.

To enable precision castings at a pressure of 10^{-5} bar, it is necessary for the selection of the refractory material to know its vapor pressure at the temperature in question in order to avoid contamination of the melt and operational disturbances. Figure 3.30 gives the vapor pressures of common simple oxides. These form gaseous suboxides such as CrO_3, CrO_2, CrO, SiO, AlO, and metal vapors, such as Mg and Cr. The respective oxygen partial pressure is calculated from the stoichiometry of the reaction equation, e.g. in the case of quartz $2\langle SiO_2\rangle = 2\{SiO\} + \{O_2\}$ then $p_{O_2} = 0.5 p_{SiO} = 5 \times 10^{-7}$ bar at about 1600 °C. Cr_2O_3 and SiO_2 are not suitable as refractory materials in vacuum operation because they are not sufficiently stable at service temperatures in the pressure range from 10^{-5} to 10^{-6} bar.

Even the use of picrochromite (MgO–Cr_2O_3) instead of the eskolaite (Cr_2O_3) does not change this situation, as the comparison of Figs. 3.30 and 3.31 **proves**. Moreover, especially in the presence of alkalis, the environmentally harmful Cr^{+6} can be formed as CrO_3.

In Fig. 3.31, the partial pressures of some oxide compounds frequently seen in the refractory industry are calculated and shown. The partial pressures of the suboxides of the oxides and their compounds differ only slightly from each other. The decomposition of dolomite leads to such high partial pressures of CO_2 (approx. 0.6 bar)

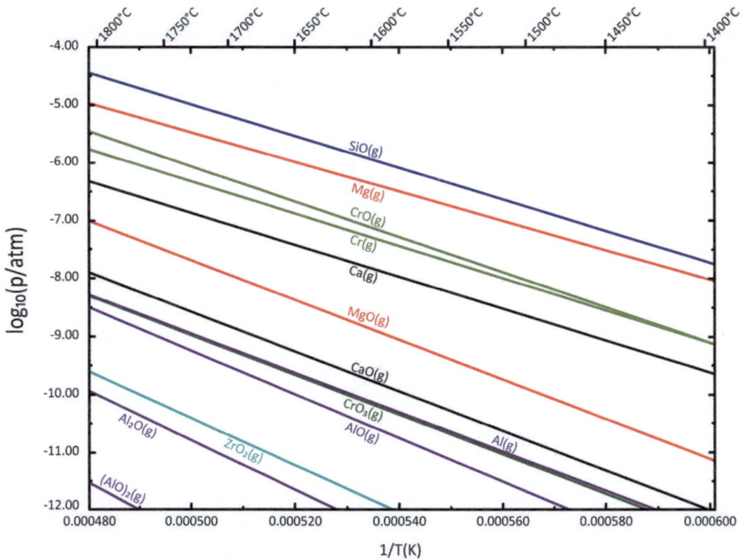

Fig. 3.30 Steam pressures of the suboxides of Cr_2O_3, SiO_2, Al_2O_3, MgO, CaO and ZrO_2 [7]

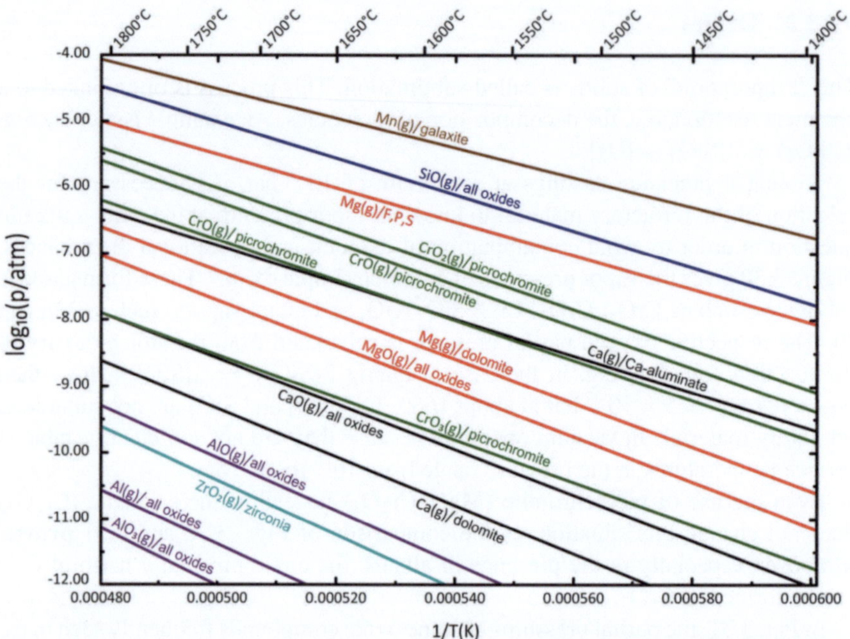

Fig. 3.31 Steam pressures of suboxides of dolomite ($MgCaC_2O_6$), spinel ($MgAl_2O_4$), Ca-aluminate ($CaAl_2O_4$), forsterite (Mg_2SiO_4), galaxite ($MnAl_2O_4$), andalusite (Al_2SiO_5), picrochromite ($MgCr_2O_4$) and zircon silicate ($ZrSiO_4$), $P_{CO_2} \approx 0.6$ bar [7]

and CO (approx. 0.001–0.01 bar) that its use in a vacuum furnace is prohibited. Free oxygen in the furnace atmosphere reduces the risk of suboxide formation (see above).

The estimation of the evaporation rate is possible with the help of equation [69]:

$$j = 44 \cdot \alpha \cdot P_i \cdot \sqrt{\frac{M_i}{T_K}} \; \left[g/(cm^2\,s) \right] \qquad (3.108)$$

pressure of $p_{total} \approx$
(refer to Sect. 3.2.3).

Since all the oxides used evaporate and, therefore, an average steam pressure of $P_{ges} \approx 5 \times 10^{-7}$ bar occurs in the recipient at 1650 °C, this value is used for the calculation. $M = 152$ g/mol of Cr_2O_3 is chosen as the molecular weight. Let the evaporation coefficient be $\alpha = 1$. The following results

$$j_{Cr_2O_3} = 44 \times 1 \times 5 \times 10^{-7} \cdot \sqrt{\frac{152}{1923}} = 6 \times 10^{-6} \left[g/cm^2\,s \right]. \qquad (3.109)$$

$MgO \cdot Cr_2O_3$

That is 216 g/m^2 h. A cube of Cr_2O_3 with the geometric surface 6 cm^2 and initial weight of 3.5 g loses 0.13 g/h. This is 3.7% of its weight. This value agrees with measurements of Koch [65]. Also, at temperatures up to 1600 °C and MgO–Cr_2O_3 sublime into a vacuum twice as fast as in flowing air [65].

Finally, Koch [65] observed that at 1600 °C the evaporation speed starts to become independent of pressure and slows down. He rightly concludes that Knudsen diffusion sets in and determines the sublimation of the chromium oxide.

Knudsen diffusion occurs when the mean free path length λ [cm] of the molecules in the gas phase equals or exceeds the dimension of the enclosing space d [cm]. The so-called Knudsen number $Kn \equiv \lambda/d$ can be used to distinguish the cases: $Kn \ll 1$ stands for the case of normal Fick's diffusion of a gas of viscosity η [g/cm s]. $Kn \gg 1$ stands for a molecular flow in which the gas has no viscosity because its molecules no longer collide with each other but only touch the wall. For $Kn \geq 1$, Knudsen diffusion is present.

The molar mass (material) transport in the gas is $j = D \cdot \Delta n/L \left[\mathrm{mol/cm^2\ s}\right]$, where the diffusion coefficient is $D = 0.33 \cdot \lambda \cdot u \left[\mathrm{cm^2/s}\right]$ and the driving *degree* $n \equiv \Delta n/L$. u [cm/s] is the (thermal) speed of the molecules and λ [cm] their mean free path length. For $Kn \ll 1$, i.e. in the Fick range, the flow speed in a tube is proportional to pressure according to Hagen–Poiseuille, which is why $D_F \sim p$. With decreasing pressure the diffusion coefficient decreases. The particle velocity of a molecular flow, on the other hand, depends on $\sqrt{T/M}$ and not on the pressure [28]. For the diffusion coefficient with $R = 8.31 \times 10^7$ erg/K mol and the tube diameter d [cm], which now substitutes the free path length λ, the following applies:

$$D_{Kn} = \frac{1}{3} \cdot d \cdot \sqrt{\frac{8 \cdot R \cdot T}{\pi \cdot M}} = 4850 \cdot d \cdot \sqrt{\frac{T}{M}} \ \left[\mathrm{cm^2/s}\right]. \tag{3.110}$$

In our example, with $d = 1$ cm and $M = 152$ g/mol at $T = 1923$ K (1650 °C), the value $D_{Kn} = 1.7 \times 10^4$ cm^2/s. D_{Kn} increases with dimension and temperature. However, with decreasing pressure, the drive **degree n** *decreases*, which is why the observed mass (material) transfer decreases.

The average diffusion coefficient in air at room temperature is about 0.2 cm^2/s. It is significantly lower because the mean free path length is only

$$\lambda_L \approx 7 \times 10^{-6} \ [\mathrm{cm}]/p \ [\mathrm{bar}]$$

Refractories have a wide pore size distribution, approximately between 10^{-6} and 10^{+3} μm. This raises the question as to at what pressure the transition occurs, i.e. where $Kn \sim 1$ is. The mean free path length is [72]

$$\lambda = \frac{k \cdot T}{\sqrt{2} \cdot \pi \cdot d_M^2 \cdot p} = 2.2 \times 10^{-8} \cdot \frac{T}{p} \ [\mathrm{cm}]. \tag{3.111}$$

$k = 1.38 \times 10^{-23}$ [N m/K] is the Boltzmann constant and $d_M = 3.74 \times 10^{-10}$ [m] is the so-called kinetic diameter of a nitrogen molecule. For the assumed air pressure of $p = 10^{-5}$ bar (10^{-2} mbar, 1 N/m^2) and $T = 1923$ K, one obtains $\lambda = 4.3$ cm, i.e. Knudsen diffusion is present in all pores. Conversely, if the mean pore diameter $d \equiv \lambda = 10\ \mu\mathrm{m} = 10^{-3}$ cm, a pressure of $p = 4{-}10^{-2}$ bar (42 mbar) is calculated at which molecular flow is present. The further the pressure drops, the larger the pores are involved in Knudsen diffusion.

In practice, gas transport occurs preferentially in the large pores, where instead of (faster) molecular flow, (slower) Fick's diffusion carries the main load. The mass (material) throughput is nonetheless larger because the cross-sectional area increases with d^2. The average diffusion coefficient in the transition region is $\tilde{D}^{-1} = D_F^{-1} + D_{Kn}^{-1}$. In the pore system, the effective diffusion coefficient is calculated by taking into account the porosity ε and the labyrinth factor γ to $D_{eff} = \tilde{D} \cdot \varepsilon \cdot \gamma$. It is very difficult to determine.

The labyrinth factor γ is the reciprocal of the so-called linear tortuosity τ. It can be derived theoretically by considering the pore structure of a body of length L as traversed by N tortuous cylindrical pores of length l. The diameter of these pores, which characterizes the transport, is d_o. Their diameter, which characterizes the transport, is d_o.

If the outer total volume is $V = F \cdot L$ and ε [–] is the porosity, then the total pore volume is $V_p = V \cdot \varepsilon = N \cdot L \cdot \pi \cdot d_o^2/4$. The total internal surface area is $F_p = V \cdot A_o = N \cdot \pi \cdot d_o \cdot l$ where $A_o\ \left[\frac{cm^2}{cm^3} \right]$ is the specific internal surface area. With the definition of the labyrinth factor $\gamma \equiv \frac{L}{l}$, one obtains $\gamma = \frac{4 \cdot \varepsilon}{A_0 \cdot d_0}$ [–].

In addition to the porosity ε, in order to apply this equation, the internal specific surface area A_o and the pore diameter d_o characterizing the transport must be known. Since this is usually not the case in practice, the approximate equation based on laboratory measurements will be used in the future.

$$\gamma = 1.84 \cdot \sqrt{\varepsilon} - 0.3 \tag{3.112}$$

used (3.62).

3.2.4 Condensation

Vapors penetrating the pore system of the refractory lining can pose a risk leading to its failure due to corrosion. The problem is that the lining's diminishing strength is often not visible from the outside. This applies to both metal and oxide vapors. In the range of their usual melting temperatures, the steam pressures of liquid metals are not always higher than those of their monoxides. Approximately, the following ratios of their steam pressures apply (Table 3.3) [73].

Table 3.3 shows on the left side metals whose steam pressures are higher than those of their oxides. On the right-hand side, the oxides have the higher steam pressure,

Table 3.3 Ratio of steam pressures of liquid metals and their monoxides

$P(Me)/P(MeO)$		$P(MeO)/P(Me)$	
Ti/TiO	1E0	MoO/Mo	1E0.5
V/VO	1E2	WO/W	1E2
Be/BeO	1E3	ZrO/Zr	1E2
Cu/CuO	1E3	ThO/Th	1E3
Cr/CrO	1E4	HfO/Hf	1E4
Mn/MnO	1E5	TaO/Ta	1E5
Fe/FeO	1E6	YO/Y	1E5
Ni/NiO	1E7	SiO/Si	1E5

i.e. in vacuum operation, deoxidation of the melt could occur via the gas phase. In practice, however, if melting superalloys, there is a risk that these alloying elements will volatilize and be will missing for the desired microstructure [74].

Magnesium and **calcium** are used for the deoxidation of steel melts.

Their solubility in liquid iron is low and their steam pressure high. At 1600 °C, the deoxidation equilibria [7] apply:

$$MgO = [Mg] + [O],$$
$$\Delta G^{o} = 348\,kJ/mol\,K = [Mg] \cdot [O] = 2 \times 10^{-10} \tag{3.113}$$

$$CaO = [Ca] + [O],$$
$$\Delta G^{o} = 425\,kJ/mol\,K = [Ca] \cdot [O] = 1.4 \times 10^{-12}. \tag{3.114}$$

(The two calculated values are reported in the literature with a scatter of 10^{-6}–10^{-15} because of experimental uncertainties [52, 75].)

The steam pressures at 1600 °C are [52]:

$$\{Mg\} = [Mg], \ \Delta G^{o} = 59\,kJ, \ K = \frac{[Mg]}{\{Mg\}} = 0023, \ P_{Mg,sat} = 17\,bar \tag{3.115}$$

$$\{Ca\} = [Ca], \ \Delta G^{o} = 55\,kJ, \ K = \frac{[Ca]}{\{Ca\}} = 0.03, \ P_{Ca,sat} = 1.2\,bar. \tag{3.116}$$

Deoxidation under vacuum is, consequently, ineffective. Therefore, alloys, such as CaSi or CaC_2, are used or a submerged bell is employed. After treatment, the deoxidizing agents still present in the melt can be easily removed under vacuum.

The vapors react with the refractory material and form new compounds. In the process, metal vapors oxidize by reducing existing oxides. For example, MgO and corundum form the spinel MgO–Al_2O_3. The free enthalpy of formation at 1600 °C is $\Delta G^{0} = 36{,}304\,J/mol$. The associated volume increase generates a high crystallization pressure of 51×10^{3} atm, which can destroy the microstructure if it is too dense (refer to Sect. 10.1).

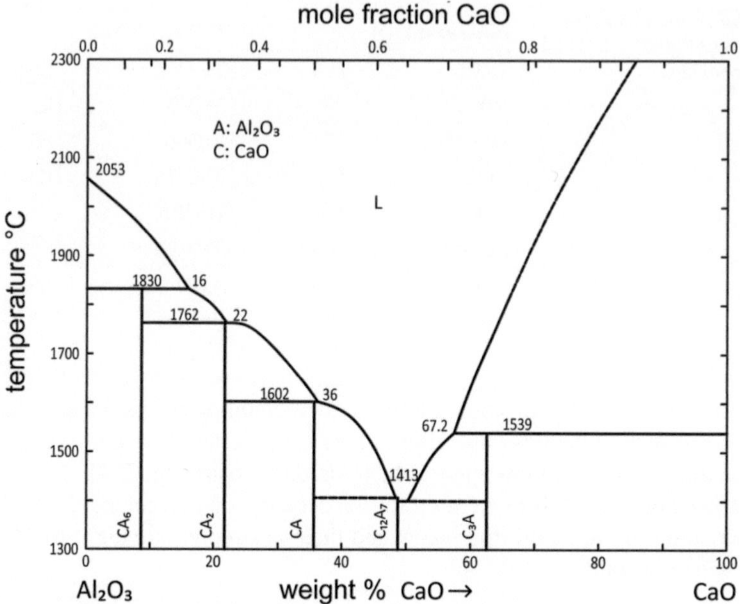

Fig. 3.32 Phase diagram Al$_2$O$_3$–CaO [43]

Corundum and CaO give rise to a series of compounds whose melting points initially decrease with increasing lime content. The compound 12CaO·7Al$_2$O$_3$ (C$_{12}$A$_7$) melts at 1413 °C. The melting points of the eutectics surrounding it are still a few degrees lower (refer to Fig. 3.32). This can result in the lining melting down.

Deoxidation of the steel with aluminum can attack the magnesia-based refractory lining and vice versa. In both cases, spinel is formed as the surface layer. The two reaction equations apply: [7]

$$\langle \text{MgO} \rangle_{\text{Ref}} + 2[\text{Al}] + 3[\text{O}] = \langle \text{MgO} \cdot \text{Al}_2\text{O}_3 \rangle,$$
$$\Delta G^0 \, (1600\,^\circ\text{C}) = -507,190 \, \text{J/mol} \tag{3.117}$$

$$\langle \text{Al}_2\text{O}_3 \rangle_{\text{Ref}} + \left[\text{Mg}\right] + [\text{O}] = \langle \text{MgO} \cdot \text{Al}_2\text{O}_3 \rangle,$$
$$\Delta G^0 \, (1600\,^\circ\text{C}) = -300,980 \, \text{J/mol}. \tag{3.118}$$

The existence fields are shown in Fig. 3.33. If 10 ppm oxygen [O] is dissolved in iron, periclase $\langle$MgO$\rangle$ exists at 1.5 ppm dissolved magnesium [Mg], until the content of dissolved aluminum [Al] exceeds 30 ppm and spinel $\langle$MgO·Al$_2$O$_3\rangle$ emerges. Above [Al] = 100 ppm, only corundum $\langle$Al$_2$O$_3\rangle$ exists. Spinel is a thermochemically very stable oxide and is well suited as a refractory material. However, spinel does not adhere firmly to the surface because its formation is connected to an increase in molar volume and crystallization pressure of about 51 × 10^3 atm may occur during

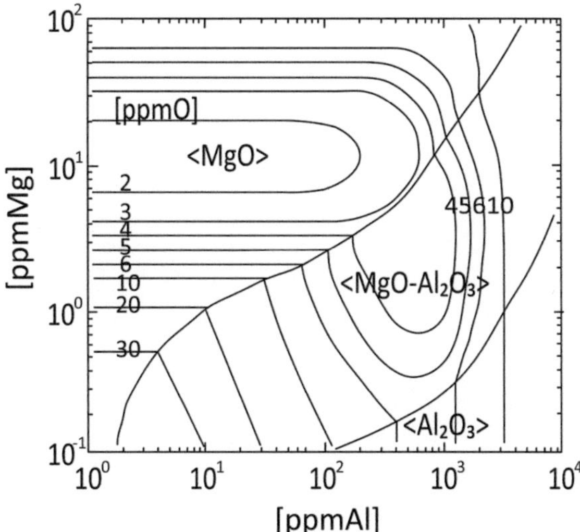

Fig. 3.33 Stability diagram of the system Fe–O–Al–Mg at 1600 °C [75]

its formation. Especially in the production of wire grades, the hard, non-deformable spinel inclusions are an undesirable disturbance of the microstructure [54].

By comparing the metal contents with each other at the same oxygen content, it can also be seen that magnesium is the significantly more effective deoxidizer compared with aluminum. It can also be largely removed from the melt by evacuation.

Magnesium vapors react with silicic acid to form forsterite ($2MgO \cdot SiO_2$). This material is preferably used by the glass industry in heat exchangers (regenerators). Its free enthalpy of formation at 1600 °C is $\Delta G^0 = 92$ kJ/mol [7].

For some years now, the bricks used for lining regenerators have increasingly been produced by firing a mixture of periclase (MgO) and zircon ($ZrO_2 \cdot SiO_2$). The MgO grain surrounds itself with a forsterite edge (hem) and ZrO_2 remains in the matrix. As a result, the corrosion resistance to exhaust gases from the glass tank containing sulfur and alkali oxide increases significantly. The reaction equation is [7]:

$$2\langle MgO \rangle + \langle ZrSiO_4 \rangle = \langle Mg_2SiO_4 \rangle + \langle ZrO_2 \rangle_{kub},$$
$$\Delta G^0 (1600\,^\circ C) = -56 \text{ kJ/FU}. \tag{3.119}$$

The change in molar volume with respect to the formation of ZrO_2 does not matter because the film (layer) thickness is small.

Manganese vapors react with *corundum* to form the spinel galaxite $MnO \cdot Al_2O_3$. The free enthalpy of formation at 1600 °C is $\Delta G^0 = 29$ kJ/mol. The change in molar volume is negligible. Destruction of the microstructure by crystallization pressure is not expected.

Chromium vapors will react with any oxide refractory material.

With *corundum,* a gapless solid solution (mixed crystal) with high melting temperatures is formed (Fig. 3.34).

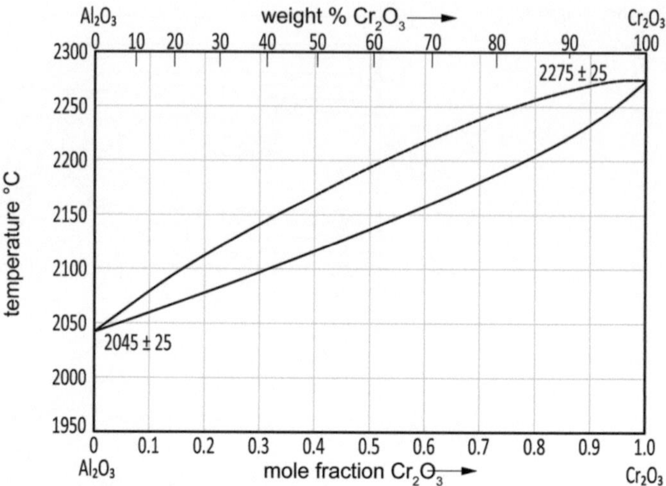

Fig. 3.34 Phase diagram Al_2O_3–Cr_2O_3 [43]

Even the smallest contents of chromium in the furnace gases lead to a permanent ruby-red coloration of the lining, which is particularly undesirable in the glass industry.

With **periclase**, chromium vapors form spinel picrochromite ($MgO \cdot Cr_2O_3$), whose chemical resistance and, compared to MgO, good thermal shock resistance are worth mentioning [27]. At 1600 °C $\Delta G^0 = -54$ kJ/mol.

Iron vapors, like chromium, have partial pressures around 10^{-4} atm at steel mill temperatures. They react with the refractory material, reducing its refractoriness. The melting temperatures drop. The compound herzynite ($FeO \cdot Al_2O_3$, $T_m = 1780$ °C, $\Delta G^0 = 16$ kJ/mol) is formed with corundum, fayalite ($FeO \cdot SiO_2$, $T_m = 1205$ °C, $\Delta G^0 = 6.4$ kJ/mol) with silica, and a solid solution (mixed crystal) [43] with periclase.

Alkali vapors above alkali oxides have partial pressures of more than 10^{-4} atm and basically reduce refractoriness. They lower the melting temperatures and form phases, e.g. ß-corundum, which lower the melting temperature and damage the microstructure by increasing the volume. A very well-known example from the glass industry is the formation of nepheline [7]:

$$\langle Na_2O \rangle + \langle Al_2O_3 \rangle + 2 \langle SiO2 \rangle = \langle Na_2O \cdot Al_2O_3 \cdot 2SiO_2 \rangle, \qquad (3.120)$$

$$\Delta G^0 \, (1400\,°C) = -80 \, \text{kJ/mol}, \quad \Delta V = 9.2 \, \text{cm}^3/\text{mol},$$

$$\Delta P = 87 \times 10^3 \, \text{atm}.$$

Thus, there is an additional risk of mechanical destruction of the microstructure, which is indicated by the high crystallization pressure (refer to Sect. 10.1.6).

In vacuum metallurgy, alkalis only occur when special slags are used, such as, for example, in secondary metallurgy.

3.2.5 *Gases in Vacuum*

3.2.5.1 Hydrogen

The solubility of hydrogen in iron (Fig. 3.35) follows Sievert's law:

$$\frac{1}{2}\{H_2\} = [H], \quad \Delta G^0\,(1600\,^\circ C) = 93\,kJ/mol,$$

$$K_H = [H]/\sqrt{P_{H_2}} = 2.5 \times 10^{-3}. \tag{3.121}$$

The mass (material) transfer takes place by diffusion [73] and can be calculated in a simplified way by the known solution of Fick's 2nd law:

$$\frac{c(t) - c_{sat}}{c_0 - c_{sat}} = \exp\left(-\frac{D \cdot F}{\delta \cdot V} \cdot t\right) \tag{3.122}$$

The diffusion coefficient of hydrogen at 1600 °C is $D = 2.4 \times 10^{-4}$ cm²/s, the thickness of the diffusion boundary layer under natural convection is $\delta = 10^{-3}$ cm [73] and the area/volume ratio of the melt is $F/V = 0.001$. Let the melt be saturated at the beginning, i.e. $c_0 = 2.5 \times 10^{-3}\%$ by weight. At the surface to the gas space,

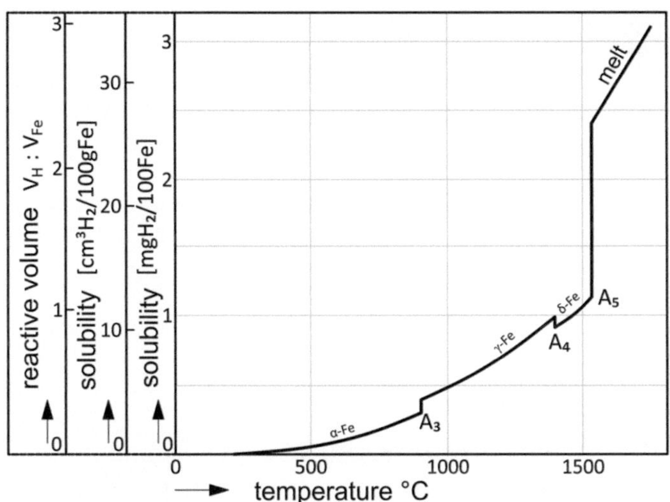

Fig. 3.35 Solubility of hydrogen in iron

the concentration $c_{sat} = 10^{-7}\%$ by weight (4×10^{-5} atm) is maintained by draining. Thus, the concentration in the entire melt drops continuously and reaches the mean value

$$c(600\,\text{s}) = 10^{-7} + \left(2.5 \times 10^{-3} - 10^{-7}\right) \cdot \exp\left(-\frac{2.4 \times 10^{-3} \times 0.001}{10^{-3}}\right) \times 600$$

$$= 6 \times 10^{-4}\,\%\,\text{by weight}. \tag{3.123}$$

after 600 s.

3.2.5.2   Oxygen

Gaseous oxygen dissolves in liquid iron according to the reaction (Sievert) [52]

$$\frac{1}{2}\{O_2\} = [O], \; K = \frac{[O]}{\sqrt{P_{O_2}}}, \; \log K = \frac{5832}{T_K} + 0.19. \tag{3.124}$$

At 1600 °C, $K = 2.0 \times 10^3$. Iron oxidule (FeO) is formed at oxygen saturation: $Fe + [O] = (FeO)$. At 1600 °C, $[O] = 0.2\%$ by weight oxygen is then dissolved in the iron and its partial pressure is $P_{O_2} = 10^{-8}$ atm. In most cases, however, oxygen is not present as a pure component but as a component of a gas mixture CO_2/CO and H_2O/H_2.

The capillary-active dissolved oxygen impairs the mass (material) transfer of gases into and out of the melt, comparable to dissolved sulfur.

3.2.5.3   Water Gas

Water gas is a mixture of the components H_2, H_2O, CO, CO_2. The following relationships apply [52]:

$$2\{H_2\} + \{O_2\} = 2\{H_2O\}, \; K = \frac{P_{\{H_2O\}}}{P_{\{H_2\}} \cdot \sqrt{P_{\{O_2\}}}},$$

$$\log K = \frac{12{,}936}{T_K} - 2.92. \tag{3.125}$$

At 1600 °C and $P_{H_2O}/P_{H_2} = 1$, $P_{O_2} = 10^{-8}$ atm [52]:

$$2\{CO\} + \{O_2\} = 2\{CO_2\}, \; K = \frac{P_{\{CO_2\}}}{P_{\{CO\}} \cdot \sqrt{P_{\{O_2\}}}},$$

$$\log K = \frac{14{,}677}{T_K} - 4.514. \tag{3.126}$$

$$CO + H_2O = CO_2 + H_2$$

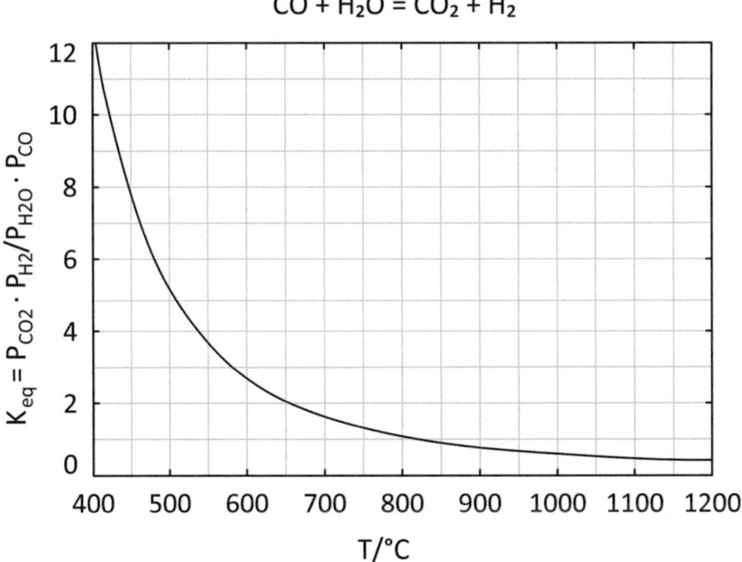

Fig. 3.36 Water gas equilibrium

At 1600 °C and $P_{CO_2}/P_{CO} = 1$, $P_{O_2} = 2 \times 10^{-7}$ atm.
The water gas reaction is [52]:

$$\{H_2O\} + \{CO\} = \{H_2\} + \{CO_2\}, \quad K = \frac{P_{H_2} \cdot P_{CO_2}}{P_{H_2O} \cdot P_{CO}},$$

$$\log K = \frac{1741}{T_K} - 1.594. \tag{3.127}$$

The progression of the equilibrium constant via temperature is shown by Fig. 3.36 with increasing temperature, the equilibrium shifts in the direction of increasing stability of the gases $CO + H_2O$. At low temperature, there is a tendency for CO to react with excretion of graphite and CO_2 (refer to Sect. 3.1.1.1). The water gas equilibrium is independent of the pressure.

3.2.5.4 Carbon Monoxide

Carbon monoxide is formed from the reaction of graphite with oxygen. In liquid iron, this is the most important deoxidation reaction in steel production [52]:

$$[C] + [O] = \{CO\}, \quad K = \frac{P_{\{CO\}}}{[C] \cdot [O]}, \quad \log K = \frac{1160}{T_K} + 2.0 \tag{3.128.1}$$

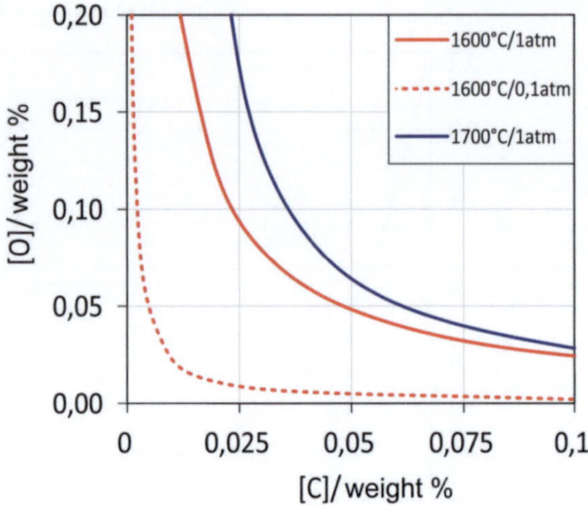

Fig. 3.37 Vacher–Hamilton equilibrium in liquid iron

At 1600 °C consequently

$$[C] \cdot [O] = 2.4 \times 10^{-3} \cdot P_{CO}. \tag{3.128.2}$$

Approximately, $a_{[C]} = [C]$ and $a_{[O]} = [O]$ are set in % by weight. The lower the CO partial pressure, the more the dissolved carbon deoxidizes in the melt. This so-called Vacher–Hamilton equilibrium (Fig. 3.37) is the basis of vacuum metallurgy.

3.2.5.5   Nitrogen

Nitrogen does not usually react directly with refractory lining. However, its behavior in steel melts will be discussed because nitrogen is an important alloying element. The phase diagram in the solidification range is shown in Fig. 3.38. The solubility of nitrogen is described by Sievert's law [51]:

$$\frac{1}{2}\{N_2\} = [N], \ [N] = K_N/\sqrt{P_{N_2}} \ \left[\% \text{ by weight}\right]. \tag{3.129}$$

The equilibrium index has the value $K_N = 4.51 \times 10^{-2}$ [% by weight atm] at 1600 °C [76]. Expressed with the mass (material) transfer coefficient $\beta = 3 \times 10^{-2}$ [cm/s], the following applies to the material flow at 1600 °C

$$j_{[N]} = \frac{\beta_N \cdot \rho_{Fe}}{100 \cdot M_{N_2}} \cdot (\% \text{ by weight} [N_S] - \% \text{ by weight} [N]) \ \left[\text{mol/cm}^2 \text{ s}\right] \tag{3.130}$$

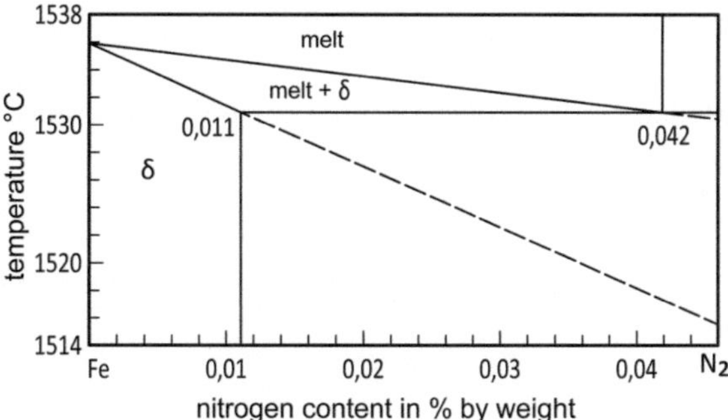

Fig. 3.38 Solubility of nitrogen in iron

% by weight $[N_S]$ is the saturation concentration depending on the respective partial pressure of nitrogen and % by weight $[N]$ depends on time. The molecular weight of nitrogen is $M_{N_2} = 28$ g/mol and $\rho_{Fe} \sim 7$ g/cm³.

In very pure iron, its absorption takes place according to the time law [76].

$$j_{[N]} = k_f \cdot P_{N_2} \left[\mathrm{mol}/\left(\mathrm{cm}^2\,\mathrm{s\,atm}\right)\right], \quad \log k_f = -6340/T - 1.38. \tag{3.131}$$

k_f [mol/cm² s] is the temperature-dependent rate constant. It is independent of the carbon content of the melt.

If the iron contains capillary-active substances, such as sulfur (S) or oxygen (O), in solution, the mass (material) transfer of the nitrogen takes place only after dissociative chemisorption of the N_2 molecules, i.e. with a time delay. The rate constants k_a [mol/cm² s] then depend on the respective chemical activity of the capillary-active element in the melt:

$$k_{a_s} = 1.7 \times 10^{-5}/(1 + 130 \cdot a_S) \left[\mathrm{mol/cm}^2\,\mathrm{s\,atm}\right] a_S \text{ and}$$
$$k_{a_o} = 1.7 \times 10^{-5}/(1 + 220 \cdot a_S) \left[\mathrm{mol/cm}^2\,\mathrm{s\,atm}\right]. \tag{3.132}$$

The same time laws apply for desorption.

Nitrogen can be removed from the melt by purging with argon. This method is particularly successful under negative pressure [77].

In titanium alloyed steels, in addition to titanium carbide, the undesirable titanium nitride is formed in the form of gold-colored inclusions. The nitrogen content drops to low values in the process. For example, in the typical Cr-Ti steel 1.4516, in % by weight (11 % Cr, 1 % Ni, 0.05–0.35 % Ti), for the chemical reaction $[Ti]+[N] = \langle TiN \rangle$ the equilibrium index in the temperature range 1550–1650 °C is [7]

$$\log K = \log([Ti] \cdot [N]) = 5.8 - 14{,}977/T \tag{3.133}$$

At 1873 K, consequently, $[Ti] \cdot [N] = 0.0064$. Excretions of TiN signal to the steelworker the formation of undesirable gas inclusions in the microstructure because the nitrides act as nuclei.

3.2.5.6 Methane

Methane (CH_4) is the essential component of natural gas. It is burned for heating. Above 620 °C, methane reacts with steam to form a hydrogen-rich gas that contains little carbon monoxide:

$$\{CH_4\} + \{H_2O\} = 3\{H_2\} + \{CO\}, \quad K = \frac{P^3_{\{H_2\}} \cdot P_{\{CO\}}}{P_{\{CH_4\}} \cdot P_{\{H_2O\}}},$$

$$\log K = \frac{11{,}796}{T_K} - 13.19. \tag{3.134}$$

Catalysts, e.g. nickel, are used to activate this reaction. The resulting gas is used for the "**direct reduction**" (3.82) of iron oxide to sponge iron at approx. 900 °C:

$$[FeO] + \{H_2\} = [Fe] + \{H_2O\}, \quad \Delta G^0 \, (900 \, ^\circ C) = 1361 \, \text{cal/mol} \tag{3.135}$$

$$[FeO] + \{CO\} = [Fe] + \{CO_2\}, \quad \Delta G^0 \, (900 \, ^\circ C) = 1890 \, \text{cal/mol} \tag{3.136}$$

As the temperature rises, the sintering tendency of the sponge iron increases and, subsequently, the risk of an operational malfunction. It is further processed in the electric arc furnace (refer to Sect. 3.1.1.4).

3.2.6 Measurement of the Leakage Rate

The leakage rate Q_L is defined as the change in the energy content of the gas trapped in the vessel with time:

$$Q_L = \frac{\partial (P \cdot V)}{\partial t} \left[\frac{\text{Pa} \, \text{m}^3}{\text{s}} \right] \tag{3.137}$$

Two equivalent results are obtained from this:
With constant volume $V = V_K$ is

$$Q_L = \theta \frac{\Delta P \cdot V_K}{\Delta t} \left[\frac{\text{Pa} \, \text{m}^3}{\text{s}} \right] \tag{3.138}$$

and at constant pressure

$$Q_L = \theta \frac{P_A \cdot \Delta V}{\Delta t} \left[\frac{\text{Pa} \, \text{m}^3}{\text{s}} \right]. \tag{3.139}$$

V_K [m^3] is the volume of the boiler. The piping and pumps are included. P_A [Pa] is the outlet pressure, and θ is a correction factor.

In practice, Eq. (3.138) is used. The prerequisite is a constant temperature. However, if, for example, air from the surroundings penetrates into a heated vacuum vessel, the penetrating gas is heated and the correction of the measured value is necessary:

$$P_E(T_A) = P_E(T_M) \cdot \frac{T_A}{T_M} \left(\text{or } V_E(T_A) = V_E(T_M) \cdot \frac{T_A}{T_M} \right). \qquad (3.140)$$

If the outside temperature is $T_A = 300$ K and the measured mean inside temperature of the gas is $T_M = 1200$ K, then the correction factor is $\theta = \frac{P_E(T_A)}{P_E(T_M)} = \frac{T_A}{T_M} = 0.25$. The final pressure $P_E(T_A)$ (or the corresponding final volume) to be used, which is related to the outside temperature T_A, is, therefore, only 25% of the measured value $P_E(T_M)$.

For a boiler volume of 5 m^3 and a pressure rise of $\Delta P_E(T_M) = 0.1$ Pa/s, from this, and taking into account the conversion $\left[\frac{\text{mbar l}}{\text{h}} \right] = 360 \left[\frac{\text{Pa m}^3}{\text{s}} \right]$, the leakage rate is calculated from Eqs. (3.138) and (3.140)

$$Q_L = 0.25 \frac{0.1 \times 5}{1} = 0.125 \left[\frac{\text{Pa m}^3}{\text{s}} \right] = 45 \left[\frac{\text{mbar l}}{\text{h}} \right]. \qquad (3.141)$$

3.3 Direct Reduction

Steel is most commonly produced worldwide by reducing the iron ore in a blast furnace (refer to Sect. 3.1.2) and the liquid molten pig iron is tapped at approx. 1500 °C. The liquid molten pig iron is then blown in a converter (refer to Sect. 3.1.4) into crude steel in the converter. Here the dissolved carbon (C) oxidizes with the blown oxygen (O_2) to form carbon monoxide (CO). This blast furnace/converter route produces approx. 1900 tons of carbon dioxide (CO_2)/1 ton of crude steel (refer to [78]). Of this, approx. 1300 t CO_2 is produced in the blast furnace process.

Due to the threat of climate change, it is desirable to reduce the CO_2 load of the process. The blast furnace process is the main option here as this is where most of the carbon dioxide is formed, on the one hand, and, on the other hand, approx. 5.7% by weight carbon is dissolved in the pig iron. This must subsequently, be removed to the largest extent possible in order to produce steel. Steel contains far less than 2% by weight carbon.

The obvious solution is to replace the carbon with a carbon-free reducing agent, such as hydrogen (H_2). However, this is only possible to a very limited extent in the blast furnace as the coke contained in the burden forms a structure in the stack and in the hearth that ensures the undisturbed passage of the gases and the liquid components (slag/metal).

The ore is not reduced by solid carbon but by carbon monoxide (CO). This is produced at high temperatures due to the oxidation of the coke by the oxygen blown into the bosh. Then it flows into the stack where the reduction takes place (refer to Sect. 3.1.2.).

However, the development of direct reduction processes in shaft furnaces, based, on the counter-current principle began as early as the beginning of the 1970s. In Germany, these were the companies PUROFER (Thyssen-Niederrhein, Oberhausen) and MIDREX (Hamburger Stahlwerke, today Arcelor-Mittal). The pelletized ore is reduced at approx. 900 °C by a mixture of carbon monoxide and hydrogen to solid sponge iron which, in addition to the pace, contains only approx. 2% carbon. This is de-charged hot and, therefore, oxidizes very quickly in air which is why it is kept under protective (inert) gas (nitrogen) and immediately compacted into briquettes. These are melted in an electric arc furnace (refer to Sect. 3.1.5) and processed into crude steel. This so-called direct reduction/electric arc route only releases approx. 1000 kg CO_2/t of pig iron, especially as the electrical energy has not yet come from renewable sources. This is set to change.

The reducing gas is obtained by the endothermic, catalytic (nickel) conversion of the natural gas methane (CH_4) with steam above 620 °C (FactSage):

$$CH_4 + H_2O = CO + 3H_2; \ \Delta G^\circ = 54{,}275 - 60.7 \cdot T_K. \tag{3.142}$$

The reduction of the iron ore takes place at approx. 900 °C and is formulated as follows starting from hematite (Fe_2O_3) (FactSage):

$$Fe_2O_3 + 3CO = 2Fe + 3CO_2; \ \Delta G^\circ = 180{,}865 - 61.58 \cdot T_K \tag{3.143}$$

and

$$Fe_2O_3 + 3H_2 = 2Fe + 3H_2O; \ \Delta G^\circ = 204{,}582 - 83.15 \cdot T_K \tag{3.144}$$

This process route produces approx. 1000 kg CO_2/1 t of crude steel as the electrical energy is predominantly obtained from fossil fuels. That is already half the CO_2 emissions of the blast furnace/converter route.

If a sufficient amount of "green" hydrogen is available for the entire process, only Eq. (3.144) applies and the CO_2 load essentially depends on the wear of the graphite electrodes which is approx. 4 kg graphite/t crude steel. It will drop to much less than 50 kg CO_2/t crude steel for the entire process.

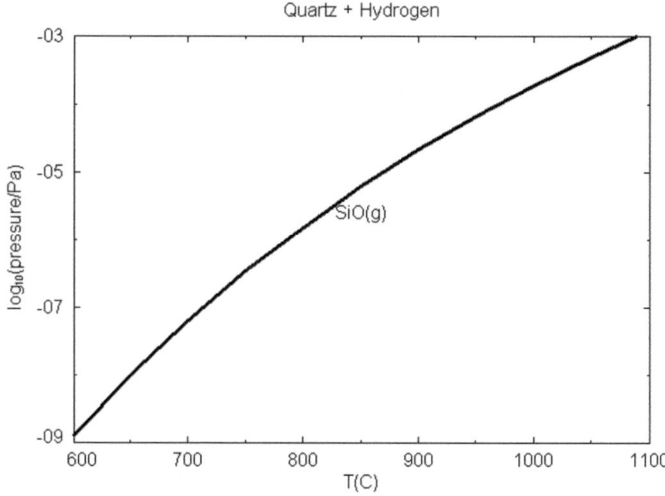

Fig. 3.39 Formation of {SiO} gas from quartz and hydrogen as a function of temperature (FactSage) [78]

The question now arises as to the appropriate lining for the shaft furnace. As in the blast furnace, high-fired, low-iron oxide fireclay with low porosity and high hardness is usually used. It is found in the system SiO_2/Al_2O_3 whose only compound is mullite (3 Al_2O_3·2 SiO_2) (refer to Fig. 4.7). Its proportion ($\leq$ 50% by weight) should be as large as possible in order to increase the refractoriness. In addition, quartz/cristobalite and glass form the structure. The latter react with the hydrogen gas at high temperatures. In the temperature range under consideration up to 1000 °C, however, this is negligible as Fig. 3.39 shows. The partial pressure of the formed suboxide {SiO} is less than 10^{-8} atm (1 Pa = 10^{-5} atm). The evaporation rate is, therefore, lower than the abrasion of the lining or its reactions with the iron oxide. Suboxides of the corundum, e.g. {Al_2O}, also play no role since their partial pressure is many times the power of ten lower than that of {SiO}.

A reaction model for calculating ore reduction in a shaft furnace was developed as early as 1971 by Förster and Pötschke [79]. It shows that continuous direct reduction is clearly superior to batchwise reduction in the retort process (HYL at the time). At 900 °C and a metallization level of 80% the gas utilization, i.e. reduction capacity, increases with increasing superficial velocity of the gas u [m/s] in relation to the reactor height H [m] and reaches approx. 0.45 t_{Fe}/m^3 h at $u/H = 5\,s^{-1}$. In comparison, a retort only achieves 0.4 t_{Fe}/m^3 h, not included the dead times (filling, emptying, heating) which are negative (Fig. 3.40). For this reason, only shaft furnaces are used today.

Fig. 3.40 Reduction
performance of a retort
(lower, blue line) and a shaft
furnace (upper, red line)
under the same conditions
[79]

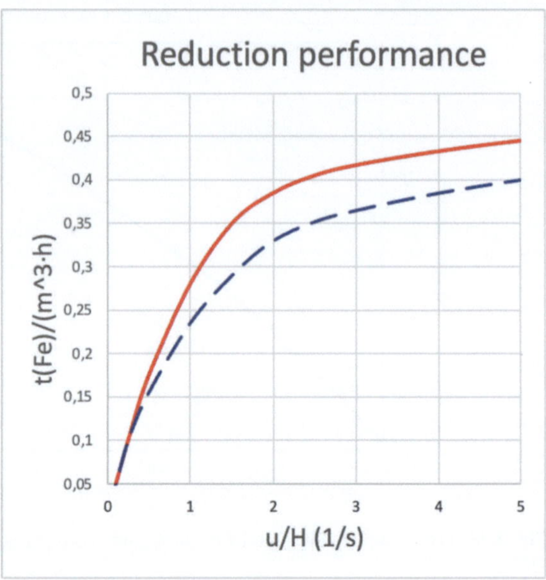

References

1. Eisenhütte: Verlag Stahl Eisen, 5th edn., p. 94 (1961)
2. Ivanov, O.: Metallurgical basic for optimization of blast furnace slags with reference to alkaline capacity. Dr.-Ing. dissertation, TUBA Freiberg in Freiberg, Germany, 2002
3. Knacke, O., Kubaschewski, O., Hesselmann, K.: Thermochemical Properties of Inorganic Substances, II. Springer (1991)
4. Baukloh, R., Knacke, O., Löscher, W.: The growth pressure of carbon. Arch. Eisenhütten. **27**(2), 95–99 (1956)
5. Krause, O., Pötschke, J.: The predictability of CO resistance of refractory material by state-of-the-art test methods—a technical and scientific approach. Interceram **57**(3), 176–180 (2008)
6. Walker, P.L., Rakszawski, J.F., Imperial, G.R.: The properties of carbon formed from CO/H_2 decomposition. J. Phys. Chem. **63**, 140–149 (1959)
7. Hack, K.: FACTSage. Thermfact & GTT Technologies (2007). www.factsage.com
8. Dierich, J.C., Bauer, W., Amirzadeh-Azl, D., Fünders, D.: Properties of synthetic titanium-containing materials to reduce wear in blast furnaces. Stahl Eisen **119**, 85–90 (1999)
9. Shinotake, A., Ootsuka, H., Sasaki, N., Ichida, M.: Blast furnace service life relating to the productivity. La Rev. Métall. CIT 203–209 (2004)
10. Raipala, K.: On hearth phenomena and hot metal carbon content in blast furnaces. Doctoral thesis, Helsinki University of Technology, Helsinki, Finland. Publications in Materials Science and Metallurgy (2003). TKK-MK-151
11. Fünders, D., Harmuth, H., Pötschke, J.: Premature wear in the hearth of a blast furnace and its reduction by the use of titanium oxide. In: Proceedings of UNITECR, Vienna, Austria (2015)
12. Levich, V.G.: Physicochemical Hydrodynamics, pp. 373–390. Prentice Hall Inc. (1962)
13. Kalvelage, L., Markert, J., Pötschke, J.: Measurement of the dissolution of graphite in liquid iron by the tracking of buoyancy. Arch. Eisenhütten. **50**, 107–110 (1979)
14. Markert, J., Pötschke, J.: Determination of interfacial tension between graphite and carbon-saturated iron melt from buoyancy. Arch. Eisenhütten. **50**(2), 53–56 (1979)
15. Borgmann, F.O.: Influencing the surface tension of carbon-saturated iron melts by additional elements. Dr.-Ing. dissertation, TU—Berlin, Germany (1971)

16. Borgmann, F.O., Frohberg, M.G.: Examinations for the determination of the surface tension of iron-carbon-melts at 1,500°C [Untersuchungen zur Ermittlung der Oberflächenspannung von Eisen—Kohlenstoff—Schmelzen bei 1500°C]. Arch. Eisenhütten. **5**(44), 337–340 (1973)

17. Park, C., Gaye, H., Lee, H.G.: Interfacial tension between molten iron and CaO–SiO_2–MgO–Al_2O_3–FeO slag system. Iron Steelmak. **36**(1), 3–11 (2009)

18. Oeters, F.: Metallurgy in Steelmaking. Verlag Stahleisen/Springer (1989)

19. Feiterna, A., Huin, D., Oeters, F., Riboud, P.-V., Roth, J.L.: Iron drop injection into slags by bursting gas bubbles. Steel Res. **71**(3), 61–69 (2000)

20. Holappa, L., Neuschütz, D., Müller, K.T.: Control of ejections caused by bubble bursting in secondary steelmaking processes. ECCR-Research Final Technical Rep. 7210-CC/123-96-C2.02a, 30.6.99

21. Hampel, M., Anezieris, C.G.: Thermal shock performance of carbon bonded materials. In: 48th International Colloquium on Refractory Materials, Aachen, Germany, pp. 28–32 (2005)

22. Brochen, E., Brückmann, C., Pötschke, J., Mittler, G., Vollenberg, I., Mudersbach, D.: Thermomechanical characterization of magnesia-based refractory products by means of wedge splitting test above 1,000°C. In: 53rd International Colloquium on Refractories, pp. 82–85 (2010)

23. Deo, B., Boom, R.: Fundamentals of Steelmaking Metallurgy, 1st edn. Prentice Hall International (1993)

24. Odenthal, H.-J., Falkenreck, U., Schlüter, J.: CFD simulation of multiphase melt flows in steelmaking converters. In: European Conference on Computational Fluid Dynamics. TU Delft (NL) (2006)

25. Ebneth, G., Pluschkell, W.: Dimensional analysis of the vertical heterogeneous buoyancy plume. Steel Res. **56**, 513–518 (1985)

26. Schürmann, E., Mahn, G., Resch, W.: Operational examinations of the progression of dephosphorization and granules formation in the slag during the basic oxygen process. Stahl Eisen **97**(23), 1069–1074 (1977)

27. Routschke, G., Wuthnow, H.: Practice Manual Refractory Materials, 5th edn. Vulkan Verlag, Essen, Germany (2011)

28. Meschede, D.: Gerthsen Physik. Springer (2005)

29. Wei, J.-H., Zhu, H.-L., Jiang, Q.-Y., Shi, G.-M., Chi, H.-B., Wang, H.-J.: Physical modeling study on back attack phenomenon and its influence on erosion and wear of refractory lining. ISIJ Inst. **50**(10), 1347–1356 (2010)

30. Park, J.M.: MgO solubility in BOF slag equilibrated with ambient air. Steel Res. **4**(72), 141–145 (2001)

31. VDEh: Slag Atlas, 2nd edn. Verlag Stahleisen GmbH Düsseldorf, Germany (1995)

32. Mudersbach, D., Geiseler, J., Kühn, M.: Optimized mathematical model to calculate the viscosity of slags. Iron Steel Slags FG Ironworks Slags Heft **8**, 213–221 (1974–2000)

33. Schmeiduch, G., Oeters, F.: Influence of some operational parameters on the wear of refractory materials in the electric arc furnace. Stahl Eisen **100**, 1188–1194 (1980)

34. Drissen, P., Engell, H.-J., Janke, D.: Viscosity of steelmaking slags at elevated MgO—content. Iron Steel Slags FG Ironworks Slags Heft **8**, 223–234 (1974–2000)

35. Ottmar, H., Oerter, A., Ameling, D.: The connection between the electrical engineering and thermal engineering basics in high performance electrical arcs. Radex Rundschau **28**, 519–527 (1973)

36. Bird, R.B., Stewart, W.E., Lightfoot, E.M.: Transport Phenomena. Wiley (1960)

37. Bannenberg, N.: Interactions between refractory material and steel and their impact on the purity of the steel. Stahl Eisen **9**(115), 79–86 (1995)

38. Reisinger, P., Presslinger, H., Hiebler, H., Zednicek, W.: MgO-solubility in steel plant slags. BHM Berg Hüttenmännische Monatsh. **144**(5), 196–203 (1999)

39. Park, J.M., Lee, K.K.: Reaction equilibria between liquid iron and CaO–Al_2O_3–$MgO_{sat.}$–SiO_2–Fe_tO–MnO–P_2O_5 slag. In: Proceedings of the 79th Steelmaking Conference, Pittsburgh (Pennsylvania), USA, 24 Mar 1996

40. Brüggmann, C., Pötschke, J.: MgO-saturation in secondary metallurgical lime-aluminate and lime-silicate slags. Steel Res. Int. **4**(82), 422–427 (2011)
41. Frohberg, M.G., Kapoor, M.L.: The application of a new basicity measure of metallurgical reactions. Stahl Eisen **4**(91), 182–188 (1971)
42. Duffy, J.A., Ingram, M.D.: Improvement of a bulk optical basicity table for oxidic systems. J. Non-Cryst. Solids **21**, 373–410 (1976)
43. Slag Atlas. Verlag Stahleisen, Düsseldorf, Germany (1995)
44. Smeets, S., Parada, S., Weytjens, J., Heylen, G., Jones, P.T., Guo, M., Blanpain, B., Wollants, P.: Behavior of magnesia-carbon refractories in VOD-ladle linings. Iron Steelmak. **4**(30), 293–300 (2003)
45. Wöhrmeyer, C., Elorza-Ricard, E., Jolly, R., Guichard, S., Brüggmann, C., Sax, A.: The impact of synthetic slags on steel ladle refractory life time. In: 51st International Colloquium on Refractories, Aachen, Germany, pp. 80–83 (2008)
46. Pötschke, J.: The influence of electrical forces on the corrosion of refractory materials. In: Refractories Manual, pp. 10–26 (2008)
47. Pötschke, J., Brüggmann, C.: Some corrosion phenomena in an induction furnace. In: Proceedings of UNITECR, Paper No. 200, Salvador, Brazil, 13–16 Oct 2009
48. Koch, R.: High temperature behavior of refractory materials in vacuum. Dr.-Mont dissertation, Montanuniversität Leoben, Austria, 1993
49. Granitzki, K.-E.: Refractories in the foundry industry. In: Ceramic News—Special Refractories, pp. 24–34. Verlag Schmid (2000)
50. Saarinen, A.: Refractories for Steelmaking. International Iron and Steel Institute. Committee on Technology, Brussels, Belgium (1992)
51. Frohberg, M.G.: Thermodynamics for Materials Engineers and Metallurgists, 2nd edn. Deutscher Verlag Grundstoffindustrie, Leipzig, Stuttgart, Germany (1993)
52. Oeters, F.: Metallurgy of Steelmaking. Verlag Stahleisen, Berlin, Germany (1989)
53. Hohlfeld, J., Janke, D.: Layer ceramic sensors for supervision of metallurgical melting and refining processes. Veitsch-Radex Rundschau **1**, 40–56 (1999)
54. Bernsmann, G.P.: Inclusion control in steel for tire cord. In: Proceedings of Conference on Metallurgy Processing and Applications of Metal Wires, Cincinnati, Ohio, USA, 6–10 Oct 1969, pp. 123–133
55. Ender, A., van den Boom, H., Kwast, H., Lindenberg, H.-U.: Metallurgical development in steel plant internal multi-injection hot metal desulphurization. Steel Res. Int. **76**(8), 562–572 (2005)
56. Asai, S., Muchi, I.: Mass transfer rate in ladle refining process. In: International Conference on Refining of Iron and Steel by Powder Injection 3. (1983), part 1, pp. 12.1–12.29, Luleå, Mefos. [S.l.: s.ed.] (1983)
57. Hauck, F.G., Pötschke, J.: Wear of submerged nozzles during continuous casting of steel. Arch. Eisenhütten. **53**, 133–138 (1982)
58. Deinet, T., Pötschke, J.: Corrosion of AMC ladle bricks. In: Proceedings 45th International Colloquium on Refractories, pp. 36–39. Aachen, Germany (2002)
59. Schubert, H.G., Schwerdtfeger, W.: Solubility of carbon in ESU slags. Arch. Eisenhütten. **45**, 437–439 (1974)
60. Granitzki, K.-E.: ASTM Metals Handbook 15 (1992)
61. Neumann, F.: (BBC): Technology of Melting for Iron and Steel Casting Special Printing of Manual of Production Technology, vol. 1. Carl Hanser Verlag (1981)
62. Volmer, M.: Kinetics of Phase Formation. Th. Steinkopf, Dresden, Leipzig, Germany (1939)
63. Köhler, M.: Bubble dynamics and erosion with acoustical cavitation. Dr.-Ing. dissertation, Univ. Stuttgart, Germany, 1999
64. Wei, J.-H., Zhu, H.-L., Jiang, Q.-Y., Shi, G.-M., Chi, H.-B., Wang, H.-J.: Physical modelling study on back attack phenomenon and its influence on erosion and wear of refractory lining. ISIJ Int. **50**(10), 1347–1356 (2010)
65. Koch, R.: High temperature of refractory materials in vacuum. Dr.-Mont dissertation, Montanuniversität Leoben, Austria, 1993

66. Baake, E., Langejürgen, M., Kirpo, M., Jakovics, A.: Analysis of transient melt flow and temperature distribution in induction channel furnaces. In: COST Action P 17 "EPM", WG V1 "Liquid Metals and Semiconductors", Workshop, Grenoble, France, 25 Mar 2009

67. Langejürgen, M.: Development and Application of a New Process for Calculation of Turbulent Flows in Induction Channel Furnaces. Series "Elektrotechnik", Bd. 12. Sierke Verlag (2009)

68. Baake, E., Jakovics, A., Pavlovs, S., Kirpo, M.: Influence of the channel design on the heat and mass exchange of induction channel furnaces. COMPEL **30**(5), 1637–1650 (2011). ISSN 0332-1649

69. Wurtz, M.: Theory and Practice of Vacuum Technology. Vieweg Verlag, Braunschweig, Germany (1965)

70. Bird, R.B., Steward, W.E., Lightfoot, E.N.: Transport Phenomena. Wiley, New York (1965)

71. Pötschke, J.: A new model for drying refractory concrete. Refractories Worldforum (2), 99–106 (2010)

72. Kuchling, H.: Handbook of Physics. H. Deutsch Verlag, Frankfurt/Main, Germany (1978)

73. Pawlek, F.: Metallurgy. W. de Gruyter, Berlin, New York (1983)

74. Winkler, O.: Theory and practice of vacuum melting. Reprint Metall. Rev. **5**(17), 1–33 (1960)

75. Seo, J.-D., Kim, S.-H.: Thermodynamic assessment of Mg-deoxidation reaction of liquid iron and equilibria of Mg–Al–O and Mg–S–O. Steel Res. **4**(71), 101–106 (2000)

76. Byrne, M., Belton, G.R.: Studies of the interfacial kinetics of the reaction of nitrogen with liquid iron. Metall. Trans. **14B**, 441–449 (1983)

77. Xie, X., Chen, C., Shi, W., Pu, H., Jin, J.: Nitrogen removal by argon bubbling during superalloy melting in a vacuum induction furnace

78. Lüngen, H.B.: Wege zur Minderung von CO_2—Emissionen in der Stahl- und Eisenindustrie in Europa. Sonderdruck VDEh, 8/2020

79. Förster, E., Pötschke, J.: Reaction-technical treatment of gas—solid material implementations in iron metallurgy. Arch. Eisenhütten. **43**(2), 135–139 (1972)

80. Jeschke, P.: Texture analysis of basic refractory brick. J. Am. Ceram. Soc. **49**(7), 360–363 (1966)

81. Schumann, H.: Metallography. VEB Deutscher Verlag für die Grundstoffindustrie, Leipzig, Germany (1962)

82. R. Telle, H. Salmang, H. Scholze: "Ceramik" (2006) Springer

83. Dubbels: Pocket Book for Machine Manufacturer, 12 (1966) 409

Chapter 4
Reactions of Refractory Materials with Steel and Slag

4.1 Comparison of Test Methods

4.1.1 Introduction

The wear of refractory material is usually subdivided in the mechanisms

- chemical dissolution,
- erosion and
- thermomechanical failure.

The chemical dissolution of the refractory material takes place both at the outer contact zone with liquid steel and its slag and "internally" by way of infiltrated melts. The infiltrated melts, especially the slag, loosen the grain structure of the microstructure so that large grains can fall out. This erosion process is promoted by a high flow rate of the melt (refer to Sect. 3.1.1.5.3.2.1).

Since the infiltrated melt has different thermomechanical properties after solidification and cooling than the surrounding structure of the refractory material, thermo-mechanically induced spalling of large areas of infiltrated material may occur while the refractory material is in service.

Here, the chemical dissolution is the main focus. The mass dissolved in time is according to Nernst (1904) [1]

$$\dot{m} = \frac{D}{\delta_N} \frac{F \cdot \rho}{100} (C_s - C_0) \; [\text{g/s}] \tag{4.1}$$

D [cm^2/s] is the mean diffusion coefficient of the refractory component concerned, e.g. (MgO) in the slag, and δ_N [cm] the thickness of the diffusion boundary layer. The quotient $\frac{D}{\delta_N}$ is designated as the mass transfer coefficient β [cm/s]. ρ [g/cm^3] is the average density of the liquid slag or steel, respectively. Diffusion is driven by the concentration difference $(C_s - C_0)$ [% by weight]. The linear corrosion progression

© The Author(s), under exclusive license to Springer Nature Switzerland AG 2024

J. Pötschke, *Refractory Fundamentals in Metallurgical Practice*,

https://doi.org/10.1007/978-3-031-63709-4_4

is as follows:

$$\frac{\Delta x}{\Delta t} = \frac{\dot{m}}{F \cdot \rho_{FF}} \ [\text{cm/s}],$$

(4.2)

i.e., the reaction area F [cm^2] can be cancelled out if Eqs. (4.1) and (4.2) are combined. Summarized it as follows:

$$\frac{\Delta x}{\Delta t} = 360 \cdot \beta \cdot \frac{\rho}{\rho_R} \cdot (C_s - C_0) \ [\text{mm/h}]$$

(4.3)

The value of the material transition coefficient upon forced laminar flow depends on the mean flow speed of the melt [1]:

$$\beta = \sqrt{\frac{4 \cdot D \cdot u}{\pi \cdot h}} \cdot \left(\frac{D \cdot \rho}{\eta}\right)^{\frac{1}{6}} \ [\text{cm/s}]$$

(4.4)

The expression in brackets takes into account the wall friction, which reduces the mass (material) transfer. It can be neglected for very slow flow (glass, slag) but not for steel [2]. There this correction factor has approximately the value 0.5 [2]. η [g/cm s] is the dynamic toughness (viscosity) and $\eta/\rho = \nu$ [cm^2/s] is the kinematic toughness (viscosity) of the fluid. The comparison of the informative value of different test methods is made by comparing the effect of the respective existing flow speed of the melts steel and/or slag on the corrosion rate.

4.1.2   Test Results

Figure 4.1 compares the most commonly used crucible test with the complex rotary drum test and the induction furnace test using the example of a refractory castable for lining a blast furnace trough (HO) [3]. The characteristic data are given in Table 4.1. The third column gives the composition of the slag saturated with CA6 (CaO·6Al$_2$O$_3$) calculated with FactSage [4] from the first two columns. Alumina, in particular, will dissolve from the lining into the slag. Pig iron and blast furnace slag were used uniformly.

As a result, the corrosion rate in the crucible test (left column) is lower than that in the rotary drum (middle column), which in turn is lower than that in the induction furnace (right column).

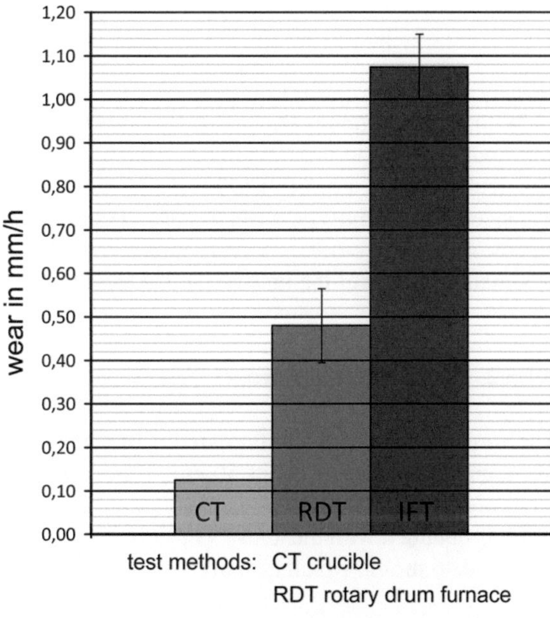

Fig. 4.1 Results of the corrosion tests on the refractory castable of a blast furnace trough (channel)

Table 4.1 Characteristic data of the refractory castable of a blast furnace trough (channel)

Weight %	Lining	Furnace slag	C_{sat}
Al_2O_3	69	12	40
SiO_2	7	35	28
CaO	1	44	29
MgO	0	7	3
SiC	18	0	18

4.1.3 Interpretation

4.1.3.1 Crucible Test

The **crucible test** is static. The specimen (80×80 mm^2 $\times 65$ mm) contains a hole (40 mm $\varnothing \times 35$ mm) in which 70 g of the slag is heated to 1500 °C in a resistance furnace at 2.5 °C/min and then held for 4 h. The temperature in the crucible is set at 0.5 °C/min. The convection u_T in the crucible [1, 2] is very low and results from the buoyancy, i.e., the characteristic length $h/2 = 0.2$ cm, the cubic thermal expansion coefficient of the slag $\beta_T = -7.5 \times 10^{-5}$ [K^{-1}], the estimated temperature gradient grad $T \approx 0.5$ [K/cm], and the kinematic toughness (viscosity) of the slag $\nu_{Sl} = 1.2$ cm^2/s [5, 6]:

$$u_T = -\frac{g \cdot \beta_T \cdot \left(\frac{h}{2}\right)^3}{\nu_{Sl}} \cdot \mathrm{grad}\, T = 2.5 \times 10^{-4}\,[\mathrm{cm/s}]. \qquad (4.5)$$

With the usual values $D_{Sl} = 2 \times 10^{-7}$ cm^2/s [5], $\rho_{Sl} = 2.9$ g/cm^3 [5], $\rho_R = 2.9$ g/cm^3 and the characteristic length $h = 0.4$ cm, the mass (material) transfer coefficient is calculated from Eq. (4.4) to be $\beta_{Sl} = 1.3 \times 10^{-5}$ cm/s. With the equilibrium distance of the alumina between lining and saturation slag $\Delta C = 28\%$ by weight, the corrosion rate (speed) $V_{corr} = 0.13$ mm/h is obtained from Eq. (4.3). This confirms well with the wear of the material of the blast furnace trough (channel) (Fig. 4.1).

4.1.3.2   Rotary Drum Furnace Test

The furnace body of the **rotary drum furnace** makes $n = 6$ revolutions/min and has a circumference of about 50 cm. This is $u = 5$ cm/s. The permanent contact circumference of the bath $h = 3$ cm is entered as the characteristic length in Eq. (4.4), and it is obtained with the above values $\beta_{Sl} = 4.6 \times 10^{-5}$ cm/s and from Eq. (1.1.3) $V_{corr} = 0.46$ mm/h, which also fits well (Fig. 4.1).

4.1.3.3   Induction Crucible Test

In the **induction crucible** (16 cm Ø, 8 cm filling height, 10 kHz), 10 kg of iron and 1.5 kg of slag are melted and held at 1500 °C for 2 h under inert gas. The recorded power in continuous operation is $N_i \approx 15$ kW. Half the filling height of the induction crucible $h = 4$ cm is chosen as the characteristic length for calculating the mean flow because of the two counter-rotating convection rolls of the melt. The flow velocity (speed) of the cast iron melt is estimated from the induced power to be $u \approx 40$ cm/s, which is very plausible [7]. Equation (4.4) is to be calculated taking into account the correction in parentheses, thus giving $\beta = 1.2 \times 10^{-4}$ cm/s as the mass transfer coefficient. From Eq. (4.3), the corrosion rate $V_{corr} = 1.2$ mm/h is obtained (Fig. 4.1).

These three experimental results are largely in agreement with the calculation because slag infiltration was insignificant. They show that the test in the induction furnace leads to the heaviest wear and is, therefore, most meaningful or rather informative in many cases.

4.2   Dissolution of Refractory Materials in Confined Vessels

Until now, the calculation of the resolution was based on the assumption that the mass of the melt is arbitrarily large compared to that of the body under consideration. At least in laboratory experiments, this is often a too strong oversimplification. Therefore, some frequently occurring examples will be discussed.

4.2.1 Dissolution of the Refractory Material in the Crucible Test (1500 °C) (Refer to Sect. 4.1.3.1)

The wall of the above crucible wetted by the slag has an area of $F = 0.44$ cm^2 if neglecting the bottom. The volume of the melt is $V = 0.044$ cm^3. The mass (material) transfer coefficient is $\beta_{Sl} = 1.3 \times 10^{-5}$ cm/s (see above). The temporal increase in concentration in the slag is (refer Fundamentals, Sect. 2.2.4.2, Eq. 2.2.37)

$$i \cdot F = V \cdot \frac{dC}{dt} \ [g/s]. \tag{4.6}$$

With the boundary conditions $t = 0$; $C = C_0$ and $t = \infty$; $C = C_S$, $C_S > C_0$ and the solution follows:

$$\frac{C(t) - C_0}{C_S - C_0} = 1 - \exp\left(-\beta \cdot \frac{F}{V} \cdot t\right). \tag{4.7}$$

Considering the alumina content $C(t)$ [% by weight] as the parameter characterizing the dissolution process, then $C_S = 40\%$ by weight and $C_0 = 12\%$ by weight. The regarding time increasing progression of the Al$_2$O$_3$ concentration in the slag is shown in Fig. 4.2.

The change in the thickness of the wall layer removed by mass (material) transfer is given by the initial condition $t = 0$, $x(t) = x_0$

$$x(t) - x_0 = -\frac{V \cdot \rho_{Sl}}{100 \cdot F \cdot \rho_R} \cdot (C_s - C_0) \cdot \left(1 - \exp\left(-\beta_{Sl} \cdot \frac{F}{V} \cdot t\right)\right) \ [cm]. \tag{4.8}$$

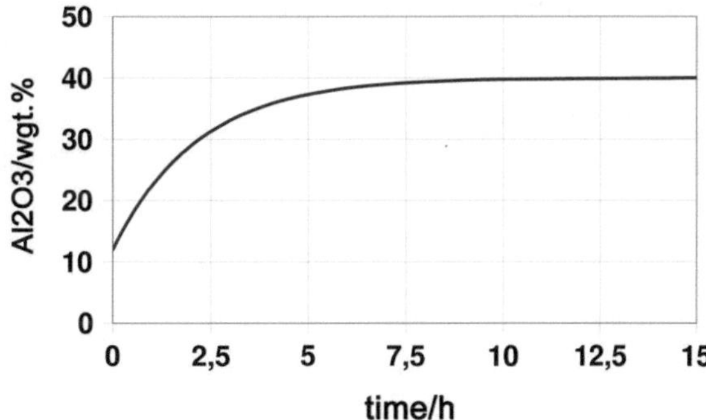

Fig. 4.2 Time progression of the Al$_2$O$_3$ concentration in the crucible

The maximum reduction in wall thickness is $(x(\infty) - x_0) = 2.8 \times 10^{-2}$ cm. After one hour, $\Delta x = 1.1 \times 10^{-2}$ cm are corroded which confirms quite well with the simplified calculation according to Eq. (4.3) and the experimental result. For the plate geometry (design) the result is correct. The corrosion of a rod in the melt can also be calculated approximately in this way. The decrease of its diameter is $\Delta d = 2\Delta x$.

The corrosion rate (speed) is $u_{corr}(t) = dx/dt$, i.e.

$$u_{corr}(t) = 360 \cdot \beta \frac{\rho_{Sl}}{\rho_R} \cdot (C_S - C_0) \cdot \exp\left(-\beta_{Sl} \cdot \frac{F}{V} \cdot t\right) [mm/h] \qquad (4.9)$$

The concentration in the slag $C(t)$ and the wear $(x(t) - x_0)$ increase with time, while the corrosion rate (speed) $u_{corr}(t)$ decreases with time. After one hour ($t = 3600$ s), the corrosion rate (speed) in our example is only $u_{corr}(t) = 8.2 \times 10^{-2}$ mm/h.

4.2.2   Dissolution of a Rod Out of Corundum in a Glass Melt (1500 °C)

In the previous calculation, the geometry (design) was not taken into account. For large dimensions (converter, ladle) this is not important in many cases. For corrosion tests in the laboratory, however, it does play a role, especially if an exact result is to be obtained (refer to Sect. 2.2.4.2.2). The most frequent case in the laboratory, the dissolution of an immersed cylinder, is dealt with here. The time required for the complete dissolution is what is desired.

In total 350 g of container glass ($\rho = 2.3$ g/cm^3, $V = 152$ cm^3) is held in a platinum crucible at 1500 °C under inert gas. A cylindrical rod out of $\alpha\beta$-fused alumina (FAB) ($R = 0.6$ cm, $\rho_R = 3.5$ g/cm^3) is immersed 3 cm deep in the melt ($F_0 = 11$ cm^2) and rotates about its longitudinal axis at $n = 50$ rpm. In the process, it dissolves completely, with $m_0 = 12$ g Al$_2$O$_3$ go in solution. This is $\Delta C = 3.3\%$. The initial concentration of alumina in the glass melt is $C_0 = 1.5\%$ by weight and the saturation concentration $C_S = 29\%$ by weight [8]. The mass required for complete saturation would be $m_S = 96$ g. Consequently, the dissolution ends well before saturation is reached. The mean diffusion coefficient in the melt is $D = 2 \times 10^{-7}$ cm^2/s [8] and the viscosity is $\eta = 76$ g/cm s [8]. Both values change little upon dissolution.

The increase of the Al$_2$O$_3$ concentration in the glass melt over time is shown graphically in Fig. 4.3. In eleven days, the concentration increases from 1.5% by weight to just 5% by weight. Then the rod is completely dissolved.

Fig. 4.3 Time progression of the Al_2O_3 concentration in glass

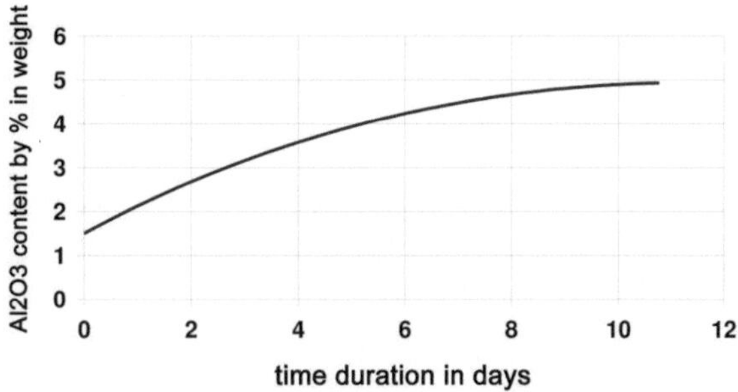

The mass transfer coefficient is calculated from the circumferential velocity u_R of the cylinder wall and the specified material values. The following experimentally confirmed dependencies apply in the glass [9]: The circumferential velocity is given by $u_R = 2\pi \cdot R \cdot n/60 = 3.14\,\text{cm/s}$ and the Reynolds number is defined as $\text{Re} = (u_R \cdot 2 \cdot R/v) = 0.114$ (laminar flow). $v = \eta/\rho = 33\,\text{cm}^2/\text{s}$ is the kinematic toughness (viscosity). From this, according to Prandtl, the thickness (viscosity) of the flow boundary layer [9] is calculated to $\delta_{Pr} = 12.64 \cdot \text{Re}^{-0.7} = 57.8\,\text{cm}$. With the Schmidt number $Sc = v/D = 1.65 \times 10^8$, the thickness of the diffusion boundary layer named after Nernst [9] is calculated for the rotating cylinder to $\delta_N = 0.5 \cdot \delta_{Pr} \cdot S_C^{-1/3} = 5.5 \times 10^{-2}\,\text{cm}$. The mass transfer coefficient is by definition [1, 2]

$$\beta = D/\delta_N = 3.8 \times 10^{-6}\,\text{cm/s}.$$

According to Friedrichs and Knacke [10] (refer to Sect. 2.2.4.2.2) the related radius ratio is

$$\varphi \equiv \left(\frac{R_0}{R_S}\right)^2 = \frac{m_0}{m_S} = \frac{12}{96} = 0.125, \tag{4.10.1}$$

the generalized time

$$\tau \equiv \frac{D \cdot F_0}{\delta \cdot V} \cdot t = 3.8 \times 10^{-6} \cdot \frac{11}{152} \cdot t = 2.8 \times 10^{-7} \cdot t \tag{4.10.2}$$

and the reduced saturation ratio:

$$\gamma \equiv \frac{C - C_0}{C_S - C_0} = \frac{5 - 1.5}{29 - 1.5} = 0.125. \tag{4.10.3}$$

The solution of the **Differential Equation** (2.2.47) $d\tau = \sqrt{\varphi} \cdot \dfrac{d\gamma}{(1-\gamma) \cdot \sqrt{\varphi - \gamma}}$ is, therefore, [10]:

$$\tau = 2 \cdot \sqrt{\frac{\varphi}{1-\varphi}} \cdot \left(\arctan \sqrt{\frac{\varphi}{1-\varphi}} - \arctan \sqrt{\frac{\varphi - \gamma}{1-\varphi}} \right) \text{ (for } \varphi < 1; \; 0 \leq \gamma \leq \varphi)$$

$$(4.11)$$

Since the dissolution is completed before saturation is reached, the reduced saturation ratio $\gamma = 0.125$ equals the referenced radius ratio $\varphi = 0.125$. Thus, Eq. (4.11) simplifies to

$$\tau = 2 \cdot \sqrt{\frac{\varphi}{1-\varphi}} \cdot \arctan \sqrt{\frac{\varphi}{1-\varphi}} \text{ (for } \varphi < 1; \; \gamma = \varphi). \qquad (4.12)$$

With $\varphi = 0.125$, the generalized time of the total process is $\tau = 0.273$. The total time required for dissolution follows from Eq. (4.10.2) at $t = 2.8 \times 10^5$ s corresponding to 11 days, which is in broad agreement with laboratory measurements [9].

4.2.3 Measurement of the Mass Transfer Coefficient During the Dissolution of Graphite in Liquid Iron (1600 °C)

The solutions of Friedrichs and Knacke [10] (Sect. 2.2.4.2.2) are specifically suitable for the determination of mass (material) transfer coefficients because already small deviations of the experimental result from the mathematical description show up clearly and, thus, indicate inadequacies in the experiment. As an example for the applicability of this method, the dissolution of dense spectral graphite in an iron melt is observed [11].

In an alumina crucible, technically pure iron, $m_{Fe} = 1300$ g, is inductively melted (10 kHz) under inert gas and held at 1600 °C. $C_0 = 3.5\%$ by weight carbon is initially added to the melt. A cylindrical rod out of spectral graphite ($R_0 = 1$ cm, $\rho_R = 1.7$ g/cm^3) is then immersed $l = 7$ cm deep into the melt and its dissolution is observed gravimetrically for about 20 min (Fig. 4.4).

The test evaluation is conducted according to Friedrichs and Knacke [10] (refer to Sect. 2.2.4.2.2). The saturation of the melt in carbon ($C_S = 5.41\%$ by weight) is described by the equation $C_{sat} = 1.3 + 2.57 \times 10^{-3} \, T_{°C}$ [% by weight] and is reached before the rod is completely dissolved. Then, the rod has the radius

$$R_S = \left(\frac{(C_S - C_0) \cdot m_{Fe}}{100 \cdot \rho_R \cdot \pi \cdot l} \right)^{1/2} = \left(\frac{(5.41 - 3.5) \times 1300}{100 \times 1.7 \times \pi \times 7} \right)^{1/2} = 0.815 \, [\text{cm}].$$

$$(4.13.1)$$

With the related radius ratio

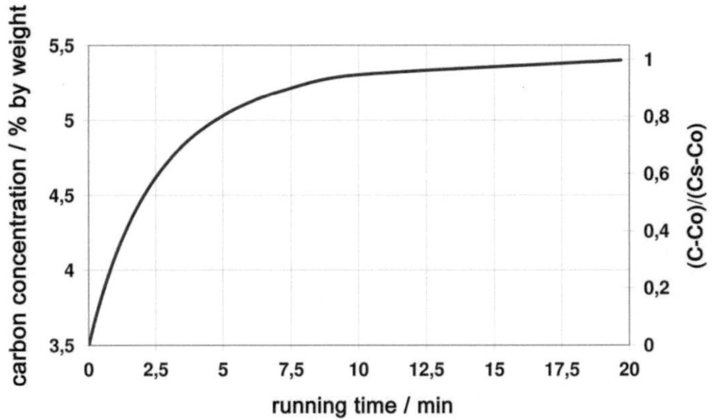

Fig. 4.4 Dissolution of a graphite rod in pure iron at 1600 °C

$$\varphi \equiv \left(\frac{R_0}{R_S}\right)^2 = \left(\frac{1}{0.815}\right)^2 = 1.51, \tag{4.13.2}$$

the reduced saturation ratio:

$$\gamma \equiv \frac{C - C_0}{C_S - C_0} = \frac{C(t) - 3.5}{5.41 - 3.5} \tag{4.13.3}$$

and the generalized time

$$\tau \equiv \frac{D \cdot F_0}{\delta \cdot V} \cdot t = \beta \cdot \frac{2\pi \cdot R_0 \cdot l}{m_{Fe}/\rho_{Fe(C)}} \cdot t = \beta \cdot \frac{2\pi \times 1 \times 7}{1300/6.65} \cdot t = \beta \cdot 0.225 \cdot t \tag{4.13.4}$$

the solution of the differential equation $d\tau = \sqrt{\varphi} \cdot \frac{d\gamma}{(1-\gamma)\cdot\sqrt{\varphi-\gamma}}$ (2.2.47) [10] is

$$\tau = \frac{\sqrt{\varphi}}{\sqrt{\varphi-1}}\left(\ln\frac{\sqrt{\varphi}-\sqrt{\varphi-1}}{\sqrt{\varphi}+\sqrt{\varphi-1}} - \ln\frac{\sqrt{\varphi-\gamma}-\sqrt{\varphi-1}}{\sqrt{\varphi-\gamma}+\sqrt{\varphi-1}}\right) \text{ (für } \varphi > 1; 0 \leq \gamma \leq 1).$$

$$\tag{4.14}$$

The face area of the specimen is not considered, only the immersed shell area $F = 2\pi R\,h$ [cm^2]. The thickness of the diffusion boundary layer is much smaller than the sample radius: $\delta \ll R$ [cm].

At the beginning, R_S (Eq. 4.13.1) and φ (Eq. 4.13.2) are calculated. Then, γ (Eq. 4.13.3) is determined next from the representation of the measured concentration progression over the duration of the experiment (Fig. 4.4). This is followed by the calculation of $\tau(C(t))$ (Eq. 4.13.4). Now, each number $\tau(C(t))$ is assigned to the measured time t/s and displayed graphically. If the measurement was without error, a straight line of slope τ/t [s^{-1}] is obtained (Fig. 4.5). From it, one calculates the mass (material) transfer coefficient:

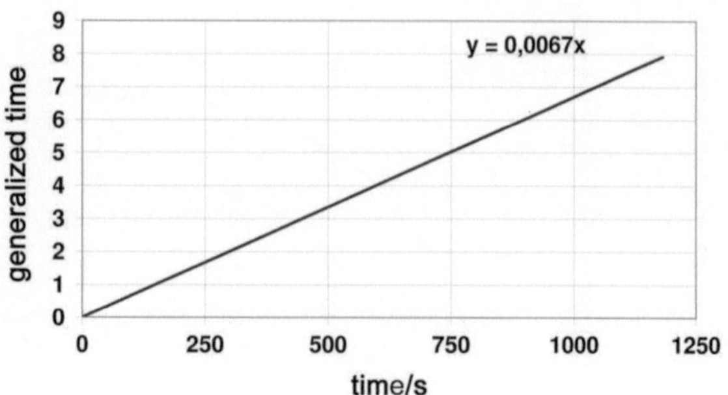

Fig. 4.5 Generalized time as a function of test duration/sec

$$\beta = \frac{V}{F_0} \cdot \frac{\tau}{t} = \frac{1}{0.225} \times 0.0067 = 0.03 \,\text{cm/s}. \tag{4.15}$$

The temperature dependence of the mass transfer coefficient of carbon in the range 1550–1650 °C has been shown in these experiments to be [11].

$$\beta = 5.21 \cdot \exp\left(-\frac{80{,}863}{R \cdot T}\right) \,[\text{cm/s}] \tag{4.16}$$

($R = 8.31$ J/K mol). Taking into account, the mean thickness of the diffusion boundary layer $\delta_N = 3.7 \times 10^{-3}$ [cm], which was also measured, we obtain for the diffusion coefficient of carbon in this temperature and concentration range [11]

$$D = 1.52 \times 10^{-3} \cdot \exp\left(-\frac{41{,}575}{R \cdot T}\right) \,[\text{cm}^2/\text{s}]. \tag{4.17}$$

At 1600 °C (1873 K), $D = 1.05 \times 10^{-4}$ [cm^2/s]. This method is very accurate and very well applicable for all refractory systems.

4.2.4 Dissolution of a MgO Rod in a CaO–Al$_2$O$_3$ Slag (1650 °C)

In a platinum crucible 350 g of calcium aluminate slag (50% by weight CaO, 35% by weight Al$_2$O$_3$, 15% by weight SiO$_2$) is kept at 1650 °C under argon. A cylindrical rod ($R = 0.9$ cm) out of densely sintered technically pure MgO ($\rho_R = 3.6$ g/cm^3) is immersed in the slag ($\rho_S = 2.7$ g/cm^3) to a depth of 3.3 cm until it is completely dissolved. At this moment, the melt has just become saturated and contains $m_S = 10.5\%$ by weight MgO dissolved. Available is, therefore, also $m_0 = 37$ g MgO, i.e. it is the special case of $m_S = m_0$. The dissolution is observed gravimetrically [12]. The temporal increase of the MgO concentration is seen from Fig. 4.6.

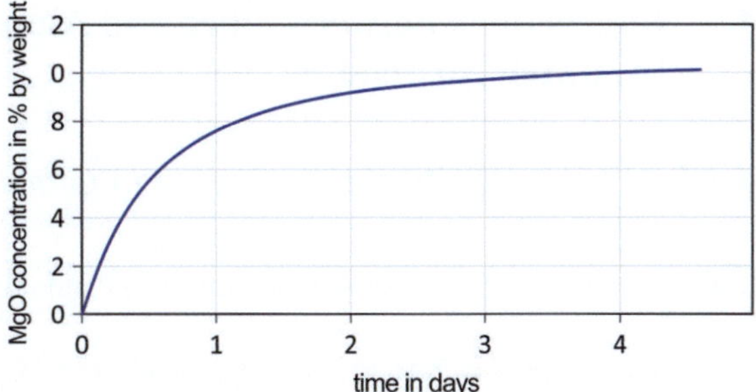

Fig. 4.6 Dissolution of a MgO rod in CaO–Al$_2$O$_3$ slag

With basicity $B = \text{CaO}/(\text{Al}_2\text{O}_3 + \text{SiO}_2) = 1$, the saturation concentration of the slag in MgO is calculated from the equation [9]

$$C_{\text{sätt}}(\text{MgO}) = \frac{9.5}{B} + 0.02(T - 1600\,^\circ\text{C})\left[\%\text{ by weight}\right] = 15.5\,\%\text{ by weight} \tag{4.18}$$

The mass transfer coefficient is experimentally found to be $\beta = 1.3 \times 10^{-4}$ cm^2/s [12].

According to Friedrichs and Knacke [10] (refer to Sect. 2.2.4.2.2), the related radii ratio

$$\varphi \equiv \left(\frac{R_0}{R_S}\right)^2 = \frac{m_0}{m_S} = 1 \tag{4.19.1}$$

is the generalized time

$$\tau \equiv \frac{D \cdot F_0}{\delta \cdot V} \cdot t = 2.1 \times 10^{-5} \cdot t \tag{4.19.2}$$

and the reduced saturation ratio:

$$\gamma \equiv \frac{C - C_0}{C_S - C_0} \to 1. \tag{4.19.3}$$

The solution of the differential equation (2.2.47) for $\varphi = 1$ and $0 < \gamma < 1$ [10] is

$$\tau = \frac{2}{\sqrt{1 - \gamma}} - 2. \tag{4.20}$$

In this case, full dissolution with simultaneous saturation ($\gamma = 1$) cannot be achieved in finite time.

4.3   Interactions Between Silica and Alumina in Deoxidized Iron Melts at 1600 °C

4.3.1   Phase Equilibria According to Gibbs

In the simplest case, the two components SiO_2 and Al_2O_3 react to form mullite [4]:

$$3Al_2O_3 + 2SiO_2 = 3Al_2O_3 \cdot 2SiO_2, \quad \Delta G^0(1600\,°C) = -31.2\,kJ/mol \quad (4.21.1)$$

Figure 4.7 shows the phase diagram of the system. It is subdivided into two systems by the peritectic compound mullite ($3Al_2O_3 \cdot 2SiO_2$). Consequently, in the (partially)-solidified state, the edge oxides are never in direct equilibrium with each other but each can only be in equilibrium with mullite on its own. The reaction (4.21.1) continues until one of the components Al_2O_3 or SiO_2 is completely consumed.

According to Gibbs, the following applies in thermodynamic equilibrium (refer to Sect. 2.1.1):

$$K = 2(Al_2O_3 \text{ or } SiO_2, \ 3Al_2O_3 \cdot 2SiO_2)$$
$$P = 2(Al_2O_3 \text{ or } SiO_2, \ 3Al_2O_3 \cdot 2SiO_2)$$

$F = K + 2 - P - R = 2 + 2 - 2 - 0 = 2$ (pressure, temperature). Since the existence of all involved phases is practically independent of the pressure, the only degree of freedom left is the temperature. For the equilibrium alumina/mullite, this applies below 1810 °C and for silica/mullite below 1595 °C (refer to Fig. 4.7).

If, for example, the molten steel deoxidized with carbon is added, further reaction equations occur [2]:

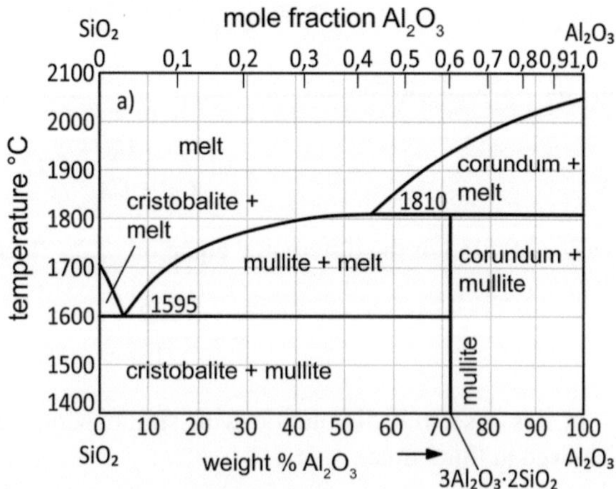

Fig. 4.7 Phase diagram of the system Al_2O_3/SiO_2 [5]

$$Al_2O_3 = 2[Al] + 3[O], \quad \Delta G^0(1600\,°C) = +470\,kJ/mol. \tag{4.21.2}$$

$$SiO_2 = [Si] + 2[O], \quad \Delta G^0(1600\,°C) = +167\,kJ/mol. \tag{4.21.3}$$

$$\{CO\} = [C] + [O], \quad \Delta G^0(1600\,°C) = +93\,kJ/mol. \tag{4.21.4}$$

$K = 9\,(Al_2O_3,\ SiO_2,\ 3Al_2O_3 \cdot 2SiO_2,\ Fe,\ [O],\ [C],\ [Al],\ [Si],\ \{CO\})$

$P = 4\,(Al_2O_3\ or\ SiO_2,\ 3Al_2O_3 \cdot 2SiO_2,\ melt,\ gas)$

$R = 4\,(4.21.1 - 4.21.4)$

$F = K + 2 - P - R = 3$ (temperature, CO partial pressure and one concentration). Deoxidation with manganese, for example, produces a liquid slag which contains all the oxide components. In addition, there is another reaction [2]:

$$(MnO)_{FeO-MnO-Slag} = [Mn] + [O], \quad \Delta G^0(1600\,°C) = +42\,kJ/mol. \tag{4.21.5}$$

It follows:

$K = 11\,(Al_2O_3,\ SiO_2,\ 3Al_2O_3 \cdot 2SiO_2,\ melt,\ [O],\ [C],\ [Al],\ [Si],\ [Mn]\,\{CO\})$

$P = 5\,(Al_2O_3\ or\ SiO_2,\ 3Al_2O_3 \cdot 2SiO_2,\ melt,\ slag,\ gas)$

$R = 5\,(4.21.1 - 4.21.5)$

$F = K + 2 - P - R = 3$ (temperature, CO partial pressure and one concentration). The addition of other components does not change the number of freedoms.

4.3.2 Experiments [13–16]

About 1700 g of technically pure iron are melted in a crucible made of sintered corundum under an argon atmosphere and held at 1600 °C for up to 3 h.

After deoxidation with high-purity graphite at a pressure of approx. 0.04 atm to an oxygen content of approx. 0.001% by weight (10 ppm), the gas is substituted by high-purity argon and, if necessary, a second, metallic deoxidizing agent is added. A rod out of fused silica (1.7 cm Ø) is then immersed in the melt. The change in weight of the immersed rod is continuously monitored via its buoyancy. The temperature and oxygen content are also monitored continuously by means of an electrochemical oxygen measuring cell (EMF) installed in the bottom of the crucible (refer to Sect. 9. 4). At the beginning and during the test, melt samples are taken with the aid of vacuum pipettes and chemically analyzed. At the end of the experiment, the specimen rod is withdrawn from the melt and subjected to phase analysis. The same applies to the slags, which are formed in particular after the addition of manganese and collect at the edge of the crucible.

4.3.3 Results and Discussion

4.3.3.1 Deoxidation by Carbon [14]

Figure 4.8 shows the surprising appearance of the fused silica rod immersed in the melt for three hours. It is covered by a light gray crystalline layer, which turns out to be 2/1—mullite (right half of the picture). Underneath, the glass is transformed into α-cristobalite in a very thin layer, next to traces of α-corundum. The mullite layer does not adhere reliably to the fused silica, so that the "seal" does not provide reliable protection against corrosion.

Figure 4.9 shows the concentration curves over time and the change in the mass of the specimen. Here, the concentration is not plotted in % in weight but in mol/ 1000 g molten iron in order to clarify stoichiometric relationships. The conversion is:

$$x_A \left[\text{mol}/1000\,\text{g} \right] \times 0.1 M_A = y_A \left[\% \text{ by weight} \right]. \tag{4.22}$$

For example, 0.06 [mol/1000 g] of silicon correspond to $0.1 - 28 = 0.168$ [% by weight].

The carbon content decreases almost linearly from 0.14 [mol/1000 g] to 0.02 [mol/ 1000 g]. In parallel, the silicon content increases linearly to 0.06 [mol/1000 g]. The carbon consequently reduces the silica stoichiometrically according to $2[C] + \langle SiO_2 \rangle = 2\{CO\} + [Si]$. After 2.5 h, the reaction comes to its apparent stop. The

Fig. 4.8 Silica glass after three hours of reaction in iron deoxidized with carbon with an average oxygen content of 0.0032% by weight in a crucible out of sintered corundum at 1600 °C. Right: mullite crystals [14]

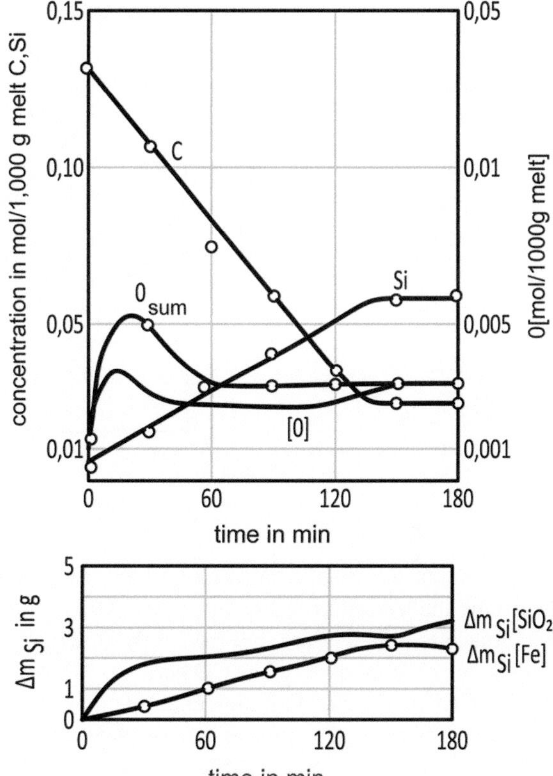

Fig. 4.9 Reaction between silica glass and liquid iron deoxidized to 0.12% by weight carbon initial content in a crucible out of sintered corundum at 1600 °C [14]

reaction time, resulting from the decrease in weight of the specimen − Δm (SiO$_2$) calculated increase of the silicon concentration in the bath Δm_{Si} (Fe), is almost equal. The chemically analyzed total oxygen content O_{total} is only slightly above the electrochemically determined value (30–40 ppm). Suspended oxides are present as impurities in the melt, especially at the beginning.

The thermodynamic explanation of the result can be taken from the comparison of the free enthalpies ΔG^0 in Eqs. (4.21.1, 4.21.2, 4.21.3, 4.21.4) and is shown in the deoxidation diagram (Fig. 4.10). During the deoxidation of the melt, oxide is formed and, at the same time, part of the metallic deoxidizing agent remains dissolved in the steel. The oxygen content [O]$_{Fe}$ in % by weight, dissolved at equilibrium at 1600 °C after the addition of a deoxidizing agent, is dotted and the linear progressing oxygen activity a[O]$_{Fe}$ is drawn as a continuous line, indicating the oxide in question. It can be seen that at high metal contents the solubility of oxygen increases again, although its activity continues to decrease. This is a consequence of the thermodynamic interaction of both components, metal and oxygen, with each other (refer to Sect. 2.1.3.3 and Fig. 2.4). The same applies for deoxidation with carbon. Its effect as a function of the CO partial pressure is also illustrated. The lower the CO partial pressure, the greater the effect.

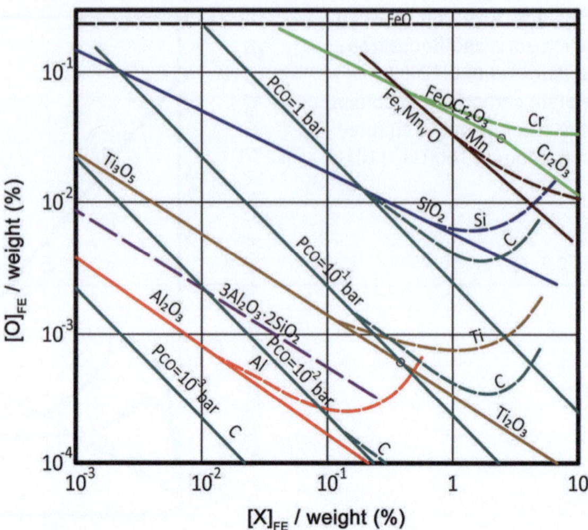

Fig. 4.10 Deoxidation diagram in liquid iron at 1600 °C [2] (MnO, SiO$_2$, 3Al$_2$O$_3$·2SiO$_2$, Al$_2$O$_3$)

As is often the case in metallurgy, here, too, local thermodynamic conditions determine the overall process. Since the CO partial pressure in the atmosphere of the furnace is initially zero and rises only slowly, the silicic acid rod and alumina crucible are reduced at very low CO partial pressure by the carbon already dissolved in the melt. Silicon and aluminum enter the melt.

The contents immediately at the surfaces of both oxides are calculated with stoichiometric dissociation according to Eqs. (4.21.2 and 4.21.3) to:

$$K_{SiO_2} = [Si] \cdot [O]^2 = [Si] \cdot \left[\frac{28}{2 \times 16} \cdot [Si]\right]^2 = 0.766 \cdot [Si]^3 = 2.2 \times 10^{-5}$$

$$[Si] = 3.0 \times 10^{-2} \, wt\% \quad and \quad [O] = 2.7 \times 10^{-2} \, wt\%; \tag{4.23.1}$$

$$K_{Al_2O_3} = [Al]^2 \cdot [O]^3 = [Al]^2 \left[\frac{2 \times 27}{3 \times 16} \cdot [Al]\right]^3 = 1.42 \cdot [Al]^5 = 8 \times 10^{-14}$$

$$[Al] = 2.2 \times 10^{-3} \, wt\% \quad and \quad [O] = 2.5 \times 10^{-3} \, wt\%. \tag{4.23.2}$$

At the fused silica surface, the oxygen activity in the iron is consequently so high that the dissolved aluminum reacts there to form Al$_2$O$_3$ or for the most part mullite [5]:

$$6[Al] + 2[Si] + 13[O] = \langle 3Al_2O_3 \cdot 2SiO_2 \rangle; \quad \Delta G^0(1600\,^\circ C) = +1827 \, kJ/mol \tag{4.24}$$

Figure 4.10 shows the mullite reaction (dotted-purple). It lies between the deoxidation effect of silicon (blue) and aluminum (red).

The reaction apparently comes to a halt when one of the partial reactions approaches its thermodynamic equilibrium. In fact, silicic acid reduction continues until one of the reaction partners is consumed.

The kinetics of silica glass dissolution, i.e. the change in concentration over time, follows a reaction order of "zero" and remains unchanged for about 2 h. The speed is independent of the concentration during this time. The formation of the dense top layer x_0 of mullite is responsible. If one first assumes for the layer formation the parabolic time law to be expected for a diffusion-determined dissolution, then k_1 [cm^2/s] applies:

$$\frac{dx}{dt} = \frac{k_1}{x_0 + x} \, [\text{cm/s}]; \quad x_0 > 0; \quad t = 0, \; x = 0. \tag{4.25}$$

After integration with the integration constant $C = -x_0{}^2$ follows

$$x(x + 2x_0) = 2k_1 \cdot t \, [\text{cm}^2]. \tag{4.26.1}$$

The layer of thickness x_0 acts as a diffusion barrier, i.e. as if $x_0 \gg x$. Therefore, in the bracket x is negligibly small over against x_0. For the layer growth x, the linear growth law results consequence

$$x = \frac{k_1}{x_0} \cdot t \, [\text{cm}]. \tag{4.26.2}$$

Since the change in layer thickness cannot be followed directly, the layer thickness $x(t)$ is approximated by using the silicon concentration [Si](t) in the bath and converted to degraded silicic acid. The concentration of silicon increases from 0.01 to 0.06 mol$_{Si}$/1000 g$_{Fe}$, that is $\Delta C = 0.05$ mol$_{Si}$/1000 g$_{Fe}$. Consequently, $j = 0.05/150 = 3.3 \times 10^{-4}$ mol$_{si}$/1000 g min.

The mass flow density is according to Nernst:

$$j \left[\frac{\text{mol}}{\text{cm}^2 \, \text{s}} \right] = \beta \left[\frac{\text{cm}}{\text{s}} \right] \cdot (C_S - C_0) \left[\frac{\text{mol}}{\text{cm}^3} \right]. \tag{4.27}$$

β is the mass (material) transfer coefficient. With the molecular weights $M_{Si} = 28$ g/mol and $M_{SiO_2} = 60$ g/mol and the density of the melt $\rho_{Fe} = 7$ g/cm^3, the conversion yields the dissolved amount of silicic acid to:

$$\Delta C \left[\frac{\text{mol}_{Si}}{1000 \, g_{Fe}} \right] = \Delta C \times 10^{-3} \cdot \frac{\rho_{Fe} \cdot M_{SiO_2}}{M_{Si}} = \Delta C \left[\frac{\text{mol}_{SiO_2}}{\text{cm}^3} \right] \tag{4.28}$$

$$\text{i.e. } 0.05 \left[\frac{\text{mol}_{Si}}{1000 \, g_{Fe}} \right] = 0.05 \times 10^{-3} \times \frac{7 \times 60}{28} = 7.5 \times 10^{-4} \left[\frac{\text{mol}_{SiO_2}}{\text{cm}^3} \right].$$

The reaction surface, without considering its change, is $F = 40$ cm^2 and the melting volume is $V = 240$ cm^3. Consequently, the mass flow density is as follows:

$$j\left[\frac{\text{mol}_{\text{Si}}}{1000\,\text{g}_{\text{Fe}} \cdot \text{min}}\right] = j \times 10^{-3} \cdot \frac{V \cdot \rho_{\text{Fe}} \cdot M_{\text{SiO}_2}}{60 \cdot F \cdot M_{\text{Si}}} = j\left[\frac{\text{mol}_{\text{SiO}_2}}{\text{cm}^2\,\text{s}}\right] \qquad (4.29)$$

i.e. $3.3 \times 10^{-4}\left[\frac{\text{mol}_{\text{Si}}}{1000\,\text{g}_{\text{Fe}}\cdot\text{min}}\right] = 3.3 \times 10^{-7} \times \frac{240 \times 7 \times 60}{60 \times 40 \times 28} = 5 \times 10^{-7}\left[\frac{\text{mol}_{\text{SiO}_2}}{\text{cm}^2\,\text{s}}\right]$.

From this, the mass (material) transfer coefficient follows directly from Eq. (4.27) to:

$$\beta = j/\Delta C = 5 \times 10^{-7}/7.5 \times 10^{-4} = 7 \times 10^{-4}\,\text{cm/s}. \qquad (4.30)$$

This value is about one power of ten smaller than that of a not impaired dissolution of silica acid in liquid iron [17–19] and proves that the mullite layer hinders the reactions. The reaction constant k is calculated as follows:

$$k = \frac{F}{V} \cdot \beta = 1.2 \times 10^{-4}\,\text{s}^{-1}. \qquad (4.31)$$

This experiment makes the interaction between the refractory material and the formation of oxides in the melt particularly clear: the lining has a significant share in the reactions within the melt. It is worth recalling the excretion of spinels as a result of the deoxidation of steel with aluminum in a magnesia lining [20].

4.3.3.2   Deoxidation by Silicon [15]

Deoxidation with silicon leads to a dense layer of mullite. Dissolution of the fused silica occurs almost stoichiometrically by the dissolved carbon (Fig. 4.11). By subtraction of the reactions (4.21.3) and (4.21.4) one gets:

$$\langle \text{SiO}_2 \rangle + 2[\text{C}] = [\text{Si}] + 2\{\text{CO}\} \quad \Delta G^0(1600\,^\circ\text{C}) = -18\,\text{kJ/mol} \qquad (4.32)$$

In the first half hour, a constant oxygen content of approx. 30 ppm is established. The kinetics follow a reaction with zero order (Fig. 4.11). The reaction constant is $k_1 \approx 10\,\text{s}^{-4-1}$. Consequently, the mass (material) transfer coefficient is $\beta \approx 6$–10^{-4} cm/s. The reduction of silica glass in iron deoxidized with carbon usually shows mass (material) transfer coefficients which are at least one power of ten higher [17–19], from which it is also concluded here that the mullite layer delays the dissolution of the silica glass.

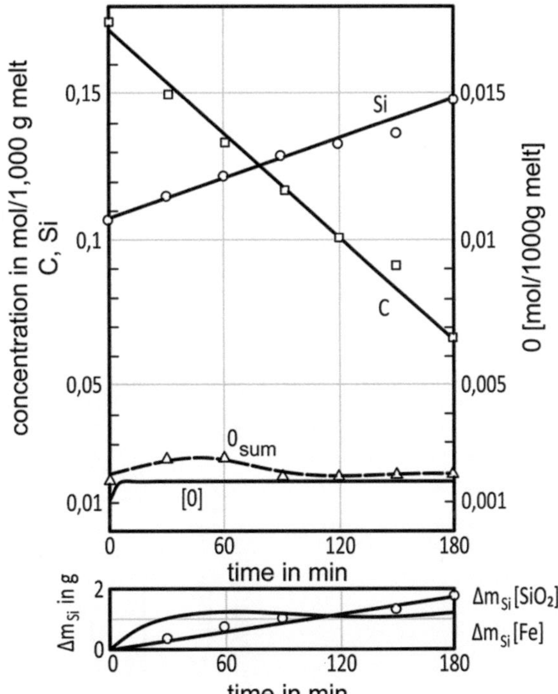

Fig. 4.11 Reaction between silica glass and liquid iron containing C and 0.3% by weight Si deoxidized at 1600 °C [15]

4.3.3.3 Deoxidation by Aluminum [15]

Deoxidation with aluminum leads to excretion of α-corundum on the fused silica with low contents of iron oxide and mullite in the transition layer directly at the rod surface. The excretion processes depend heavily on the initial concentration of aluminum, whereas the dissolution speed depends only slightly on it.

Figure 4.12 shows the progression of a particularly striking result: carbon reduces the silicic acid. The oxygen content is approx. 30 ppm. It is clearly evident from the time progressions of the contents of silicon and carbon that a reaction with zero order is also observed here. Other experiments [13, 15] show that an increasing initial content of aluminum only slightly reduces the reaction constant : $k_1 \approx 10^{-4}$ s^{-1}. The excretion form of the alumina behaves differently: With increasing aluminum content (max. 0.065%), the corundum layer formed from dendrites becomes increasingly more dense (Fig. 4.13).

However, the result of the buoyancy measurement in the lower partial image should be emphasized. Immediately upon immersion of the silica glass, an oxide layer grows. After half an hour, it bursts off and starts to form again (Fig. 4.14). Although this process is not the rule, it shows that the "protective top layer" cannot be relied upon.

From the difference between the total oxygen content and the dissolved share, it can be seen in Fig. 4.12 that Al deoxidation has produced a strong suspension of

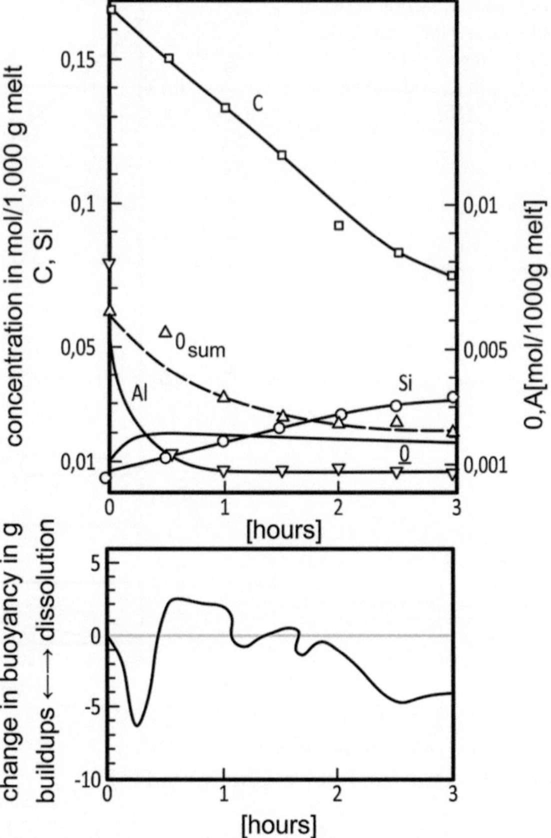

Fig. 4.12 Reaction between silica glass and liquid iron with 0.02% by weight Al at 1600 °C [13]

oxides in the melt, which decreases in time. If the steel matrix is dissolved chemically with bromine methanol and the inclusions are examined microscopically, α-corundum with a wide variety of growth forms is found. Figure 4.15 shows a so-called "alumina nest." It is formed at high mass (material) flow density [21] and is a branched single crystal. This type of inclusion is feared because it is very difficult to separate from the melt. It consists mainly of the melt itself.

4.3.3.4   Deoxidation by Manganese [13, 15]

The progression of slagging of silica in iron deoxidized with manganese at 1600 °C is shown in Fig. 4.16. After deoxidation with manganese, a fused slag film immediately forms on the immersed fused silica:

$$3\langle SiO_2\rangle_{rod} + 2[Mn] = 2(SiO_2,\ MnO)_{Sl} + [Si] \tag{4.33}$$

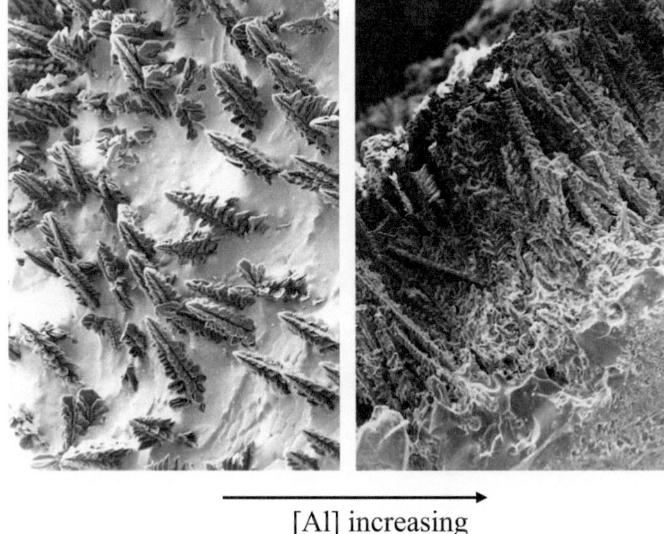

Fig. 4.13 Corundum dendrites of the surface layer on fused silica grown in an Al-deoxidized iron melt at 1600 °C [15]

Fig. 4.14 Spalled top layer of alumina formed on a base of quartz glass in Al-deoxidized iron at 1600 °C [15]

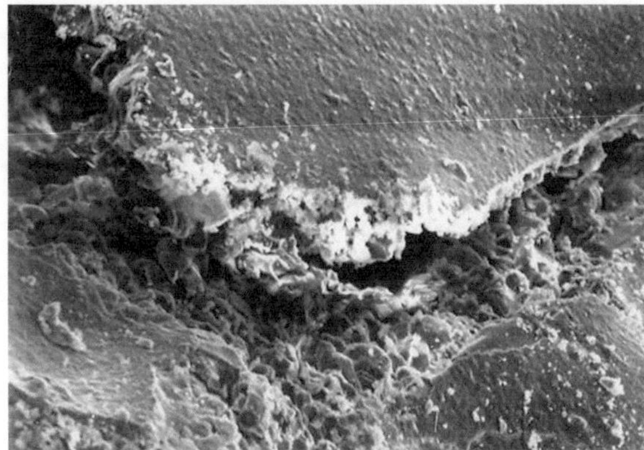

This slag consists of about half each of both oxides, but already contains small amounts of Al_2O_3 dissolved. A large part of the slag collects at the edge of the corundum crucible, where it reaches saturation with alumina (30% by weight SiO_2, 30% by weight MnO, 40% by weight Al_2O_3). The crucible is visibly attacked. The manganese content changes only slightly because manganese oxide from the slag (refer to dark brown line in Fig. 4.10) is at least partially reduced as manganese back into the bath by all metallic components dissolved in the iron.

Here, too, the reactions at the rod and at the crucible are completely different and neither of them is in equilibrium with the melt which is further away from the

Fig. 4.15 "Alumina nest" formed during the deoxidation of iron by aluminum at 1600 °C [15]

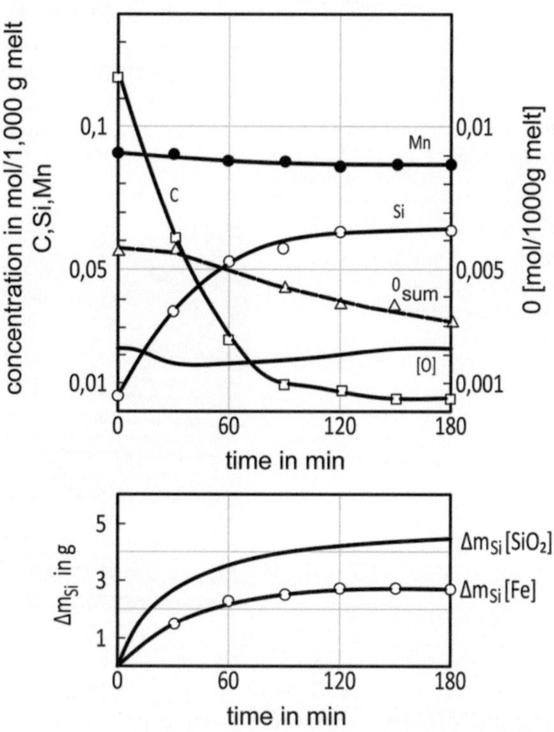

Fig. 4.16 Reaction between fused silica and manganese-containing iron in the sintered corundum crucible at 1600 °C [15]

reaction site. In the lower part of Fig. 4.16, it is clear that, although the dissolved silicon content tends toward a final value, the dissolution of the glass rod does not. Theoretically, equilibrium is never reached but the reactions continue until either the glass or the crucible is completely dissolved. The content of oxygen dissolved in the iron is almost constant and is about 30 ppm. A considerable amount of oxides is emulsified in the steel which decreases in time.

Overall, this is a reaction of first order. The speed constant is on average 4×10^{-4} s^{-1} and the mass (material) transfer coefficient is consequently $\beta = 2.4 \times 10^{-3}$ cm/s. This value agrees with data in literature [22] and confirms that the mass (material) transfer in the slag layer formed on the rod determines the reaction speed.

In summary, it should be noted that different refractory materials react with each other in melts even if they are separated locally. For this reason, it is recommended, for example, to always place two bars of the same material next to each other in the crucible test in the induction furnace and only then to install a pair of another material next to it. In this way, different materials can be tested at comparable conditions.

It should also be noted that the formation of cover layers exerts a strong influence on the reactions between the oxides of the refractory materials and the melt. This is also confirmed by studies on the reaction of magnesite-chrome bricks in deoxidized iron melts [16]. A problem encountered in practice is the formation of spinels in the production of aluminum-deoxidized wire grades [20] using refractories containing MgO.

References

1. Bird, R.B., Stewart, W.E., Lightfoot, E.N.: Transport Phenomena. Wiley, New York (1965)
2. Oeters, F.: Metallurgy in Steelmaking. Verlag Stahleisen/Springer (1989)
3. Brüggmann, C., Pötschke, J.: MgO saturation in secondary lime-aluminate and lime-silicate slags. Steel Res. **4**(82), 422–427 (2011)
4. Hack, K.: FACTSage, Thermfact & GTT Technologies (2007). www.factsage.com
5. VDEh: Slag Atlas, 2nd edn. Verlag Stahleisen GmbH Düsseldorf, Germany (1995)
6. Mudersbach, D., Drissen, P.M., Kühn, M., Geiseler, J.: Viscosity of slags. Steel Res. **72**, 86–90 (2001)
7. Neumann, F.: Technology of melting for iron and steel casting. In: Special Printing Handbook of Manufacturing Technique, Band 1. Carl Hanser Verlag (1981)
8. Dunkl, M., Brückner, R.: Corrosion of refractory material by a container glass melt. Glastechn. Ber. **62**(1), 10–19 (1989)
9. Brückner, R.: Interactions between glass melt and refractory material. Glastechn. Ber. **53**(4), 77–88 (1980)
10. Friedrichs, H.-A., Knacke, O.: Process Engineering Basic Types of Isothermal Reactions in Blown Through and Overflown Fills. Research Reports of the State of North Rhein-Westphalia, Nr. 2240 (1973)
11. Kalvelage, L., Markert, J., Pötschke, J.: Measurement of the dissolution of graphite in liquid iron. Arch. Eisenhüttenwesen. **50**(3), 107–110 (1979)
12. Brüggmann, C.: A contribution to slagging of MgO in secondary metallurgical slags. Dr.-Ing. dissertation, TU Bergakademie Freiberg/Saxony, Germany, 2011
13. Knacke, O., Markert, J., Pötschke, J.: Gravimetric and Electrochemical Studies of the Reaction of Silica with Deoxidized Iron Melts. Metallurgical Transactions A, Iron and Steel Institute, Pittsburgh, Pennsylvania, USA, p. 9 (1978)
14. Hauck, F., Markert, J., Pötschke, J.: The reaction between silica and alumina in carbon deoxidized iron melt. Arch. Eisenhüttenwesen. **50**(5), 189–193 (1979)
15. Markert, J., Pötschke, J.: The influence of alumina on the reduction of silicic acid in iron melts containing manganese, silicon and aluminum. Arch. Eisenhüttenwesen. **50**(6), 231–236 (1979)
16. Hauck, F., Pötschke, J.: Reaction of refractory magnesite-chrome bricks in deoxidized iron melts. Arch. Eisenhüttenwesen. **50**(4), 145–150 (1979)

17. Frohberg, M.G., Leygraf, H.: Contribution to kinetics of silicic acid reduction. Arch. Eisenhüttenwesen. **41**, 501–508 (1970)
18. Oeters, F., Heyer, K., Vardag, S.G.K.: About the dissolution of suspended oxide particles in liquid iron. Arch. Eisenhüttenwesen. **38**, 83–90 (1967)
19. Kienow, S., Knüppel, R., Oeters, F.: Examination results regarding corrosion kinetics of refractory materials with the assistance of the rotating disc. Arch. Eisenhüttenwesen. **46**, 57–64 (1975)
20. Bernsmann, G.P.: Inclusion control in steel for tire cord. In: Proceedings of Conference on Metallurgy Processing and Applications of Metal Wires, Cincinnati, Ohio, USA, 6–10 Oct 1969, pp. 123–133
21. Steinmetz, E., Lindenberg, H.U., Moersdorf, W., Hammerschmid, P.: Growth forms of aluminum oxides in steels. Arch. Eisenhüttenwesen. **48**, 569–574 (1977)
22. Abratis, H.: Reduction of silicic acid and silicic acid-containing liquid iron-manganese alloys. Arch. Eisenhüttenwesen. **41**, 909–915 (1970)

Chapter 5
Infiltration

5.1 Basic Equations

In general, the following applies to the calculation of a capillary equilibrium (refer to Sect. 2.1.3.4, Fig. 2.10)

$$\left[P_{\text{atm}} + \rho_S \cdot g \cdot h_u\right]_{\text{surrounding}} = \left[P_{G,K} - \rho_S \cdot g \cdot h_K + 4 \cdot \sigma_{SG} \cdot \cos\theta_{SR}/d\right]_{\text{capillary}} \tag{5.1}$$

The explanation makes four steps easier:

1. A large-diameter ceramic tube d [m] open at both ends is immersed in a large quantity of melt. Atmospheric pressure P_{atm} and gas pressure P_{GK} in the tube are equal. The filling level h_K in the tube equals that of the surroundings h_U, i.e. $(h_U - h_K) = 0$. The principle of communicating vessels applies here.
2. The diameter d of the tube is greatly reduced, so that the capillary pressure, which is dependent on the wetting, becomes noticeable

$$P_K = \rho_S \cdot g \cdot h_K = \frac{4 \cdot \sigma_{SG} \cdot \cos\theta_{SR}}{d} \left[\text{N/m}^2\right] \tag{5.2}$$

h_K [m] is the filling height of the capillary, due to the capillary force alone. σ_{SG} is the interfacial (boundary) surface tension between the melt (S) and the atmosphere (G). θ_{SR} is the wetting angle between the melt (S) and the ceramic material (R). For $\theta_{SR} < 90°$, i.e. wetting, the melt rises upward in the tube against gravity ($h_K > 0$). For $\theta_{SR} > 90°$, i.e. non-wetting, on the other hand, the melt is shifted downward, i.e. displaced from the capillary ($h_K < 0$).

In case of wetting of the wall by the melt ($\theta_{SR} < 90°$), infiltration of the pore space is inevitable. An example is slag in oxidic refractories. On the other hand, if the refractory material is not wetted by the melt, such as, e.g. by steel, the pressure P_U

© The Author(s), under exclusive license to Springer Nature Switzerland AG 2024
J. Pötschke, *Refractory Fundamentals in Metallurgical Practice*,
https://doi.org/10.1007/978-3-031-63709-4_5

applied from the surroundings must overcome the capillary pressure, i.e. be greater, to achieve infiltration: $(P_U - P_K) > 0$.

3. This pressure may result from the change of the gas atmosphere P_{atm} and/or from the change of the original height of the melt to the value h_U. For example, if the ferrostatic height increases when a ladle is filled, infiltration into the refractory material is enhanced. For the surroundings

$$P_U = P_{atm} + \rho_S \cdot g \cdot h_U \tag{5.3}$$

must be taken into account. The gas pressure P_{atm} above the melt and in the pore space of the refractory material P_{GK} do not have to be the same, e.g. this applies in the case of pressure fluctuations in the recipient. Then, this pressure difference must also be taken into account. However, this is usually waived due to ignorance of P_{GK} and only P_{atm} is used for the calculation.

The critical diameter d of one capillary, upon its exceedance the infiltration occurs, is given by Eqs. (5.2) and (5.3) to be:

$$|d| \geq \frac{4 \cdot \sigma_{SG} \cdot \cos \theta_{SR}}{P_{atm} - P_{GK} + \rho_S \cdot g \cdot (h_U + h_K)}. \tag{5.4}$$

For $P_{atm} = P_{GK}$, the pressure loses its influence. If also $h_U = 0$, the filling height of the capillary follows, as is known,

$$h_K = \frac{4 \cdot \sigma_{SG} \cdot \cos \theta_{SR}}{d \cdot \rho_S \cdot g} \tag{5.5}$$

4. In porous refractory material, the pores point in all directions. Equations (5.1)–(5.4) describe the usually considered case of the capillary pointing vertically upwards. In order to consider the direction, the angle α formed by the capillary and, consequently, the filling level with the horizontal are considered. The product $h_K \times \sin \alpha$ is considered. For $\alpha = 90°$, h_K remains unchanged. For $\alpha = 180°$ $(0°)$, h_K is discarded. If the capillary points downward from the bottom, $\alpha = 270°$, i.e. in the direction of gravitation (pores in the bottom brick), the vectors of gravity pressure and capillary pressure are equally directed. One calculates the critical pore diameter with the equation (refer to Eq. 5.1):

$$|d| \geq \frac{4 \cdot \sigma_{SG} \cdot \cos \theta_{SR}}{P_{atm} - P_{GK} + \rho_S \cdot g \cdot (h_U + h_K \cdot \sin \alpha)} \tag{5.6}$$

The flow speed of the melt in a pore of diameter d [m] is obtained from the equation of Hagen–Poiseuilles:

$$\frac{dl}{dt} = \frac{d^2 \cdot (P_1 - P_2)}{32 \cdot \eta \cdot l} \ [\text{m/s}], \ (P_1 \geq P_2) \tag{5.7}$$

The driving pressure difference can be, among others, the capillary pressure (Eq. 5.2).

5.2 Slag Infiltration

Because of the low slag height, the static pressure is neglected. From Eqs. (5.2) and (5.7), the flow velocity (speed) of the liquid slag with its dynamic viscosity η [g/m s] in a pore having a diameter d [m] is given by the following:

$$\frac{dl}{dt} = \frac{d \cdot \sigma \cdot \cos\theta}{8 \cdot \eta \cdot l} \ [\text{m/s}]. \tag{5.8}$$

The distance l covered in the pore is not identical with the geometric penetration depth x because the pores have windings (turns). The conversion succeeds with the so-called labyrinth factor λ, which is linked to the porosity ε [1, 2]:

$$\lambda = \frac{x}{l} = 1.84 \cdot \sqrt{\varepsilon} - 0.3 \ [-] \tag{5.9}$$

Summarizing both equations, we get

$$\frac{dx}{dt} = \lambda^2 \frac{d \cdot \sigma \cdot \cos\theta}{8 \cdot \eta \cdot x} \ [\text{m/s}] \tag{5.10}$$

Separation of variables and integration leads to

$$x = \lambda \cdot \left(\frac{d \cdot \sigma \cdot \cos\theta}{4 \cdot \eta} \cdot t \right)^{0.5} \ [\text{m}] \tag{5.11}$$

and by switching after the time to

$$t = \frac{4 \cdot \eta \cdot x^2}{\lambda^2 \cdot d \cdot \sigma_{\text{Schlacke}} \cdot \cos\theta} \ [\text{s}]. \tag{5.12}$$

Example [3] A slag unsaturated with MgO of the composition 50% by weight CaO, 35% by weight Al_2O_3 and 15% by weight SiO_2 infiltrates a (decarburized) MgO–C ladle brick at 1600 °C. The following values apply: apparent porosity $\varepsilon = 0.2$, labyrinth factor $\lambda = 0.52$, mean pore diameter $d = 8 \ \mu\text{m} = 8 \times 10^{-6}$ m, geometric (design) infiltration depth $x = 9 \times 10^{-3}$ m, dynamic viscosity of the slag: $\eta = 0.17$ Pa s, surface tension of the slag: $\sigma_S = 0.57$ N/m and wetting angle: $\theta = 34°$. Putting all values into Eq. (5.12) gives for the infiltration time, $t = 54$ s ≈ 1 min.

Conversely, e.g. Eq. (5.11) answers the question how far the infiltration penetrates in 10 min (600 s) if the wetting angle changes. Then, the above numerical values

Table 5.1 Infiltration depth of refractory material by a CAS slag in 10 min

θ [°]	x [cm]
90	0
60	2.3
45	2.8
30	3.1
0	3.3

(Table 5.1) apply:

$$x = 1.34 \times 10^{-3} \cdot \sqrt{\cos\theta \cdot t}\,[\text{m}]. \tag{5.13}$$

The wetting angle θ must be less than 90° as a prerequisite for infiltration.

5.3   Steel Infiltration

Usually, the wetting angle between liquid steel and refractory material is greater than 90°. The refractory material is not wetted. The capillary pressure (Eq. 5.2) is negative, i.e. counteracts infiltration. Therefore, it must now be calculated with the assistance of Eqs. (5.1 or rather 5.6) above which pore diameter d [m] the molten steel infiltrates.

The external pressure is $P_{atm} = 1 \times 10^5$ Pa (1 bar). The pressure in the pores is $P_{GK} = 1 \times 10^3$ Pa. The density of the melt is $\rho_S = 7000$ kg/m^3 and the acceleration due to gravity $g = 9.81$ m/s^2. The $h_U = 1.45$ m high steel bath corresponds to an external pressure of 1 bar. Let the capillary length be $h_K = -\,0.2$ m and the inclination $\alpha = 90°$. The surface tension of the steel is $\sigma_{SG} = 1.5$ N/m and the wetting angle to the refractory material is $\theta_{SR} = 120°$. With these typical numbers, $d \geq 1.6 \times 10^{-5}$ m, i.e. pores larger than 0.016 mm will contain steel. Replacing the difference $(h_U + h_K) > 0$ with the ferrostatic height $h_S = 1.45$ m in Eq. (5.4), the capillary diameter is calculated to be $d = 1.5 \times 10^{-5}$ m because $h_U \gg h_K$. Furthermore, the influence of the inclination angle α is (then) of minor importance. If one additionally substitutes the difference of the gas pressures only by the gas pressure in the recipient, i.e. $P_{atm} = 10^5$ Pa (1 bar), the result remains unchanged $d = 1.5 \times 10^{-5}$ m because $P_{atm} \gg P_{GK}$. Consequently, for the evaluation of the infiltration behavior of the melts, it is sufficient in practice for $\theta_{SR} > 90°$ to approximately utilize the equation

$$d \geq \frac{-4 \cdot \sigma_{SG} \cdot \cos\theta_{SR}}{P_{atm} + \rho_S \cdot g \cdot h_S} \tag{5.14}$$

(refer to Sect. 3.2.1.2).

Using the typical viscosity of liquid steel $\eta_S = 0.003$ Pa s, the infiltration depth x and infiltration time t can be calculated analogously to Eqs. (5.11) and (5.12) [4]:

$$x = \lambda \cdot \sqrt{\left(\frac{d \cdot \sigma_S \cdot \cos\theta}{4 \cdot \eta_S} + \frac{d^2 \cdot \rho_S \cdot g \cdot h}{16 \cdot \eta_S} \right) \cdot t} \ [m], \qquad (5.15)$$

$$t = \frac{16 \cdot \eta \cdot x^2}{\lambda^2 \cdot \left(d^2 \cdot \rho_s \cdot g \cdot h + d \cdot \sigma_{S/G} \cdot \cos\theta \right)} \ [s]. \qquad (5.16)$$

With the given numerical values, it can be seen that liquid steel having a height h = 1.45 m (1 atm) is able to penetrate $x = 0.18$ m deep into a pore of diameter $d = 5 \times 10^{-5}$ m in one minute. It is, therefore, due to the temperature gradient in the lining that no steel penetrates through the wall to the outside; it solidifies beforehand.

It should be noted that in a crucible induction furnace the bath superelevation must be included in the ferrostatic height [4, 5].

The wetting angle of carbon-bonded refractory materials (AC, MC, AMC) by slag may exceed 90°. Then Eqs. (5.14) to (5.16) given for steel infiltration must be used to calculate the infiltration [4].

References

1. Jeschke, P.: Texture analysis of basic refractory brick. J. Am. Ceram. Soc. **49**(7), 360–363 (1966)
2. Pötschke, J., Routschka, G., Simmat, R.: The evaluation of oxidation experiments on carbon-containing refractories. In: Proceedings of 45th International FeuerFest—Koll., Aachen, 16–17 Oct 2002, pp. 36–39
3. Brüggmann, C.: Ein Beitrag zur Verschlackung von MgO in sekundärmetallurgischen Schlacken. Dr.-Ing. dissertation, TUBA—Freiberg, 2011
4. Pötschke, J., Brüggmann, C.: Some corrosion phenomena in an induction furnace. In: Proceedings of UNITECR, Paper No. 200 (2009)
5. Pötschke, J.: Corrosion of refractory in vacuum metallurgy—new aspects of wear in an induction furnace. In: Safeway—Vacuum Conference, Gabelbach, Ilmenau, 26–27 Nov 2009

Chapter 6
Slagging of MgO

6.1 Introduction

In 1970, Langhammer and Geck [1] published much noted experiments conducted gravimetrically on the slagging of MgO. According to these, the decay and the chemical dissolution of the refractory material play a reciprocal role. This interpretation was made qualitatively on the basis of microstructural examinations. In the present work, [2] comparable experiments are carried out with enhanced technology and with the aim of obtaining quantitative results that can be evaluated.

6.2 Experiments

6.2.1 Experimental Setup

An electrically heated laboratory chamber furnace (furnace volume ≈ 3 l), accessible from both top and bottom, is used, which can be operated in air up to a temperature of 1750 °C (Fig. 6.1) [3].

Temperature control by a type B thermocouple and suspension of the specimen, by means of a platinum wire connected to an Al_2O_3 tube, are performed from above. Thermocouple and specimen are flexibly positioned in the heating chamber with screw clamps outside the furnace. The chimney effect is suppressed by using an Al_2O_3 tube closed on one side.

Cylinders with a diameter of 17.85 mm and a height of 25 mm, which were drilled out of a MgO–C ladle brick (MgO content in oxidic component $\approx 97.5\%$ by weight, C content after coking $\approx 6.77\%$ by weight), serve as test specimens. They are equipped with a drill (hole) for suspension and then aged at 1100 °C for 15 h in

© The Author(s), under exclusive license to Springer Nature Switzerland AG 2024
J. Pötschke, *Refractory Fundamentals in Metallurgical Practice*,
https://doi.org/10.1007/978-3-031-63709-4_6

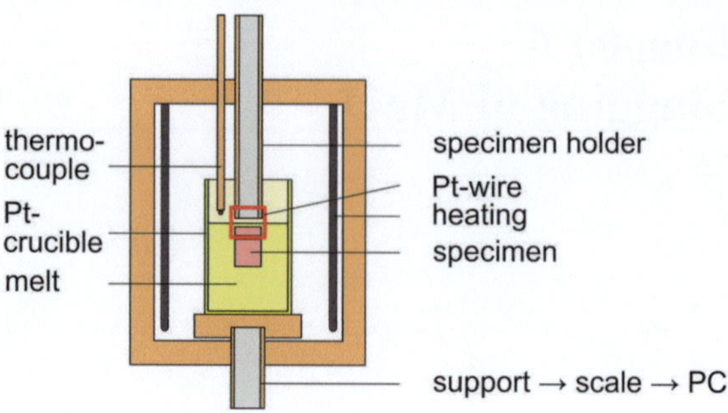

Fig. 6.1 Schematic illustration of the experimental setup [3]

air. The completely decarburized specimens then have a bulk density $\rho_R = 2.79$ g/cm^3 and a porosity $\varepsilon = 0.198$, determined according to EN 993-1 [4]. Each specimen weighs 17.45 g.

The scale support made of sintered corundum is inserted into the furnace chamber from below (refer to Fig. 6.1). A PtRh (80/20) crucible with an inner diameter of 68 mm and an inner height of 80 mm, containing 350 g of slag, is placed on it. The slag is composed of the powdered individual components, burnt lime, quartz powder, tabular clay and, if necessary, magnesium oxide, and mixed intensively. The average particle size of all individual components is in the range < 63 μm.

The test is carried out with two calcium aluminate slags at 1600 °C:

1. An MgO-saturated composition: 45.2% by weight CaO, 31.6% by weight Al_2O_3, 13.6% by weight SiO_2, and 9.6% by weight MgO.
2. A composition not saturated with MgO: 50% by weight CaO, 35% by weight Al_2O_3 and 15% by weight SiO_2. Their MgO solubility is 9.6% by weight, i.e., 37.2 g MgO dissolve in 350 g slag.

6.2.2 Experimental Procedure

- Fixation of the thermocouple in the specimen chamber and of the specimen centrally above the crucible.
- Heating at 2.5 °C/min to 1650 °C.
- Three hours holding at 1650 °C to homogenize the molten slag and sinter the test specimen.
- Cool down to 1600 °C and hold for 30 min.
- Start of gravimetric measurement data acquisition and immediate, complete immersion of the specimen.

6.3 Results and Discussion

6.3.1 Changes in Weight and Microstructure Over Time

For both slags, Fig. 6.2 shows the change in weight over time after infiltration of the microstructure had occurred within one minute.

In unsaturated slag, a rapid and approximately linear weight loss is observed. The specimen disintegrates completely after only 16 min and dissolves further in the slag.

No change in weight is observed in the saturated slag for 90 min. At time $t \approx 91$ min, however, the specimen abruptly detaches from the suspension.

To interpret the result in saturated slag, further experiments are performed: Decarburized and pre-sintered MgO small plates of dimension $25 \times 25 \times 10$ mm for 3 h at 1650 °C are then aged in pre-melted MgO-saturated slag at 1600 °C. The heating-up rate from room temperature is 10 °C/min. From the small plates, quenched in water after a holding time of 1, 24 and 48 h, polished sections are made. The microstructure is examined by means of scanning electron microscopy. A microstructure not aged with slag serves as a reference condition.

Figure 6.3 gives the back-scattered electron images of the microstructures at $300\times$ magnification.

The initial microstructure (**A**), which is exclusively thermally loaded, exhibits finely formed sinter bridges. The grains in the fine grain range are present in their original fracture form, which is characterized by pointed corners and sharp edges of the grain contour. In the slag, on the other hand, a drastic microstructural remodeling

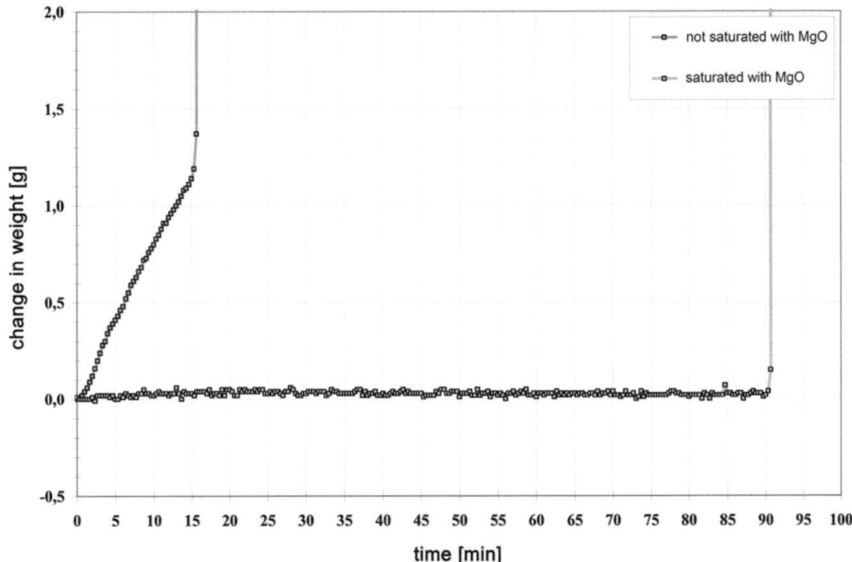

Fig. 6.2 Time progression of change in weight [3]

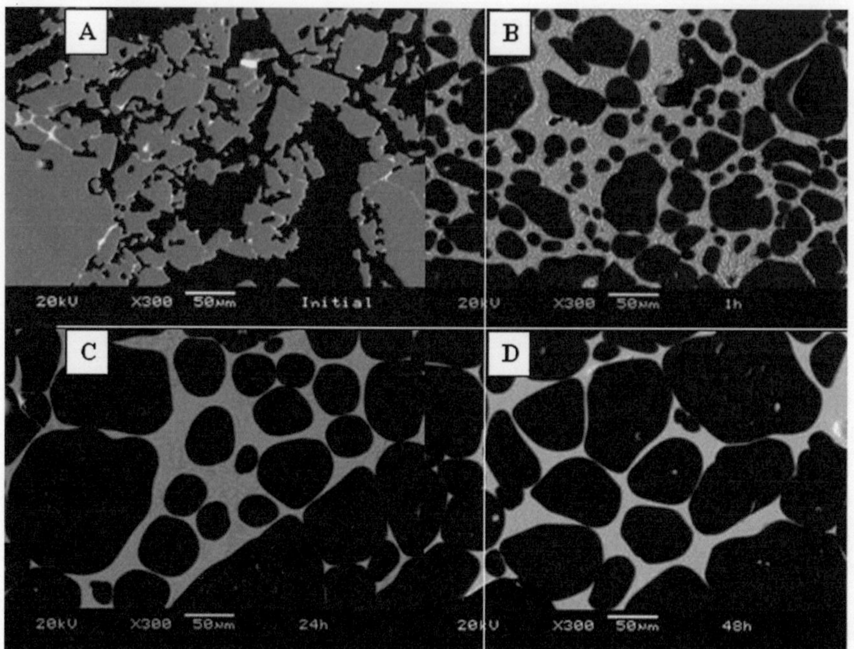

Fig. 6.3 Back scatter electron images of the binding phase [3]

takes place. With increasing time ($\mathbf{B} = 1$ h, $\mathbf{C} = 24$ h, $\mathbf{D} = 48$ h), rounded, almost spherical grains are formed, some of which are isolated in the slag, i.e., without skeleton-forming solid material bridges. The expansion of the contact surfaces is continuously reduced even further.

6.3.2 Discussion

6.3.2.1 Infiltration

The infiltration depth observed in time t [s] is x [m]. Because of the porosity ε [–], the actual distance l [m] covered by the slag is longer. Both distances are linked by the labyrinth factor λ [–] [5, 6]:

$$\lambda = \frac{x}{l} = 1.84 \cdot \sqrt{\varepsilon} - 0.3 \ [-]. \tag{6.1}$$

The flow velocity of the slag in the pores is (Hagen–Poiseuille)

$$\frac{\mathrm{d}l}{\mathrm{d}t} = \frac{d \cdot \sigma \cdot \cos\theta}{8 \cdot \eta \cdot l} \ [\mathrm{m/s}]. \tag{6.2}$$

With respect to its penetration depth x, Eq. (6.1) gives the expression

$$\frac{\mathrm{d}x}{\mathrm{d}t} = \lambda^2 \frac{d \cdot \sigma \cdot \cos\theta}{8 \cdot \eta \cdot x} \ [\mathrm{m/s}]. \tag{6.3}$$

Separation of variables, integration and rearrangement by time leads to

$$t = \frac{4 \cdot \eta \cdot x^2}{\lambda^2 \cdot d \cdot \sigma_{\mathrm{Schlacke}} \cdot \cos\theta} \ [\mathrm{s}]. \tag{6.4}$$

The following values apply to the experiment at 1600 °C:

Apparent porosity $\varepsilon = 0.198$, labyrinth factor $\lambda = 0.519$, mean pore diameter $d = 8\ \mu\mathrm{m} = 8 \times 10^{-6}$ m, geometric infiltration depth (cylinder radius) $x = 9 \times 10^{-3}$ m, dynamic viscosity of slag: $\eta = 0.17$ Pa s [7], surface tension of slag: $\sigma_{\mathrm{S}} = 0.57$ N/m [7, 8] and wetting angle: $\theta = 34°$ (measured).

Placing all values into Eq. (6.4) yields for the infiltration time, as observed, $t = 54$ s ≈ 1 min.

6.3.2.2 Interaction with MgO-Saturated Calcium Aluminate Slag

Particle disintegration in combination with Ostwald ripening offers itself as an explanation for the temporal change of the microstructure.

Particle Disintegration

Particle disintegration is the result of infiltration of the grain boundaries by slag. As a result, the grain agglomerate disintegrates and the grain size spectrum can be widened by the formation of secondary grains. The following applies as an energetic criterion for the interfacial (boundary surface) energies

$$\sigma_{\mathrm{MgO}} > 2 \cdot \sigma_{\mathrm{MgO\text{-}Schlacke}}, \tag{6.5}$$

which applies in the present case.

For the wetting angle, determined according to DIN 53170 [9], a value of $\theta = 34°$ [10] is obtained on a monocrystalline small plate out of molten MgO. The surface tension of the MgO-saturated slag σ_{S} is estimated to be 570 mN/m [7, 8]. At 1600 °C, $\sigma_{\mathrm{MgO}} = 903$ mN/m [11]. In thermodynamic equilibrium, the vector surface or rather interfacial (boundary surface) energies are related according to Young [12]:

$$\cos\theta = \frac{\sigma_{\mathrm{MgO}} - \sigma_{\mathrm{MgO\text{-}Schlacke}}}{\sigma_{\mathrm{Schlacke}}} \tag{6.6}$$

Thus, the interfacial (boundary surface) tension $\sigma_{\text{MgO-slag}} = 430\,\text{mN/m}$ is obtained. The criterion according to Eq. (6.5) is thus fulfilled. Particle disintegration takes place but does not explain the rounding of the grains.

Ostwald Ripening

Material reprecipitation by Ostwald ripening is based on the Gibbs–Thomson effect (refer to Sect. 2.1.3.4, Eq. 2.1.4.7). The equilibrium concentration $c(r)$ of a MgO grain at the interface (boundary) with the slag is proportional to its radius of curvature r. For a flat surface, $c(\infty) = c_S$ and it holds:

$$c(r) = c_S \cdot \left(1 + \frac{2 \cdot \sigma_{\text{MgO-Schlacke}} \cdot V_{M_{\text{MgO}}}}{r \cdot R \cdot T}\right) \left[\text{mol/m}^3\right] \qquad (6.7)$$

Since for $r < \infty$ always $c(r) > c_S$, small, strongly convexly curved surfaces, such as grains, grain corners and edges, dissolve. The released concentration of MgO excretes on less strongly curved surfaces. Consequently, the microstructure becomes more coarse and energy-minimized, rounded grain shapes are formed according to the crystallographic orientations of the cubic MgO [13].

If the grains are isolated from each other, the time dependence [14] applies for the mean grain radius of a distribution in the case of diffusion-determined growth or diffusion-determined dissolution.

$$\bar{r} \sim t^{1/3}. \qquad (6.8)$$

This will be reviewed for the case at hand:

For this purpose, 370 grain areas of the binding phase are measured microscopically at the times 1, 24 and 48 h and the corresponding circular surface-equivalent, averaged grain radii $\bar{r}$ are calculated. Plotting the values and subsequent potential regression provide the relationship shown in green in the double logarithmic mesh in Fig. 6.4. The conformity with Eq. (6.8) is given.

This enables us to make statements about the time progression of the grain size distribution. For this purpose, the concentration gradient in the vicinity of a curved surface with diffusion-determined mass (substance) transfer is estimated taking $\text{grad}\,c \simeq (c(r) - c)/r$ [15]. Here, $c\ [\text{mol/m}^3]$ is the current average concentration in the slag. The temporal change of the radius is then

$$\frac{\text{d}r}{\text{d}t} = -\frac{D \cdot V_{M_{\text{MgO}}}}{r} \cdot (c(r) - c)\ [\text{m/s}]. \qquad (6.9)$$

Putting Eq. (6.7) into Eq. (6.9), we obtain

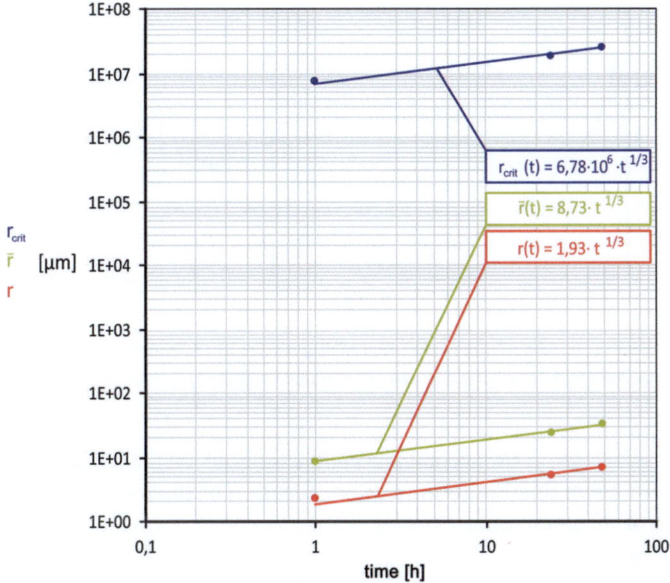

Fig. 6.4 Time evolution of critical radius r_{crit} (blue), mean radius of distribution $\bar{r}$ (green) and smallest radius of distribution r (red), coefficient of determination for all progressions is $R^2 = 0.99$ [3]

$$\frac{\mathrm{d}r}{\mathrm{d}t} = \frac{D \cdot V_{M_{MgO}}}{r} \cdot \left[\underbrace{(c - c_S)}_{A} - c_S \cdot \underbrace{\frac{2 \cdot \sigma_{MgO\text{-}Schlacke} \cdot V_{M_{MgO}}}{r \cdot R \cdot T}}_{B} \right] \text{[m/s]}. \qquad (6.10)$$

If $A > B$, the grain grows. If $A < B$, it shrinks. The critical grain radius r_{krit} is obtained for $\mathrm{d}r/\mathrm{d}t = 0$. It is obtained by equating the terms A and B:

$$r_{krit} = \frac{2 \cdot \sigma_{MgO\text{-}Schlacke} \cdot V_{M_{MgO}}}{R \cdot T} \cdot \frac{c_S}{(c - c_S)} \text{[m]} \qquad (6.11)$$

If $r < r_{krit}$, the grain perishes, but if $r > r_{krit}$, it continues to grow. In the second case, the grain size distribution shifts continuously toward higher values of r, i.e., r_{krit} also increases (Fig. 6.4, blue straight line) and the concentration difference ($c - c_S$) decreases (Fig. 6.5). For very long times, as expected, all grain radii of the microstructure decay except for a single crystal in its equilibrium form.

By separating the variables of Eq. (6.10), we get

$$\left(r^2 - r \cdot D \cdot V_{M_{MgO}} \cdot (c - c_S)\right)\mathrm{d}r = - c_S \cdot \frac{2 \cdot \sigma_{MgO\text{-}Schlacke} \cdot D \cdot V^2_{M_{MgO}}}{R \cdot T} \mathrm{d}t \text{ [m]}$$

$$(6.12)$$

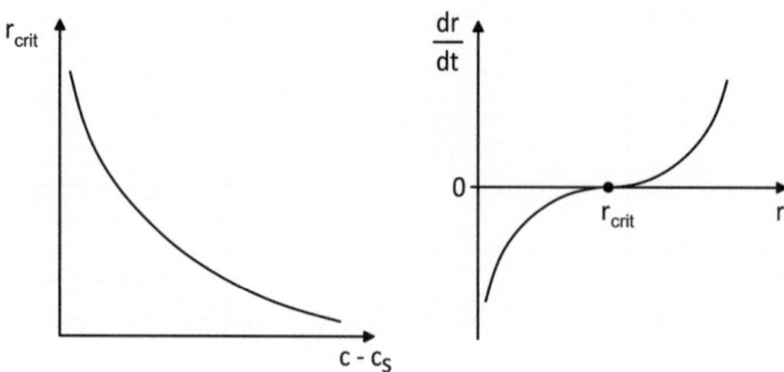

Fig. 6.5 Qualitative representation of the relationship between concentration difference and critical radius (left) and between critical radius and growth rate (right) [3]

Since $(c - c_S) \ll c_S$, the second term in the brackets on the left side is negligible and the well-known relation follows after execution of the integration for the time progression of the grain radius:

$$r = \left| -\frac{6 \cdot \sigma_{\text{MgO-Schlacke}} \cdot D \cdot V^2_{M_{\text{MgO}}} \cdot c_S}{R \cdot T} \cdot t \right|^{1/3} \quad [\text{m}] \tag{6.13}$$

(In the case where the process is determined by convection, D/r would have to be replaced in Eq. (6.10) by the mass transfer coefficient $\beta = D/\delta$ with the flow boundary layer δ, making $r \sim t^{1/2}$.)

In the present case, the MgO specimen detaches from the suspension after a time $t = 90$ min $= (5400$ s$)$. With Eq. (6.13) and σ_{MgO} slag $= 0.430$ N/m, $V_{\text{MgO}} = 2 \times 10^{-5}$ m^3/mol. $c_S = 6478.7$ mol/m^3 and the universal gas constant $R = 8.31$ J/(mol K), at this time at 1873 K the value is calculated to be $r = 2.2$ μm and $d = 4.4$ μm, respectively, which agrees well with the fine solid material bridges giving the bond shown in Fig. 6.3a.

From Eq. (6.13), the diffusion coefficient of the MgO in the slag can be estimated because the size of the smallest grain radii of the distribution over time is known in addition to the mean one and shown in red in Fig. 6.4. With $t = 3600$ s, $r = 1.93 \times 10^{-6}$ m, and the above values, $D_{\text{MgO}} \approx 1.3 \times 10^{-11}$ m^2/s $\approx 1.3 \times 10^{-7}$ cm^2/s results.

6.3.2.3   Interaction with Calcium Aluminate Slag Not Saturated with MgO [2]

Model Dissolution and Decay of Refractory Materials

In the case of the slag not saturated with MgO, an almost linear weight change is observed up to time $t \approx 15.5$ min (Fig. 6.2). This corresponds to edge decay, i.e., the detachment of the MgO grains from the specimen progressing from the outside to the inside. The already detached grains continue to dissolve in the slag by diffusion. At time $t \approx 16$ min, the specimen finally disintegrates completely.

The reason for this is that, due to infiltration of the slag not saturated with MgO, the grains providing the bond in the inner part of the specimen are weakened by dissolution. During the further course, the mechanisms of particle disintegration already described above and—if saturation of the infiltrated slag has occurred—material precipitation by Ostwald ripening are decisive for the time of complete disintegration. If one assumes, for example, that the already infiltrated slag immediately becomes saturated with MgO ($C_S = 9.6\%$ by weight), it follows from Eq. (6.13) that at the time of disintegration ($t = 960$ s) the dissolved grain size is $d \approx 2.5$ μm.

Since the solely diffusion-determined dissolution of the cylindrical specimen over its surface would far exceed the observed times [3], the model "Decay and Dissolution of Piles" developed by Gans [16] already in 1973 is followed to describe the joint action of decay and dissolution as seen in the experiment. It is presented in a simple form adapted to the wear of refractory material and then applied to the experiment.

The starting point is a cube-shaped edge region of edge length A_0 infiltrated by slag, which consists of N_0 also cube-shaped crystallites of edge length $a_{\min} \leq a_0 \leq a_{\max}$. The size distribution $f(a_0)$ of the crystallites is proportional to mass, i.e., the weight of each grain size class a_0 is equal. The finer the grain size class, the more particles it contains. Then, the following applies:

$$f(a_0) = \frac{dm(a_0)}{da_0} = \text{const.} \tag{6.14}$$

(This formal restriction has no significant effect on the accuracy of the result but facilitates the calculation and its interpretation.)

The interval of the initially present reduced edge lengths of all crystallites results with the definition $\rho = a_0/a_{\max}$ to be

$$x = \frac{a_{\max} - a_0}{a_{\max}} = 1 - \rho. \tag{6.15}$$

$t_a = a_0/2u_a$

In a monodisperse system, $x = 0$. For a polydisperse system with an increasingly broad grain spectrum, $0 < x < 1$.

The reduced running time (τ) of the process (t_Σ), consisting of decay (t_z) and dissolution (t_a), is given by the dissolution time $t_a = a/2u_{0a}$ of a crystallite of size a_0 to be

$$\tau = \frac{t_\Sigma}{t_z + t_a}. \tag{6.16}$$

t_Σ is the running time, u_a is the linear dissolution rate of the crystallites. If $\tau = 1$, the total process is concluded.

The reduced dissolution time of a (e.g., largest) crystallite ($a_{\max}$) by diffusion is

$$\alpha = \frac{t_a}{t_z + t_a}. \tag{6.17}$$

t_z is the decay time of the considered cube (A_0) and t_a is the dissolution time of its crystallites in the slag. For $\alpha \to 1$, $t_z \to 0$, i.e., the body decays immediately and the increase in concentration of the dissolved component in the melt is determined solely by the dissolution of all individual crystallites. The increase in concentration in the melt can be relatively rapid (e.g., Fig. 6.8, green curve).

The reduced decay time is as follows:

$$1 - \alpha = \frac{t_z}{t_z + t_a}. \tag{6.18}$$

For $\alpha \to 0$, $t_a \to 0$ is possible, or it is $t_z \gg t_a$. The process is solely decay-determined. In the borderline case, e.g., the blue process in Fig. 6.8. Then applies exactly, if the body does not decay at all. Its disintegration occurs exclusively via the outer surface and takes a relatively long time. For $\alpha = 0$, therefore, the grain structure of the refractory material plays no role, but only the outer geometry (design) of the product (refer to Figs. 6.7 and 6.8).

The sum of the two reduced times is

$$\alpha + (1 - \alpha) = 1. \tag{6.19}$$

One can see the symmetry: One result for $\alpha \leq 0.5$ can be identical to such a one for $(1 - \alpha)$ if $\alpha \geq 0.5$. Example: $\alpha = 0.4$ is identical in result with $(1 - \alpha)$, i.e., for $\alpha = 0.6$. Therefore, for $\alpha > 0.5$, the number $(1 - \alpha)$ must be used for the numerical calculation of the turnover conversion.

If $\alpha = 0.5$, decomposition and dissolution occur at the same rate, i.e., the increase in concentration in the melt can occur at a relatively high speed. For $\alpha \neq 0.5$, the conversion decreases in both directions, as either decay or dissolution lag behind. Invariably, the slowest partial step determines the speed. This process is likely to correspond to the behavior of refractory material in many cases and is shown in red in Fig. 6.8 as an example.

Related to the reduced dissolution time α of the largest grain a_{max} is the reduced running time of the total process (refer to Eq. 6.17)

$$\tau' = \frac{2u_a \cdot t_\Sigma}{a_{max}} = \frac{t_\Sigma}{t_{a_{max}}} = \frac{\tau}{\alpha}. \tag{6.20}$$

The degree of decay v of the observed refractory material of edge length A_0 is

$$v = 1 - \frac{N(t_\Sigma)}{N_0} = 1 - \left(\frac{A(t_\Sigma)}{A_0}\right)^3$$

$$= 1 - \left(1 - \frac{t_\Sigma}{t_z}\right)^3 = 1 - \left(1 - \frac{\tau}{1 - \alpha}\right)^3 \tag{6.21}$$

and at first independent of the grain size distribution.

The decay rate is the derivative with respect to time:

$$\frac{d\upsilon}{d\tau} = \frac{3}{1-\alpha} \cdot \left(1 - \frac{\tau}{1-\alpha}\right)^2. \tag{6.22}$$

For $\tau = (1-\alpha)$ the decay is completely finished. For $\tau = 1$, decay and dissolution are completely finished (refer to Eq. 6.16).

To calculate the degree of conversion Φ, which is the fraction (share) already dissolved in the slag and, for easier understanding, a monodisperse system is first considered, i.e., a crystal class with edge length a_0 is singled out. Analogous to Eq. (6.21), we obtain

$$\phi = 1 - \left(\frac{a(t_\Sigma)}{a_0}\right)^3 = 1 - \left(1 - \frac{t_\Sigma}{t_a}\right)^3 = 1 - \left(1 - \frac{\tau}{\alpha}\right)^3. \tag{6.23}$$

The conversion rate of all dissolving crystallites of the same size follows from the derivation to

$$\frac{d\phi}{d\tau} = \frac{3}{\alpha} \cdot \left(1 - \frac{\tau}{\alpha}\right)^2. \tag{6.24}$$

For $\tau = \alpha$, the decomposition is complete. For $\tau = 1$, decay and dissolution are completed.

In the next step, if we consider the entire grain spectrum $a_{\min} \leq a_0 \leq a_{\max}$, the conversion speed must be related to the overall distribution Eq. (6.15) and integrated via it [16]. This is equivalent to calculating the integral mean of the distribution and it follows taking Eq. (6.15):

$$\frac{d\phi_1}{d\tau'} = \frac{3}{x} \cdot \int_{1-x}^{1} \frac{1}{\rho} \cdot \left(1 - \frac{\tau'}{\rho}\right)^2 d\rho \quad \text{for } 0 \leq \tau' \leq (1-x) \tag{6.25}$$

and

$$\frac{d\phi_2}{d\tau'} = \frac{3}{x} \cdot \int_{\tau'}^{1} \frac{1}{\rho} \cdot \left(1 - \frac{\tau'}{\rho}\right)^2 d\rho \quad \text{for } (1-x) \leq \tau' \leq 1. \tag{6.26}$$

Equation (6.25) describes the initial period in which, after the refractory material has started to disintegrate, the crystallites suspended in the melt are involved in the dissolution. Equation (6.26) describes the dissolution process after some of the initially present crystallites have already completely dissolved but the disintegration is not yet completely finished. This should correspond better to what is seen in practice. By integrating the right-hand side [16], the conversion rate (speed) is given by the following:

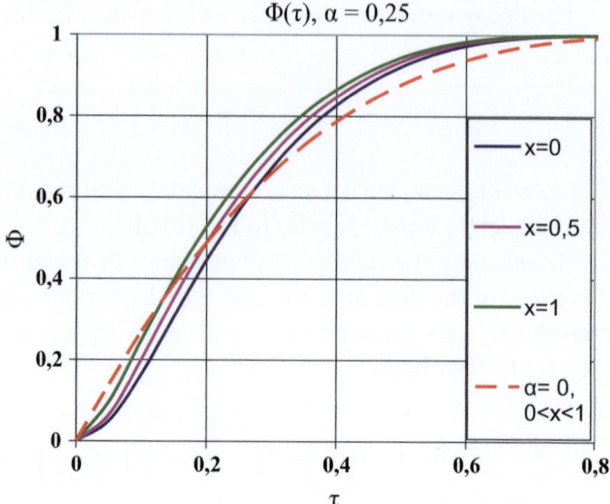

Fig. 6.6 Calculated conversion Φ as a function of the reduced dissolution time $\alpha = 0.25$ and the reduced grain size distribution x

$$\frac{d\phi_1}{d\tau'} = -\frac{3 \cdot \ln(1-x)}{x} - \frac{6}{1-x} \cdot \tau' + \frac{3 \cdot (2-x)}{2 \cdot (1-x)^2} \cdot \tau'^2 \quad \text{for } 0 \leq \tau' \leq (1-x)$$

$$(6.27)$$

and

$$\frac{d\phi_2}{d\tau'} = -\frac{3}{x} \cdot \left(-\ln \tau' - \frac{3}{2} + 2 \cdot \tau' - \frac{1}{2} \cdot \tau'^2\right) \quad \text{for } (1-x) \leq \tau' \leq 1 \quad (6.28)$$

The integration gives the degree of conversion Φ to

$$\phi_1 = -\frac{3 \cdot \ln(1-x)}{x} \cdot \tau' - \frac{3}{1-x} \cdot \tau'^2 + \frac{2-x}{2 \cdot (1-x)^2} \cdot \tau'^3 \quad \text{for } 0 \leq \tau' \leq (1-x)$$

$$(6.29)$$

and

$$\phi_2 = \frac{3}{x} \cdot \left[-\tau' \cdot \ln \tau' - \frac{1}{6} \cdot \tau'^3 + \tau'^2 - \frac{1}{2} \cdot \tau' - \frac{1-x}{3}\right] \quad \text{for } (1-x) \leq \tau' \leq 1$$

$$(6.30)$$

The result describes the conversion as a consequence of the dissolution after relatively rapid decay of the polydisperse system, i.e., for the case $t_z \to 0$ resp. $\alpha \to 1$. The closer $x \to 1$, the faster the dissolution because, relative to the same a_{max}, the share of fine grains proportional to mass increases (refer to Eq. 6.15 and Figs. 6.6 and 6.7).

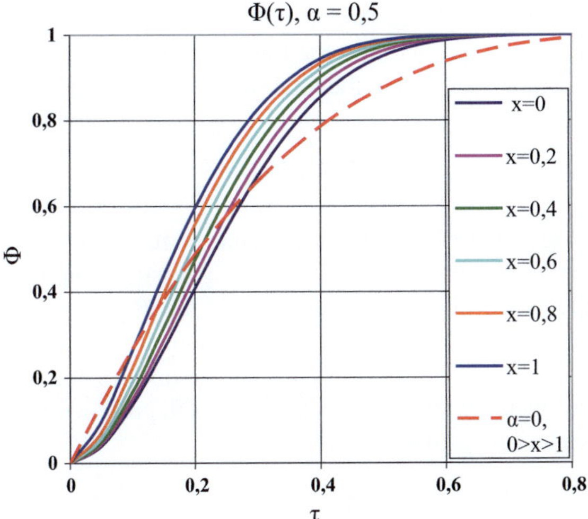

Fig. 6.7 Calculated conversion Φ as a function of the reduced dissolution time $\alpha = 0.5$ and the reduced grain size distribution x

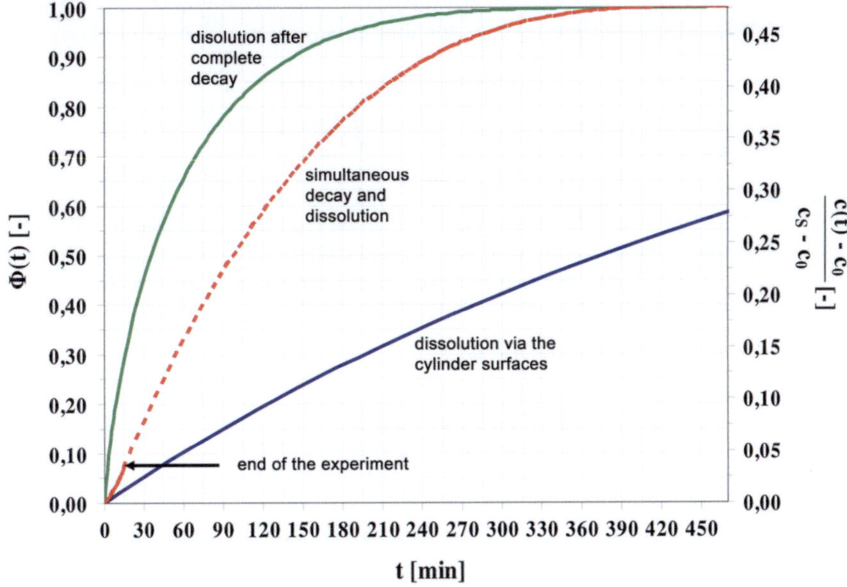

Fig. 6.8 Time progression of the experimentally measured and calculated degree of conversion (left ordinate) and the reduced saturation ratio (right ordinate)

In order to take the steadily advancing edge decay into account, the lifetime y of the crystallites already detached into the melt must be included in addition to the reduced running time τ. Related to the total time, the following applies:

$$\eta = \frac{y}{t_z + t_a} \tag{6.31}$$

The reduced lifetime η of a crystallite depends on its size a_0 and is in the range $0 \leq \eta \leq \alpha$. For $\eta = 1$, all crystallites are completely dissolved in the melt.

By linking the temporal changes connected to the degree of decay (3.2.3.1.8b), which is independent of the initial grain spectrum, to the degree of conversion Eq. (6.22) and subsequent integration over all lifetimes of the dissolving crystallites, one gets the changing total surface area of all still present crystallites $\Omega(\tau)$ with the speed of dissolution u_a, to be

$$\Omega(\tau) = \int_{\eta_1}^{\eta_2} \left(\frac{d\upsilon}{d\tau}\right)_{\tau - \eta} \cdot \frac{d\phi}{d\eta} d\eta. \tag{6.32}$$

The renewed integration of the total surface area of all remaining crystallites $\Omega(\tau)$ over the reduced running time τ yields the total progression of corrosion $\Phi(\tau)$:

$$\Phi(\tau) = 1 - \frac{V(t_\Sigma)}{V_0} = \int_0^\tau \Omega(\tau) d\tau \tag{6.33}$$

V_0 is the initial volume of the refractory material under consideration and $V(t_\Sigma)/V_0$ its time-dependent share (fraction). If Eq. (6.32) is introduced into Eq. (6.33), taking into account Eq. (6.20) and the result of Eqs. (6.27) and (6.28), then for all $x < 1$ and $\alpha \leq 0.5$ or $(1 - \alpha) \leq 0.5$, the total course (progression) of corrosion (dissolution with simultaneous decay) [16] leads to

$$\Phi_1(\tau) = \frac{3}{\alpha \cdot (1 - \alpha)} \cdot \int_0^\tau \int_{\eta_{11}}^{\eta_{12}} \left(1 - \frac{\tau - \eta}{1 - \alpha}\right)^2$$

$$\cdot \left(\frac{-3 \cdot \ln(1 - x)}{x} - \frac{6 \cdot \eta}{\alpha \cdot (1 - x)} + \frac{3 \cdot (2 - x) \cdot \eta^2}{2 \cdot (1 - x)^2 \cdot \alpha^2}\right) d\eta d\tau \tag{6.34}$$

and

$$\Phi_2(\tau) = \frac{9}{\alpha \cdot (1 - \alpha) \cdot x} \cdot \int_0^\tau \int_{\eta_{21}}^{\eta_{22}} \left(1 - \frac{\tau - \eta}{1 - \alpha}\right)^2$$

$$\cdot \left(\ln \alpha - \ln \eta - \frac{3}{2} + \frac{2 \cdot \eta}{\alpha} - \frac{\eta^2}{2 \cdot \alpha^2} \right) \mathrm{d}\eta \mathrm{d}\tau. \tag{6.35}$$

Equation (6.34) describes, comparable to Eq. (6.29), the initial period of decay and dissolution without particles going completely into solution. $\Omega(\tau)$ Eq. (6.35) describes, comparable to Eq. (6.30), the further process until finally all particles are dissolved.

For a clear distinction, the integration **variable** τ of the outer integral will be denoted by t in Eqs. (6.34) and (6.35) in future. The integration **limit** of the outer integral remains τ. The integration limit of the inner integral denoted by η is then t, while η remains as the integration variable. In addition, the following functions are defined to simplify the illustration:

$$g_1(t, \eta) = \frac{3}{\alpha \cdot (1 - \alpha)} \cdot \left(1 - \frac{t - \eta}{1 - \alpha} \right)^2$$
$$\cdot \left(\frac{-3 \cdot \ln(1 - x)}{x} - \frac{6 \cdot \eta}{\alpha \cdot (1 - x)} + \frac{3 \cdot (2 - x) \cdot \eta^2}{2 \cdot (1 - x)^2 \cdot \alpha^2} \right) \tag{6.36}$$

and

$$g_2(t, \eta) = \frac{9}{\alpha \cdot (1 - \alpha) \cdot x} \cdot \left(1 - \frac{t - \eta}{1 - \alpha} \right)^2 \cdot \left(\ln\left(\frac{\alpha}{\eta} \right) - \frac{3}{2} + \frac{2 \cdot \eta}{\alpha} - \frac{\eta^2}{2 \cdot \alpha^2} \right). \tag{6.37}$$

Decay and dissolution of the refractory material thus proceed in successive and partly overlapping temporal phases. It is, therefore, necessary to first integrate them separately within their physically justified limits and then to add them together. In this way, the process can be followed step by step:

For the conversion integral within the limits $0 \leq \tau \leq 1$, i.e., the entire process of continuous edge dissolution with simultaneous complete dissolution of all crystallites in the slag, the following applies [16]:

$$\Phi(\tau) = \int_0^{\alpha \cdot (1-x)} \left(\int_0^t g_1(\eta, t) \mathrm{d}\eta \right) \mathrm{d}t$$
$$+ \int_{\alpha \cdot (1-x)}^{\alpha} \left(\int_0^{\alpha \cdot (1-x)} g_1(\eta, t) \mathrm{d}\eta + \int_{\alpha \cdot (1-x)}^t g_2(\eta, t) \mathrm{d}\eta \right) \mathrm{d}t$$
$$+ \int_{\alpha}^{1-\alpha} \left(\int_0^{\alpha \cdot (1-x)} g_1(\eta, t) \mathrm{d}\eta + \int_{\alpha \cdot (1-x)}^{\alpha} g_2(\eta, t) \mathrm{d}\eta \right) \mathrm{d}t$$

$$+ \int_{1-\alpha}^{1-x\cdot\alpha} \left(\int_{t-(1-\alpha)}^{\alpha\cdot(1-\alpha)} g_1(\eta, t)\mathrm{d}\eta + \int_{\alpha\cdot(1-\alpha)}^{\alpha} g_2(\eta, t)\mathrm{d}\eta \right) \mathrm{d}t$$

$$+ \int_{1-x\cdot\alpha}^{\tau} \left(\int_{t-(1-\alpha)}^{\alpha} g_2(\eta, t)\mathrm{d}\eta \right) \mathrm{d}t \tag{6.38}$$

The analytical calculation of the integrals shown in Eq. (6.38) is intricate. An explicit function illustration in the form of equations would fill several pages. Therefore, the calculation is done numerically with the help of a **Maple script** [3]. (The program can be provided by the authors if the interested party already works with **Maple**.) Thus, the closed-form calculation of the degree of conversion $\Phi(\tau)$ is possible for all $0 < x < 1$ and $0 < \alpha \leq 0.5$ or $0 < (1 - \alpha) \leq 0.5$ in a straightforward way.

Figures 6.6 and 6.7 show some calculations graphically.

For $\alpha \to 0$, the process is solely determined by the (impaired) decomposition, which is why the grain size distribution x has no influence on the conversion. The decisive factor is the dissolution of the geometric surface of the refractory product by diffusion. With increasing $\alpha \to 0.5$, the conversion depends increasingly on the width of the grain size distribution and reaches the highest speed at $x \to 1$ because the large number of small grains dissolves rapidly compared to the largest grain. Therefore, in the case of a monodisperse distribution of grains ($x = 0$), the dissolution occurs relatively slowly (refer to Eq. 6.15).

It is also evident that with increasing $\alpha > 0$ the dissolution is initially increasingly delayed and later progresses much more rapidly. This process is also linked to the grain size distribution and is most intensive in the monodisperse case ($x = 0$). With an increasing share of fines ($0 < x \leq 1$), the effect weakens. The entire process finally ends with the dissolution of the largest grain.

Application of the Model to the Measurement Results

The mean curve (red) in Fig. 6.8 describes the measured progression of dissolution and decay of the specimen.

In order to describe the experiment quantitatively, besides the decay time $t_z = 16$ min the knowledge of the dissolution time t_a of the largest crystallite is required. As an approximation, the required dissolution time of a spherical MgO crystallite can be calculated with

$$t_a \approx \frac{1}{2.1^2} \cdot \frac{R^2}{4 \cdot D} \approx \frac{R^2}{17.64 \cdot D} \ [\text{s}] \tag{6.39}$$

(refer to Sect. 2.2.2, Eq. 2.2.27).

In the present case, the radius of the largest, spherically imagined crystallite is $R = 0.25$ cm. For the diffusion coefficient D, the value 1.3×10^{-7} cm^2/s measured in (6.3.2.1) is used. It comes from Eq. (6.39) for the dissolution time $t_a = 27{,}254$ s $= 454$ min. The result of the calculation does not take into account any movement of the slag so that shorter dissolution times can be expected in operational practice.

According to Eq. (6.17), the reduced dissolution time is

$$\alpha = t_a/(t_z + t_a) = 454/(16 + 454) = 0.996. \tag{6.40}$$

The calculation of the conversion integral (Eq. 6.38) is done in general for $\alpha \leq 0.5$. If, however, $\alpha \geq 0.5$, then α must be replaced by $(1 - \alpha)$ because the solutions are symmetric in α and $(1 - \alpha)$ (refer to Eq. 6.19). Consequently, the calculation in **Maple** is thus done with $\alpha = 1 - \alpha = 1 - 0.966 = 0.034$. The interval of the reduced edge length is assumed to be $x = 0.99$ according to Eq. (6.15), which is justified by the microstructure (fine grain < 10 μm, coarse grains $= 5000$ μm). Since t_z is known by way of the experiment and t_a by way of the calculation, according to Eq. (6.16) the abscissa, i.e., $0 \leq \tau \leq 1$, can be converted into a time $t_\Sigma = t$ [min].

The result of the **Maple** calculation **is** shown by the red curve in Fig. 6.8. The complete conversion of the specimen occurs in 470 min. At the time of its decay, $t_z = 16$ min or $\tau = 0.034$, about 9% of the specimen is dissolved in the slag, corresponding to a reduced saturation ratio $\gamma \approx (0.09 \times 17.7)/37.2 \approx 0.043$ or an average MgO concentration $c(t) \approx 0.043 \times 9.6\%$ by weight $\approx 0.41\%$ by weight of the slag. A gross analysis, on the other hand, e.g., in the form of slag samples drawn, would yield a considerably higher MgO concentration at the time of decay because the fragments not yet dissolved would be analyzed simultaneously. The model thus offers the possibility to follow the further dissolution of the sample in the slag by calculation, even after the specimen has disintegrated.

Approximate calculation of the measured turnover is possible with by applying Eq. (6.41) if the large crystallites are determinant for the time progression [16]:

$$\phi_{\mathrm{Exp}} = \frac{1}{20(1 - \alpha)^3} \cdot \left\{ \begin{array}{l} -15\alpha + 24\alpha^2 - 10\alpha^3 + \left(60 - 90\alpha + 36\alpha^2\right)\tau \\ + (45\alpha - 60)\tau^2 + 20\tau^3 \end{array} \right\}. \tag{6.41}$$

For example, for $t = 235$ min, $\tau = t/t_{\mathrm{ges}} = t/470$ gives $\tau = 0.5$. Further above, $\alpha = 1 - \alpha = 1 - 0.966 = 0.034$ was calculated. Thus, it follows from Eq. (6.41) $\phi_{\mathrm{Exp}} = 0.88$, which sufficiently conforms with the result in Fig. 6.8.

If the decomposition takes place immediately, i.e., $t_Z = 0$, with all crystallites already dispersed in the slag at time $\tau = 0$, the curve shown in green, which is obtained from Eq. (6.30), applies. Once again, the total process is determined by the dissolution time of the largest crystallite ($t_a = 454$ min). As expected, this is characterized by a strong rush in the initial phase. On the other hand, for $t = 375$ min the difference between the two processes has dropped to less than 2%. Therefore, and not least because of the only slight difference in the total duration of the process (454 min over against 470 min), i.e., $\alpha \rightarrow 1$, Eqs. (6.29) and (6.30) can be used approximately to describe the degree of conversion quantitatively [17]. If one additionally considers the aspect of the microstructure of refractory material and, thus, the interval of the reduced edge length with $x \to 1$, then even the entire process can be reduced to Eq. (6.30) in the case of rapid decay. One finally obtains for the degree of conversion the strongly simplified relationship

$$\phi(\tau) = 3 \cdot \left[-\tau \cdot \ln \tau - \frac{1}{6} \cdot \tau^3 + \tau^2 - \frac{1}{2} \cdot \tau \right]. \tag{6.42}$$

In the case that the decomposition of an apparent porous, grain-dispersed body takes place very slowly, i.e., the grain bonds remain very stable over a long period of time and $\alpha \to$ strives toward 0, the course of the degree of conversion or the reduced saturation ratio approaches that of a purely diffusion-determined dense body dissolved over the outer phase boundary (shell surface) (blue curve, dissolution time $t_a = 28.5$ h) [18]. Such a microstructure represents the ideal case for refractory material to resist chemical corrosion.

Since the melt is not yet saturated with MgO after the complete dissolution of the cylinder, the solution of the differential equation is [19] (refer to Sect. 2.2.4.2, Eq. 2. 2.48.2):

$$\tau_D = 2 \cdot \sqrt{\frac{\varphi}{1-\varphi}} \cdot \arctan \sqrt{\frac{\varphi}{1-\varphi}} \tag{6.43}$$

$\varphi \equiv \frac{R(t)^2}{R_S^2} = \frac{m(t)}{m_S} = \frac{C(t)-C_0}{C_S-C_0} \equiv \gamma$ represents the relative ratios of the running radius R, the dissolved mass m and the concentration C to their value at (hypothetical) achievement of saturation. The saturation radius is $R_S = 1.3$ cm and, therefore $\varphi_E = \frac{R_0^2}{R_S^2} = 0.48$. The ratio of the conversion ratio ϕ to the reduced saturation ratio γ is $\phi/\gamma = \varphi_E = 0.48$ from which Eq. (6.43) τ_{DE} gives 1.47. From this, the total dissolution time is calculated to be

$$t_a = \frac{\tau_{DE} \cdot V_S \cdot \delta}{F_0 \cdot D} \text{ [s]}. \tag{6.44}$$

The crucible contains 350 g of slag of density $\rho_S = 2.72$ g/cm^3. Its volume is therefore $V_S = 128.7$ cm^3. The thickness of the diffusion boundary layer is estimated to be to $\delta = 1 \times 10^{-3}$ cm [14]. The shell surface of the immersed specimen is ($R = 0.9$ cm, $h = 2.5$ cm) $F_0 = 14.1$ cm^2. The diffusion coefficient in the slag is $D = 1.3 \times 10^{-7}$ cm^2/s [12]. Equation (6.44) gives $t_a = 7 \times 10^4 \times \tau_{DE} = 1.03 \times 10^5$ s ≈ 28.6 h for the dissolution time, which is just like the exact result.

The model offers the possibility to make a quantitative decision in a laboratory experiment to which degree the refractory material in question corrodes by way of decay or as result of diffusion-controlled dissolution.

Disintegration Time

The disintegration time can only be roughly estimated because in most cases the statistically recorded peculiarities of the microstructure are not sufficient for the exact calculation. Only the binder phase is considered. It is veined with pores with an average diameter of $d = 400\ \mu$m. The slag penetrates these pores in about one minute. The time law [18] applies regarding the dissolution of the binding phase into the already infiltrated slag:

$$t = \frac{d^2}{\pi^2 \cdot D} \cdot \left(\ln \frac{8}{\pi^2} - \ln \frac{c_S - c(t)}{c_S - c_0} \right) \tag{6.45}$$

$D = 1.3 \times 10^{-7}\ \text{cm}^2/\text{s}$ is the diffusion coefficient in the slag, $d = 400\ \mu$m is the mean pore diameter, c_0, c_S and $c(t)$ are the concentrations of the dissolved component, e.g., MgO, at onset, upon saturation and in time, respectively. Estimating that the time of decay t_Z occurs once there is 60% saturation of the infiltrate, Eq. (3.2.3.3.1) gives $t_Z \approx 0.71 \cdot \frac{d^2}{\pi^2 x D} \le 880$ s. This corresponds to an infiltration time of 15 min.

For longer times, e.g., in a slag already strongly enriched in MgO, Ostwald ripening determines the decay. Assuming a saturation of the slag at the moment of decay of 99%, Eq. (6.45) gives $t_Z = 4.4 \cdot \frac{d^2}{\pi^2 \cdot D} = 5480$ s $= 91$ min. This also corresponds to what was observed.

6.4 Summary

The slagging process of MgO in a slag saturated with MgO and in a slag not saturated with MgO was monitored by thermogravimetric measurements at 1600 °C [3]. It is shown that even in contact with MgO-saturated slag, the microstructure decomposes after some time. Corrosion is, therefore, not completed despite the presence of MgO saturation of the slag. This is due to the overlapping mechanisms of particle disintegration and Ostwald ripening.

In contact with not saturated MgO slag, rapid edge decay and finally complete disintegration of the microstructure occurs. The complex interplay of disintegration and dissolution, which can also be attributed to the involvement of the above physical mechanisms, is discussed and quantitatively understood using a model provided by Gans [16]. The numerical evaluation is carried out with the aid of a **Maple** script. However, in the case of rapid disintegration of the microstructure and relatively slow dissolution of the grains in the slag, the process can be described with sufficient accuracy by means of the simple Eq. (6.30). If the dissolution takes place solely via the shell surface of the cylinder, Eq. (6.43) must be applied [19]. If both processes take place simultaneously in the refractory material, Eq. (6.41) can be used as an approximation [16].

Thus, the model offers the possibility of making a quantitative decision in a laboratory experiment as to the degree to which the refractory material in question corrodes by decay or as a result of diffusion-controlled dissolution. It is confirmed that the enhancement of the corrosion behavior of refractory materials in slag must be achieved primarily by a strongly formed microstructure bond. This finding can be quantitatively transferred to the examination results of Langhemmer and Geck [1, 17].

References

1. Langhammer, H.-J., Geck, H.G.: Laboratory examination of slagging resistance of refractory materials towards iron oxide-containing steel mill slags at 1,600 °C. Arch. Eisenhüttenwesen. **41**(11), 1081–1091 (1970)
2. Brüggmann, C., Pötschke, J.: Disaggregation and dissolution of refractories using the example of MgO/lime-aluminate slag. In: Proceedings 54th International Colloquium on Refractories, Aachen, Germany, pp. 139–141 (2011)
3. Brüggmann, C.: A contribution to slagging of MgO in secondary metallurgical slags. Dr.-Ing. dissertation, TUBA—Freiberg, Germany, 2011
4. DIN EN 993-1: Test Method for Densely Shaped Refractory Products—Part 1: Determination of Bulk Density, Open Porosity and Total (True) Porosity (1995)
5. Jeschke, P.: Texture analysis of basic refractory brick. J. Am. Ceram. Soc. **49**(7), 360–363 (1966)
6. Pötschke, J., Routschka, G., Simmat, R.: The evaluation of oxidation experiments on carbon-containing refractories. In: Proceedings 45th International Colloquium on Refractories, Aachen, Germany, 16–17 Oct 2002, pp. 36–39
7. Turkdogan, E.T.: Properties of Molten Slags and Glasses, pp. 163–174. Metal Society London, England (1983)
8. Verein Deutscher Eisenhüttenleute (VDEH): Slag Atlas, 2nd edn. Verlag Stahleisen GmbH Düsseldorf, Germany (1995). ISBN 3-514-00457-9
9. DIN 51730: Testing of Solid Fuels—Determination of the Ash Melting Behavior (1998)
10. Schatt, W.: Sintering Processes—Basics. VDI-Verlag GmbH Düsseldorf, Germany (1992). ISBN 3-18-4012182
11. Livey, D.T., Murray, P.: The wetting properties of solid oxides and carbides by liquid metal. Presentation given at 2nd Plansee Seminar "De re Metallica", Reutte (Tirol), Austria, 19–23 June 1955
12. Young, T.: An essay on the cohesion of fluids. Philos. Trans. R. Soc. Lond. **95**, 65–87 (1805)
13. Lee, W.E., Zhang, S.: Melt corrosion of oxide and oxide-carbon refractories. Int. Mater. Rev. **44**(3), 77–104 (1999)
14. Telle, R.: Sintering. In: Salmang/Scholze Keramik, Telle, R. (eds.) 7th edn. Springer Berlin Heidelberg, New York (2007). ISBN 3-540-63273-5
15. Kahlweit, M.: Boundary Surface Appearances, Basics of Physical Chemistry, Band 7. Steinkopff-Verlag, R. Haase, Aachen, Germany (1981)
16. Gans, W.: Collapse and Dissolution of Aggregate Material, Habilitation. RWTH Aachen, Germany (1973)
17. Pötschke, J.: Wear of Refractory Materials by Slag and Steel, Seminar Documents VDEh (2001)
18. Bird, R.B., Stewart, W.E., Lightfoot, E.M.: Transport Phenomena. Wiley (1960)
19. Friedrichs, H.A., Knacke, O.: Process Engineering Basic Types of Isothermal Reactions in Blown Through and Overflown Fills. Research Reports of the State of North Rhein-Westphalia No. 2240. Verlag Opladen (1973)
20. Pötschke, J.: Personal Communication, s. [15]

Chapter 7
Premature Wear of Refractory Material as a Result of Marangoni Convection

7.1 Introduction

Marangoni convection is an interfacial convection of two adjacent liquid phases driven by potential differences, such as gradients of concentration and/or temperature [1]. The interfacial convection occurs at high velocity, which significantly increases the mass transfer between the two phases or an adjacent solid phase, respectively [2]. Consequently, this effect plays a significant role in many technical processes. Specific mention is made of the welding of metals [3] and the breeding of single crystal semiconductors, such as silicon, according to the method of Czochalski [4, 5].

As early as 1966, Marwedel [6] pointed out that Marangoni convection has a strong influence on the corrosion of the refractory material during the production of flat glass. A year later, Brückner [7] deepened the understanding by experiments on model substances. In 1982, Hauck and Pötschke [8] proved that the premature wear of the submerged nozzle during continuous casting of steel is a result of Marangoni convection.

The quantitative calculation of the flow velocity in a simplified model by Levich [2] was an important step toward understanding this effect. In practice, however, the boundary conditions are only known imperfectly and there are a large number of experimental and theoretical examinations of Marangoni convection [9]. Particularly great efforts to understand Marangoni convection have been made since 1974 in the German μ-g-program (TEXUS, D1, D2, Spacelab, etc.) [10].

Marangoni convection starts when the system's own and experimentally determined critical Marangoni number $\mathrm{Ma_c}$ is exceeded (refer to Fundamentals Sect. 2. 2.5.2). If, for example, the surface tension σ [g/s^2] of the liquid depends on a change in concentration Δc, then [2] applies:

$$\mathrm{Ma_c} \leq \frac{\partial \sigma}{\partial c} \cdot \frac{\Delta c \cdot h}{\eta \cdot D}. \qquad (7.1)$$

© The Author(s), under exclusive license to Springer Nature Switzerland AG 2024   187
J. Pötschke, *Refractory Fundamentals in Metallurgical Practice*,
https://doi.org/10.1007/978-3-031-63709-4_7

(η [g/cm s] is the viscosity, D [cm^2/s] the diffusion coefficient and h [cm] a characteristic length, e.g. the layer thickness of the moving fluid.)

Some authors [11–13] have developed models to quantitatively describe the dissolution of solid bodies in liquids. In our experience, however, their application to the wear of refractory materials does not lead to satisfactory results. A compilation of various examinations in the metallurgical field was made in 1998 by Mukai [14].

The present work presents the own experiments regarding premature wear of refractory material carried out on glass, slag and steel melts between 1995 and 2006 as a unified, closed model and investigates its applicability to metallurgical corrosion problems.

7.2   Geometric Shape of the Corrosion Trough

Figure 7.1 illustrates the design of the corrosion trough (channel) in principle. Since liquid glass and slag usually wet the refractory wall structure, the molten metal is drawn up the wall concavely by the height h [m]. Steel, on the other hand, does not wet the lining, which is why its melt bends convexly downward by the height $- h$ [cm] (refer to Sect. 2.1.3.4). The slag is pushed between the wall and the molten steel. The geometric shape of the melt is described in both cases by Laplace's equation [15]. With twice the capillary length

$$2 \cdot \xi = 2 \cdot \sqrt{\frac{\sigma}{\rho \cdot g}} \ [\text{cm}]; \tag{7.2}$$

the vertical height y of the corrosion channel at each point x of its horizontal expansion is calculated to [15]:

$$y = 2 \cdot \xi \cdot \sin(0.5 \cdot (\psi - \theta)) \ [\text{cm}] \tag{7.3}$$

$$(\bar{x} - \bar{x}_0) = \sqrt{1 - \bar{y}^2} + 0.5 \cdot \ln \frac{\bar{y}}{1 + \sqrt{1 - \bar{y}^2}} \ [\text{cm}]. \tag{7.4}$$

The reduced lengths in x- and y-directions (Fig. 7.1) are:

$$\bar{x} = \frac{x}{2 \cdot \xi} \ [-] \quad \text{and} \quad \bar{y} = \frac{y}{2 \cdot \xi} \ [-] \tag{7.5}$$

σ [mN/m] is the interfacial tension of the liquid phase with its liquid environment, e.g. glass/air, slag/air and steel/slag.

θ [°] is the wetting angle of the liquid phase (glass, slag, steel/slag) with the refractory material. θ is calculated from the equation of Young (1803):

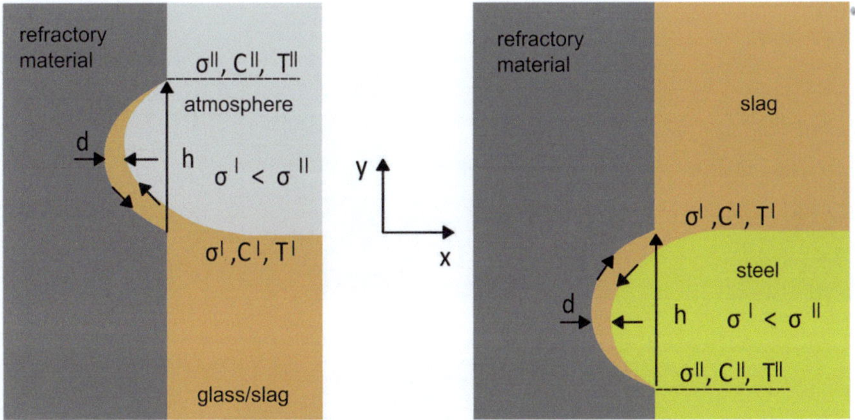

Fig. 7.1   Formation of the corrosion trough (channel) [16]

$$\sigma_{\text{solid}} = \sigma_{\text{liquid}} \cdot \cos\theta + \sigma_{\text{solid/liquid}} \left[\text{mN/m} \equiv \text{g/s}^2\right]. \tag{7.6}$$

Ψ [°] is the angle of inclination of the wetted surface to the horizontal plane of the expanded liquid. Upon reaching the stationary state of corrosion of the refractory material by liquid glass or slag then $\psi = 180°$ (compare Fig. 7.1). Then $y = h$ is the maximum height of rise of the melt. In the steel/slag/refractory material, the wetting angle θ is greater than 90°; i.e. the steel/slag phase boundary does not rise at the refractory material but bends downward. Then h is negative and is calculated with $\psi = 0°$.

Figure 7.2a–c shows three examples that the shape of the corrosion zone basically adapts to the course of the melt calculated according to Laplace.

At the beginning of corrosion, the shape of the corrosion trough (channel) is exactly described by Laplace's equation (Fig. 7.2a).

Looking at the second example (Fig. 7.2b), the shape of the corrosion trough (channel) has changed as a result of progressive corrosion: Namely, the angle of inclination of the wall ψ decreases increasingly. However, since the wetting angle θ does not change, the meniscus of the melt slides down along the curved wall with increasing corrosion.

Taking the third example (Fig. 7.2c), the shape of the corrosion trough deviates more strongly from the Laplace course of the molten steel with increasing time because gravity pulls the molten steel downward in the corrosion trough.

Just like the rise height, the average thickness of the liquid layer is calculated from the equilibrium between capillary force and gravity [17]:

$$d = 0.5 \cdot \frac{\sigma \cdot \cos\theta}{\rho \cdot g \cdot h} \ [\text{cm}] \tag{7.7}$$

The bulge of the corrosion zone is approximated by a semicircle:

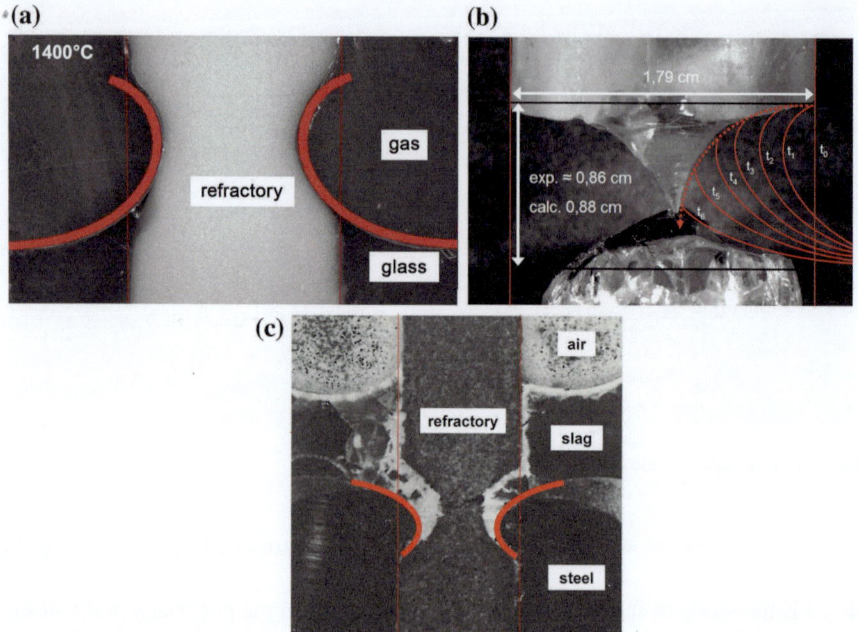

Fig. 7.2 a Corrosion of sintered corundum by a glass melt at 1400 °C [17]. **b** Corrosion of polycrystalline fused magnesia by CA slag at 1650 °C [18]. **c** Corrosion of submerged nozzle material in slag/iron at 1600 °C [19]

$$s = 0.5 \cdot \pi \cdot h \ [\text{cm}] \tag{7.8}$$

(The length s [cm] is needed for the later calculation of the gradients of concentration and temperature, as well as for the calculation of the mass transfer coefficient.)

7.3 Speed of the Marangoni Flow

Levich [2] (refer to Fundamentals Sect. 2.2.5.2) calculates the flow velocity starting from the surface of a liquid in a flat shell:

$$u_{\text{M}} = \frac{d'}{4 \cdot \eta} \cdot \left(\frac{\text{d}\sigma}{\text{d}C} \cdot \text{grad}\,C + \frac{\text{d}\sigma}{\text{d}T} \cdot \text{grad}\,T \right) \ [\text{cm/s}], \tag{7.9}$$

d' is the thickness of the flowing layer and smaller than d. An estimation shows that in slags with viscosity $\eta \approx 1$ g/cm s the flowing layer d' is $\ll 2$ cm [2].

In Fig. 7.1, the direction of flow is drawn in the corrosion trough. It is directed in such a way that one area of low interfacial energy flows toward another with high interfacial energy in order to lower the surface energy there as well. It is now assumed that a vortex is formed in the slag layer d and that approximately $d' = 0.5 \cdot d$ if the viscosity does not change much due to the intake of solute. However, in the case of corrosion by liquid glass, for example, its viscosity changes so much that its maximum value at saturation is used for η. In the metallurgical field, on the other hand, this is not so. $d\sigma/dc$ and $d\sigma/dT$ are the changes in the interfacial energy due to the component(s) dissolved out of the refractory material and the temperature change along the stretch s, respectively (refer to Eq. 7.8):

$$\frac{d\sigma}{dC} = \frac{\sigma^{\mathrm{I}} - \sigma^{\mathrm{II}}}{C^{\mathrm{I}} - C^{\mathrm{II}}} \left[\mathrm{mN/m\,\% \ by\ weight}\right] \quad \text{and} \quad \frac{d\sigma}{dT} = \frac{\sigma^{\mathrm{I}} - \sigma^{\mathrm{II}}}{T^{\mathrm{I}} - T^{\mathrm{II}}} \left[\mathrm{mN/m\,K}\right]. \quad (7.10)$$

The values with index I correspond to the initial state of the fluid and II describes the saturation state or a changed temperature as a result of heat losses. The gradients are:

$$\mathrm{grad}\,C = \frac{C^{\mathrm{I}} - C^{\mathrm{II}}}{|s|} \left[\% \ \mathrm{by\ weight/cm}\right] \quad \text{and} \quad \mathrm{grad}\,T = \frac{T^{\mathrm{I}} - T^{\mathrm{II}}}{|s|} \left[\mathrm{K/cm}\right] \quad (7.11)$$

7.4 Mass Transfer

If the slag flows along the refractory material with the calculated flow velocity u_{M} [m/s], the average mass transfer coefficient β [m/s] is [20] (refer to Sect. 2.2.4, Eq. (2.2.34)):

$$\beta = \sqrt{\frac{4 \cdot D \cdot |u_{\mathrm{M}}|}{\pi \cdot |s|}} \cdot \left(\frac{D \cdot \rho}{\eta}\right)^{\frac{1}{6}} \quad [\mathrm{cm/s}] \quad (7.12)$$

The bracket expression takes the wall friction into account which reduces the mass transfer and is neglected in the case of slow flow [17, 18]. D [cm^2/s] is the diffusion coefficient of the component dissolved in the liquid, e.g. Al_2O_3 in the glass melt or MgO in the slag.

$\eta/\rho = v$ [cm^2/s] is the kinematic viscosity of this fluid.

7.5 Corrosion Rate

The corrosion speed (rate) is calculated under the assumption of a linear boundary layer diffusion according to Nernst (1904) (refer to Sect. 2.2.4, Eq. (2.2.32)):

$$V_{\text{CORR}} = 360 \cdot \frac{\rho_{\text{S}}}{\rho_{\text{FF}}} \cdot \beta \cdot (C_{\text{sätt}} - C_0)\ [\text{mm/h}] \tag{7.13}$$

ρ_{S} [g/cm^3] is the density of the liquid, e.g. glass or slag, and ρ_{ref} is the bulk density of the refractory material. C_{sat} is the saturation concentration of the refractory material in the liquid flowing along it, and C_0 is its initial concentration in % by weight.

7.6 Examples

7.6.1 Corrosion of Fused Cast α-β-Al$_2$O$_3$ (FAB) Refractory Material by Container Glass at 1400 and 1500 °C

The thorough theoretical analysis of this corrosion process is possible because of the very carefully measured results of Dunkl and Brückner [21] regarding the critical values: corrosion rate, surface tension, viscosity, density and diffusion coefficients for glass of different compositions at 1400 and 1500 °C. Based on this and taking into account experimental experience with slag and steel melts [17–19, 21–25], it became possible to establish the somewhat simplified model described above and to compare it with laboratory practice [17, 21].

Table 7.1 shows the good conformity between the experimental result and the simplified calculation described. Figure 7.3 shows the same comparison for all laboratory experiments by Dunkl and Brückner [21] graphically and confirms the good conformity between theory and practice.

Table 7.1 Corrosion of a cylinder made of fused cast α-β-Al$_2$O$_3$ (FAB) by container glass at 1400 °C [17] (FF = refractory)

T	[°C]	1400
σ_0	[mN/m]	320
$\sigma_{Sätt}$	[mN/m]	380
θ	[°]	5
ρ_S	[g/cm^3]	2.3
ρ_{FF}	[g/cm^3]	3.5
η_0	[g/(cm s)]	160
$\eta_{Sätt}^{*}$	[g/(cm s)]	2788
D	[cm^2/s]	6×10^{-9}
C_0	[% by weight]	1.3
C_S	[% by weight]	29

Equation	Result	Value	Dimension
(1)	$2 \cdot \xi$	0.75	cm
(2)	y/h	0.75	cm// ψ 180°
(6)	d	0.079	cm
(7)	s	1.18	cm
with	d'	0.047	cm
(8)	u_M^{*}	2.1×10^{-4}	cm/s
(9)	$d\sigma/dC$	2.17	mN/(m % by weight)
(10)	grad C	-23.4	% by weight/cm
(11)	β	1.2×10^{-6}	mm/h
(12)	V_{corr}	7.7×10^{-3}	mm/h

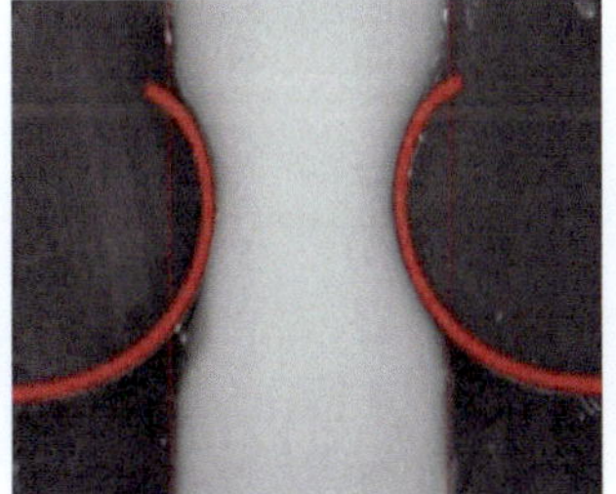

M. Dunkl and R. Brückner, 1989; $V_{exp} = 8.1 \times 10^{-3}$ mm/h

7.6.2 Dissolution of a Cylinder of Densely Sintered MgO in a Lime-Aluminate Slag at 1650 °C

The composition of the slag in % by weight is 50% CaO, 35% Al$_2$O$_3$, 15% SiO$_2$; saturation is achieved with 10.5% MgO$_{sat}$. Due to the cylindrical shape of the specimen (1.82 cm Ø) and the slight increase in MgO concentration in the slag during the experimental procedure, the increase in concentration in the slag is not linear in time but slows down with time. The comparison of experiment and calculation is made by relating half the corroded diameter (9 mm) to the total test time of 11 h. Within the accuracy of the material data, the achieved conformity is good. Table 7.2 shows an example calculation [18].

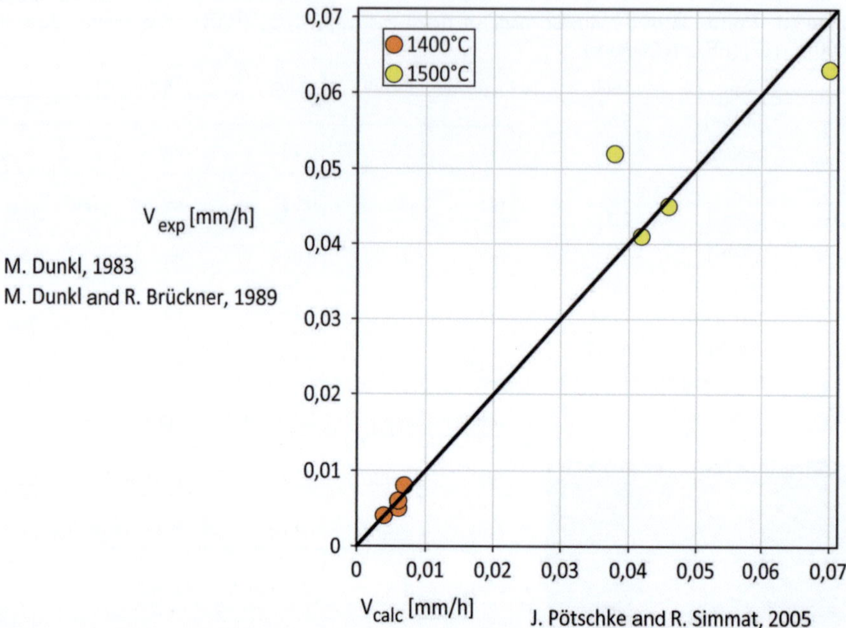

Fig. 7.3 Comparison of experimentally determined corrosion speeds of fused corundum in container glass [21] with calculated values [17]

Table 7.2 Corrosion of a cylinder of densely sintered MgO in a calcium aluminate slag at 1650 °C [18] (FF = refractory)

T	[°C]	1650
σ_0	[mN/m]	540
$\sigma_{Sätt}$	[mN/m]	565
θ	[°]	32
ρ_S	[g/cm^3]	2.7
ρ_{FF}	[g/cm^3]	3.3
η_0	[g/(cm s)]	1
D	[cm^2/s]	2.4×10^{-7}
C_0	[% by weight]	0
C_S	[% by weight]	10.5

Equation	Result	Value	Dimension
(1)	$2\,\xi$	0.9	cm
(2)	h	0.87	cm// $\psi = 180°$
(6)	d	0.1	cm
(7)	s	1.36	cm
mit	d'	0.05	cm
(8)	u_M	0.23	cm/s
(9)	$d\sigma/dC$	2.38	mN/(m % by weight)
(10)	grad C	-7.7	% by weight/cm
(11)	β	2.3×10^{-4}	mm/h
(12)	V_{corr}	0.7	mm/h

C. Brüggmann, 2010; $V_{exp} = 0.8$ mm/h

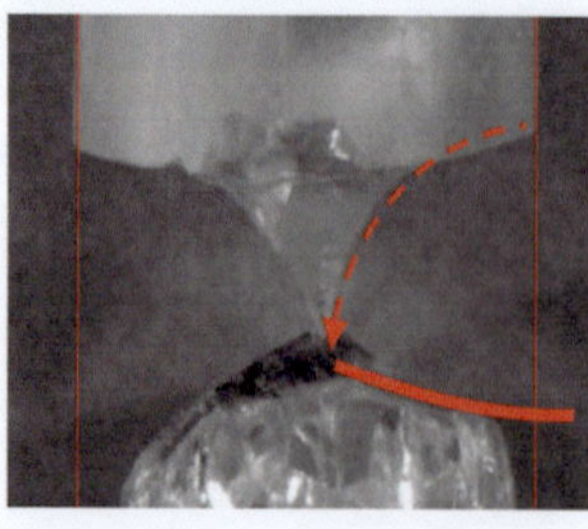

7.6.3 *Corrosion of AMC Linings of a Laboratory Induction Furnace (20 kg Steel, 2 kg Slag, 10 kHz) at 1550–1650 °C*

Over the course of 10 years, the corrosion properties of different refractory linings in alternating steel-slag combinations at steel mill temperatures were tested with a large number of corrosion tests in the induction furnace at the German Institute for Refractories and Ceramics (DIFK) in Bonn, Germany [22–24].

Figure 7.4 illustrates schematically the test setup and the observed zones of premature wear. They are always located along the three-phase lines refractory material/slag/air, or to a significantly greater extent along refractory material/slag/steel.

A cross-section through the corrosion zone is shown in Fig. 7.5. Clearly visible is the premature wear on both three-phase lines refractory material/slag/steel and refractory material/slag/atmosphere (Ar, 5% H_2).

Figure 7.6 shows the temporal change of the corrosion zone schematically. It can be seen that although the wetting angle $\theta_{Ref/St/Sl}$ determines the curvature of the steel/slag phase boundary in contact with the refractory material, the slag runs into the gap. The rapid convection of both phases is transferred to the slag at least near the gusset so that corrosion progresses most rapidly in this area. During this process, $\theta_{Ref/St/Sl}$ remains unchanged. With the assistance of gravity and electromagnetic forces, the corrosion zone curves downward over time. However, in this case, the thickness of the slag layer can only be calculated imprecisely according to Eq. (7.7). The corrosion gap between the refractory material and the steel/slag phase boundary, into which the slag runs, is better approximated by the simple geometric relationship and calculated

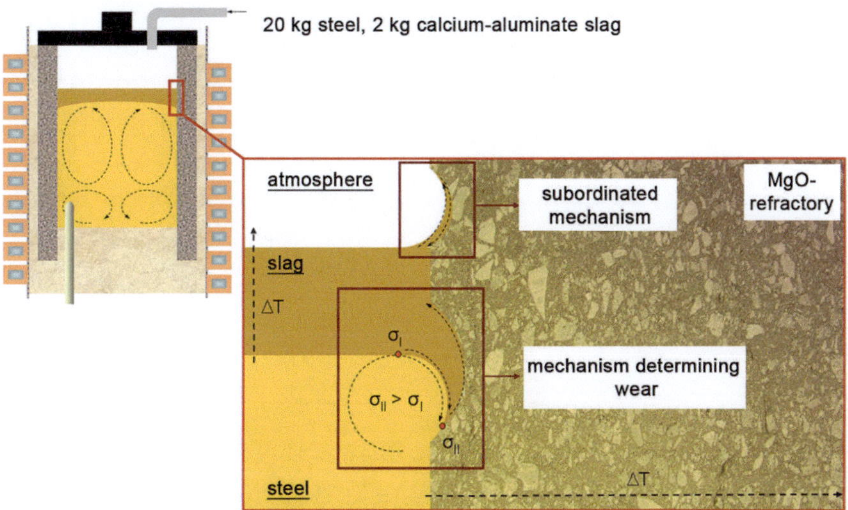

Fig. 7.4 Schematic illustration of the zones of premature wear in an induction crucible test [16]

Fig. 7.5 Partial view of a cut-open induction crucible test [16]

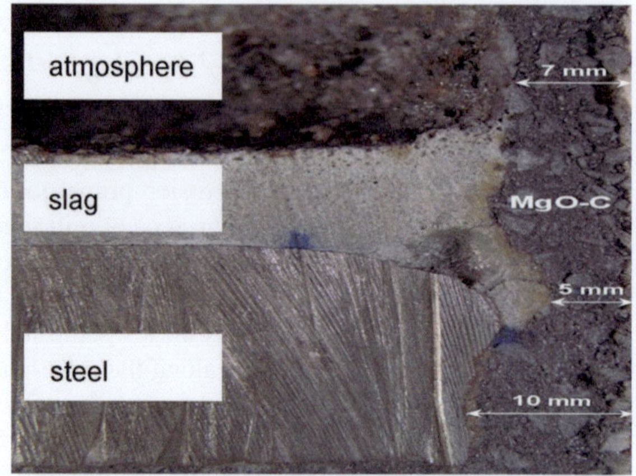

$$d = h \cdot \tan\left(180 - \theta_{\mathrm{Ref/St/Sl}}\right) \tag{7.14}$$

(refer to Fig. 7.6 for $x = d$). The mass transfer determining the corrosive rate takes place from the refractory material into the slag.

In summary, the equation for corrosion in the refractory material/slag/steel system is:

Fig. 7.6 Formation of the corrosion zone in time sequence: $t_0 < t_1 < t_2$

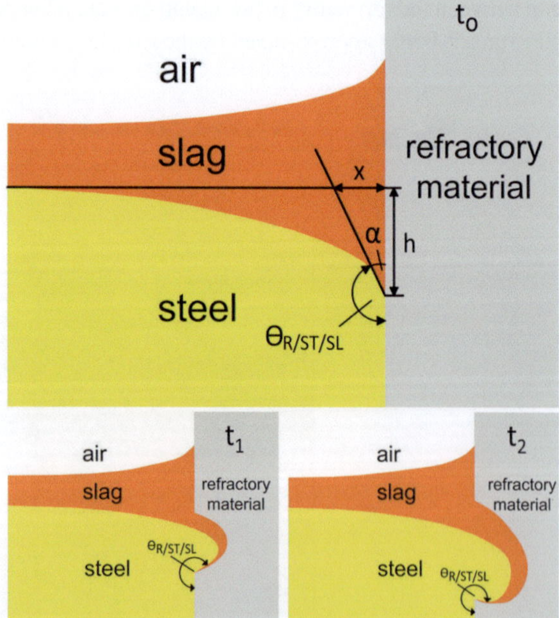

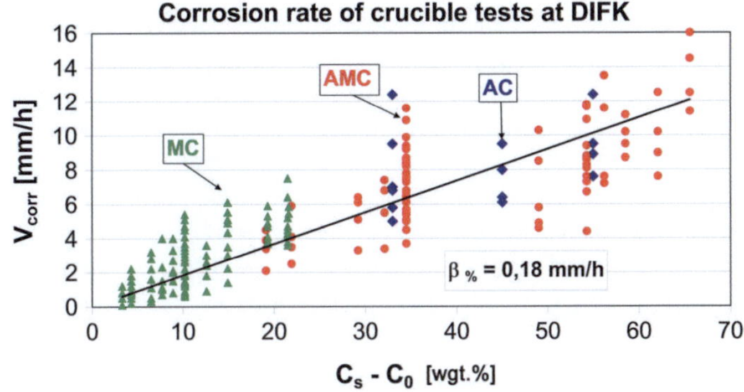

Fig. 7.7 Corrosion of AMC linings of a laboratory induction furnace (20 kg steel, 2 kg slag) at 1550–1650 °C [22–24]

$$\beta_M = 360 \cdot \frac{\rho_{Sl}}{\rho_R} \cdot \left(\frac{4 \cdot D_{Sl} \cdot \tan\left(180 - \theta_{Ref/St/Sl}\right)}{\pi^3 \cdot h \cdot \eta_{St}} \cdot \left(\frac{D \cdot \rho}{\eta} \right)_{Sl}^{1/3} \right)^{1/2}$$

$$\cdot \left(\frac{d\sigma_{St/Sl}}{dC_{[O]}} \cdot \left(C_{[O]}^{I} - C_{[O]}^{II} \right) + \frac{d\sigma_{St/Sl}}{dT} \cdot \left(T^{I} - T^{II} \right) \right)^{1/2} \quad [\text{mm/h} \, \%]. \quad (7.15)$$

h is calculated according to Eq. (7.3) for $\psi = 0°$.

The compilation of all results for carbon-bonded AC, MC and AMC materials [23] is shown in Fig. 7.7. The mean corrosion rate is plotted against the equilibrium distance of the corrosion-determining component(s) in the slag, i.e. the driving concentration difference $(C_{sat} - C_0)$ [% in weight].

The corrosion rate is first calculated separately for both components, Al_2O_3 and MgO. The result is compared with the experiment, and a decision is made from it. The main component of the refractory material, Al_2O_3 or MgO, determines the corrosion rate (speed). In the transition region ($10\% < MgO < 50\%$), however, both components must be taken into account [22, 24]. The scatter results from the different materials and test conditions [23]. The individual values can be calculated quite accurately, taking into account the conditions prevailing in the experiment [22–24].

The mean mass transfer rate $\beta_\% = 0.18$ mm/h given in Fig. 7.7 must be multiplied by the driving concentration difference $(C_{sat} - C_0)$ [% in weight] to calculate the mean corrosion rate (speed). For corrosion of MgO-C materials, the average is $\beta_\% \approx 0.24$ mm/h and for Al_2O_3-C $\beta_\% \approx 0.19$ mm/h [23].

Table 7.3 shows the calculation of an example. Here, too, a sufficient conformity between experiment and calculation is achieved [25].

The difficulty of determining the substance (material) data is discussed below. (In the original Table 7.3 the numerical values used directly for the calculation are marked with *.)

In contrast to the first two examples, if applying the simplified model, it should be noted that in the steel/slag system two liquid phases are present next to each other,

Table 7.3 Corrosion of the AMC lining of a laboratory induction furnace by steel and CA slag at 1650 °C [24] (FF = refractory)

T	[°C]	1650
$C_{[O]}^{I}$	% by weight	0.005*
$C_{[O]}^{II}$	% by weight	0.0025*
$\sigma_{St/Sl}$	[mN/m]	1135*
σ_{St}	[mN/m]	1550
σ_{Sl}	[mN/m]	550
$\theta_{St/Sl}$	[°]	40
$\sigma_{FF/St}$	[mN/m]	1900*
σ_{FF}	[mN/m]	900*
$\theta_{FF/St}$	[°]	130
$\theta_{FF/St/Sl}$	[°]	150*
ρ_{St}	[g/cm^3]	7*
ρ_{Sl}	[g/cm^3]	2.7*
ρ_{FF}	[g/cm^3]	3.3*
η_{St}	[g/(cm s)]	0.04*
η_{Sl}	[g/(cm s)]	1*
D	[cm^2/s]	2.4×10^{-7}*
ΔC	[% by weight]	34*

Equation	Result	Value	Dimension
(1)	$2\,\xi$	0.81	cm
(2)	h	-0.79	cm
(5)	θ	150	°
(6)	d	0.45	cm
(7)	s	-1.23	cm
(8)	$d' = d$	0.45	cm
Equation	Result	Value	Dimension
(15)	$d\sigma/dC$	-4.4×10^4	mN/(m % by weight)
–	$d\sigma/dT$	-0.2	mN/(m K)
(10)	grad C	-2.0×10^{-3}	% by weight/cm
(10)	grad T	$+16.2$	K/cm
(8)	u_M	244	cm/s
(11)	β	7.2×10^{-4}	cm/s
–	$\beta_\%$	0.21	mm/(h %)
(12)	V_{corr}	7.2	mm/h

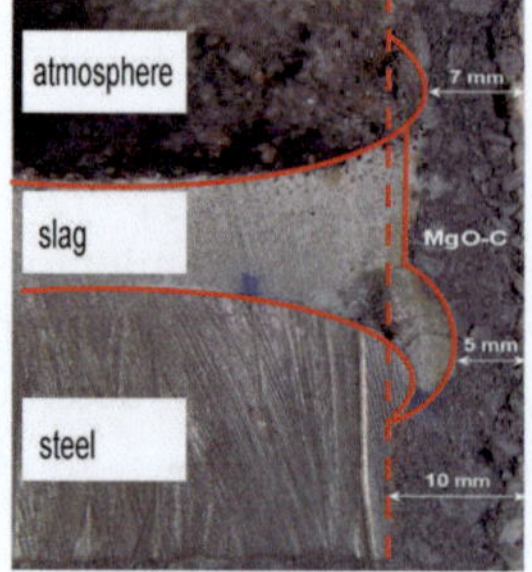

which react with each other fluid-dynamically and chemically. This is not the case in the glass/air and slag/air systems. First of all, experimental experience teaches that the Marangoni flow is primarily determined by gradients in the steel bath. Consequently, the concentration and temperature gradients in the steel are along the steel/slag phase boundary and decisive for the corrosion progress.

Oxygen dissolved in steel in particular, but also sulfur [3], has a strong capillary-active effect. In molten pure iron, its surface tension changes with the oxygen concentration as [26]:

$$\sigma_{Fe-1650°C} = 542 - 464 \cdot \log[\%O] \ [mN/m]. \tag{7.16}$$

From the contents dissolved in the iron, the following can be calculated according to Eq. (7.14)

$$C^{I}_{[O]} = 0.005\,[\%] \quad \text{and} \quad C^{II}_{[O]} = 0.0025\,[\%]$$
$$\sigma^{I}_{Fe} = 1610\,[\text{mN/m}] \quad \text{and} \quad \sigma^{II}_{Fe} = 1750\,[\text{mN/m}] \tag{7.17}$$

If $\sigma_{Fe-1650°C}$ (Eq. 7.14) is derived according to the oxygen content dissolved in the steel, the following applies [22–25]

$$\frac{\mathrm{d}\sigma}{\mathrm{d}C_{[O]}} = -\frac{220}{C_{[O]}}\,\left[\text{mN/m}\,\%\ \text{by weight}\right]. \tag{7.18}$$

The change of the surface tension with temperature is [26]

$$\mathrm{d}\sigma/\mathrm{d}T = -0.2\,\text{mN/m K}. \tag{7.19}$$

Steel is not pure iron, which is why the surface tension in practice is about $\sigma_{Fe} = 1550$ mN/m. However, the dependencies just calculated continue to apply quite well. The wetting angle $\theta_{Ref/St/Sl}$ between the steel/slag phase boundary and the refractory material is calculated from Eq. (7.16) to be $\theta_{Ref/St/Sl} = 150°$ using the values $\sigma_{Ref} = 900$ mN/m, $\sigma_{St/Sl} = 1135$ mN/m, and $\sigma_{St/Ref} = 1900$ mN/m given in Table 7.3. These are the values that are used to calculate.

Another point should be addressed here: Hauck [8] found experimentally that the mass transfer of the carbon dissolved in the slag also plays a role for the corrosion rate of submerged nozzle refractory material with approx. 30% by weight carbon. This is not observed for AMC material with a significantly lower carbon content.

7.7 Discussion

The model presented here is based on the assumption of laminar flow of the corroding liquid along the refractory material [17]. This is always fulfilled for liquid glass and slag, since the Reynolds number

$$Re = |u_M| \cdot d/v \tag{7.20}$$

is very small. This is not the case in the molten steel bath, where the Reynolds number exceeds 2500, i.e. the flow is turbulent. The mass transfer from the slag into the steel bath should not be a problem, but much more than from the refractory material into the slag.

Examinations with two mutually insoluble liquids, e.g. water/toluene, to which a third, capillary-active liquid, e.g. acetone, is added in a very small quantity, show that a strong, turbulent flow sets in, which can even shift the original position of the phase boundary [10]. Under these circumstances, the division of the layer thickness d into two equal parts is physically meaningless. Therefore, $d = d'$ is set. Besides, the wall friction must be taken into account when calculating the mass transfer coefficient $\beta_{R/SI}$ (Eq. 7.12).

In the experiments with steel and slag, the calculation of the mean thickness d [cm] of the slag gusset enclosed between refractory material and bath according to Eq. (7.14) is a very rough approximation because the geometrical conditions change over time (refer to Fig. 7.6).

Only the observation that the experiments can be described with these assumptions justifies the use of the presented model also for steel/slag/refractory material systems. It, therefore, remains to be stated that the model presented is only to be understood as a step toward a physically exact model.

Another point should also be explicitly mentioned: The used substance (material) values are mostly taken from literature and carefully selected. However, the scatter of values in literature is very large. An exact verification of the model requires exact values, as, e.g., in example 1 glass melting/refractory material [17].

7.8   Summary

On the basis of a large number of experiments, a model is developed to describe the corrosion of refractory materials by glass, slag and steel/slag melts. Simple physical relationships are used, which simplifies the application for laboratory tests and quickly leads to initial statements. The application in operational practice is only possible to a limited extent because the boundary conditions there usually change considerably from batch to batch.

Together with Dr. C. Brüggmann, this work was presented in excerpts to the Technical Committees for Physical Chemistry and Metallurgical Process Development of the Steel Institute VDEh, Düsseldorf, Germany, on March 11, 2011.

References

1. Marangoni, C.: Spread of the drops of a liquid on the surface of another surface. Ann. D. Phys. Chem. **143**, 337 (1871)
2. Levich, V.G.: Physicochemical Hydrodynamics, pp. 373–390. Prentice Hall Inc. (1962)
3. Scheller, P.R., Brooks, R.F., Mills, K.C.: Influence of sulfur and welding conditions on penetration in thin strip stainless steel. Weld. Res. Suppl. **2**, 69s–75s (1995)
4. Czochalski, J.: A new process for measuring the crystallization speed of metals. Z. Phys. Chem. **92**, 219–221 (1918)
5. Müller, G.: On the Formation of Inhomogeneities in Semi-Conductor Crystals During Their Manufacture Out of the Melt. Selisch—Fachbuchverlag (1986)
6. Marwedel, H.J.: Interpretation of the additional corrosion at the flux line of glass melts. Glastech. Ber. Ber. **39**(9), 399–402 (1966)
7. Brückner, R.: The erosion of model substances and refractory materials. Glastech. Ber. Ber. **40**(12), 451–462 (1967)
8. Hauck, F.G., Pötschke, J.: Wear of submerged nozzles during continuous casting of steel. Arch. Eisenhüttenwesen. **53**, 133–138 (1982)

9. Schwabe, D.: Marangoni effects in crystal growth melts. Phys. Chem. Hydrodyn. **2**(4), 263–280 (1981)
10. Brückner, R., Christ, H.: Diffusion-Related Boundary Surface Convection. Final Report TEXUS I–X, DLR, pp. 121–125 (1977–84)
11. Hrma, P., Kashcheyev, J.D.: Dissolution of a solid body governed by surface-free convection. Chem. Eng. Sci. **25**, 1679–1688 (1970)
12. Stanek, V., Szekely, J.: The effect of surface-driven flows on the dissolution of partially immersed solids in liquids. Chem. Eng. Sci. **25**, 699–715 (1970)
13. Harada, Hiragushi, K., Mukai, K.: Slag film movement in a local corrosion zone of solid silica at PbO-SiO$_2$ slag surface. Ceram. Bull. **69**, 1189–1193 (1990)
14. Mukai, K.: Marangoni flows and corrosion of refractory walls. Phil. Trans. R. Soc. Lond. **365**, 1015–1026 (1998)
15. Langbein, D.: Capillary Surfaces—Shape—Stability—Dynamics, in Particular Under Weightlessness. In: Springer Tracts in Modern Physics, p. 178 (2002)
16. Pötschke, J., Brüggmann, C.: Premature wear of refractories due to Marangoni Convection. Steel Res. Int. **7**, 637–644 (2012)
17. Pötschke, J., Simmat, R.: The wear of refractories by glass melts due to Marangoni Convection. In: 48th International Colloquium on Refractories, Aachen, Germany, pp. 113–117 (2005)
18. Brüggmann, C.: A contribution to slagging of MgO in secondary metallurgical slags. Dr.-Ing. dissertation, TUBA—Freiberg, Germany, 2011
19. Hauck, F.G.: Wear of submerged nozzles during continuous castings of steel. Dr.-Ing. dissertation, RWTH—Aachen, Germany, 1981
20. Bird, R.B., Stewart, W.E., Lightfoot, E.M.: Transport Phenomena. Wiley (1960)
21. Dunkl, M., Brückner, R.: Corrosion of refractory material by a container glass melt. Glastech. Ber. Ber. **62**(1), 10–19 (1989)
22. Pötschke, J., Deinet, T., Routschka, G., Simmat, R.: Properties and corrosion of AMC—refractories part II: corrosion by steel/slag. In: Proceedings of UNITECR—2003—Congress, Osaka, Japan, pp. 584–587 (2003)
23. Pötschke, J., Deinet, T.: Premature corrosion of refractories by steel and slag. In: Millennium Steel, pp. 109–113 (2005)
24. Pötschke, J., Deinet, T.: The corrosion of AMC—refractories by steel and slag. In: UNITECR—2005—Congress, Orlando, Florida, USA, pp. 75–79 (2005)
25. Pötschke, J., Deinet, T.: The Corrosion of Refractory Castables, Interceram, Refractory Manual, pp. 6–10 (2005)
26. Borgmann, F.O.: Influence of surface tension of carbon-saturated iron melts by additional elements. Dr.-Ing. dissertation, TU—Berlin, Germany, 1971
27. Richardson, F.D.: Physical Chemistry of Melts in Metallurgy. Academic Press, London, NY (1974)
28. Sun, H., Nakashima, K., Mori, K.: Influence of slag composition on slag-iron interfacial tension. ISIJ Int. **3**(46), 407–412 (2006)

Chapter 8
Formation and Behavior of Ceramic Inclusions

Refractory material reacts chemically with metallic melts, specifically steel, and it can dissolve. As the temperature decreases, the components accumulate in the melt in relation to their solubility and excretions, e.g. spinels, can be formed there. This is particularly true during solidification of the melt. The excretions, e.g. oxides and oxysulfides, are formed preferentially in the interdendritic residual melt and mark the morphology of the microstructure.

A second process, which has received little attention to date, is the displacement of already excreted particles by the solidification front itself. This can also lead to considerable accumulations of impurities in the interdendritic structure. Both processes overlap, so that their differentiation is difficult in individual cases.

Since both mechanisms, even after deoxidation has taken place, contribute significantly to the contamination of the microstructure by refractory material, they are addressed here.

8.1 Formation of Inclusions During Solidification

The section of a phase diagram is shown in Fig. 8.1. The solidification of the melt of initial concentration C_0 takes place in the temperature range ΔT. The concentration of the enriched residual melt is $C_{li} = C/k_{oc}$. k_c [–] is the ratio of the solidus/liquidus concentration and, just like the ascent of the liquidus line $m = \Delta T/\Delta C_{li}$ [K/%], is considered to be constant. It can be seen that in the case shown the dissolved component enriches before the solidification front [1]. The higher the dissolved component C_{li}, the lower the solidification temperature.

If solidification takes place at the velocity v [cm/s], the maximum accumulated (stacked) concentration C_0/k_C at the phase boundary decreases by diffusion into the melt [2]:

J. Pötschke, *Refractory Fundamentals in Metallurgical Practice*,
https://doi.org/10.1007/978-3-031-63709-4_8

Fig. 8.1 Concentration shift in an alloy during its solidification

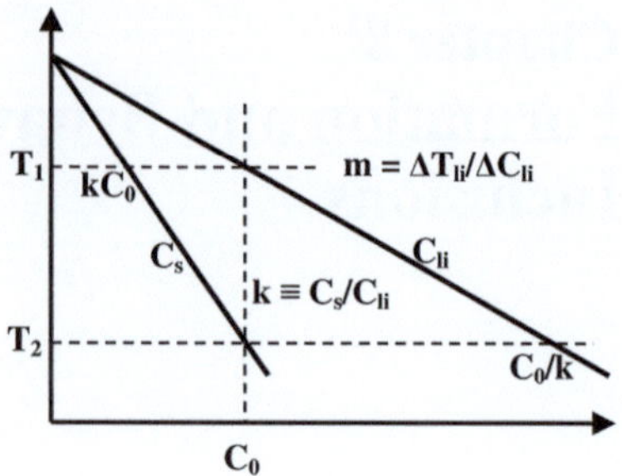

$$C_{\text{li}} = C_0 \cdot \left[1 + \frac{1 - k_c}{k_c} \cdot \exp\left(-\frac{v}{D} \cdot x \right) \right]. \tag{8.1}$$

D [cm^2/s] stands for the diffusion coefficient of the component under consideration, e.g. oxygen dissolved in the steel. Figure 8.2 shows the result of the concentration distribution according to Eq. (8.1) graphically. Starting from the solidification front, the liquidus temperature increases. At the same time, the actual temperature gradient $G = \Delta T / \Delta x$ [K/cm] is present in the melt. If its temperatures T_x are below those of the liquidus line $T_{\text{li},x} = T_{C_0} - |m_1| \cdot C_{\text{li},x}$, it is called constitutional super cooling **CU** [1]. It follows the relation $\Delta T_{\text{CU}} = T_x - T_{\text{li},x}$ and satisfies the condition [1–3].

$$\frac{G}{V \cdot C_0} \leq -\frac{m}{D} \cdot \frac{1 - k_c}{k_c}. \tag{8.2}$$

The constitutionally undercooled melt causes the destabilization of the initially flat solidification front increasingly the smaller $\frac{G}{V \cdot C_0}$ becomes, and, consequently, essentially determines the solidification morphology. Figure 8.3 schematically illustrates two examples.

Fig. 8.2 Constitutional supercooling (CU) of a solidifying alloy

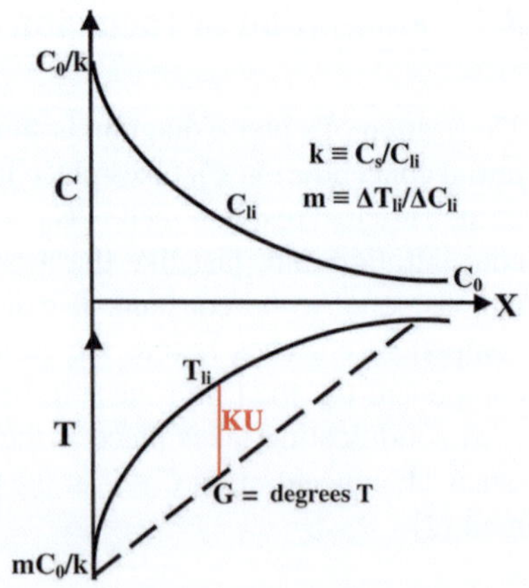

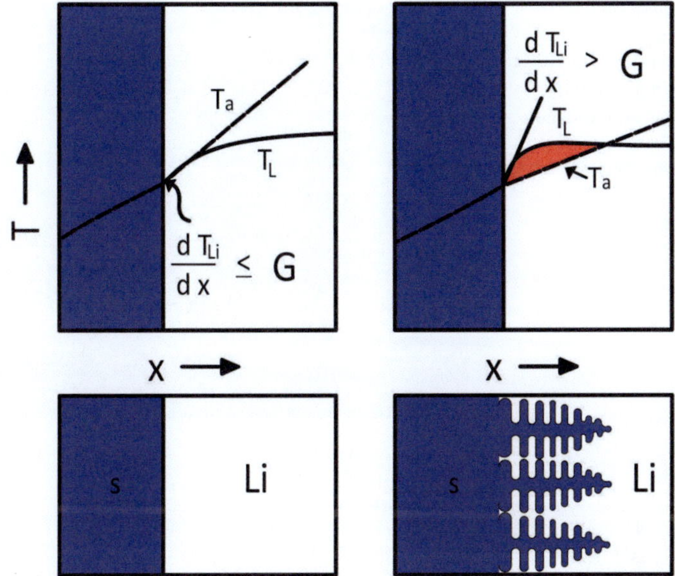

Fig. 8.3 Solidification morphology as a result of constitutional supercooling [3]

With growing constitutional supercooling, cells develop which become dendrites by the formation of side arms. This is because residual melt also accumulates (enriches) in the area between the cell walls that grow toward each other. Constitutional supercooling develops there as well. As a result of branching, it is reduced by the growth of further cells.

The calculation of this process is complex and was done approximately by Bolling and Tiller [4]. Here the simplified "pencil model" is used [5]. It is sketched in Fig. 8.4.

From simple geometric relationships, the ratio of crystallization rates in the side/main growth direction is obtained:

Fig. 8.4 "Pencil model" as simplified solidification model [5]

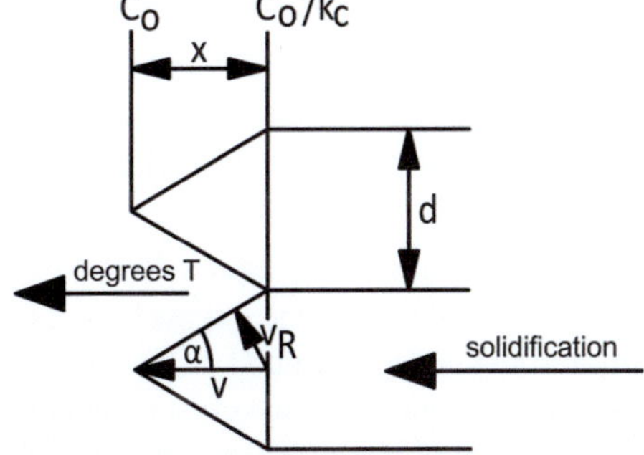

Fig. 8.5 Reaction of a solidifying steel melt (0.013% S, 0.0035% O, 0.016% Al) in an alumina crucible. Mean solidification rate $u = 0.03$ cm/s. Magnification: 500× [6]

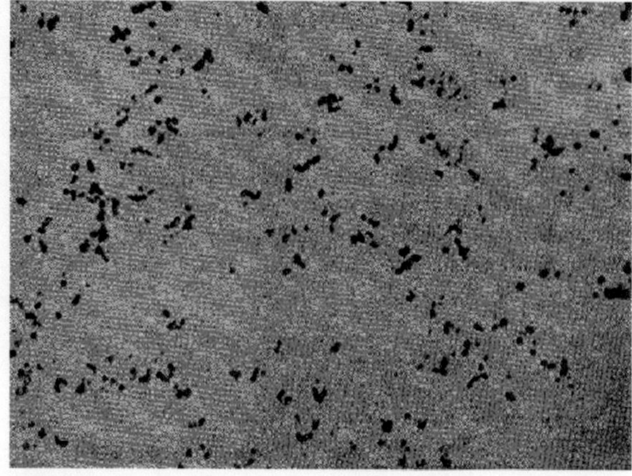

$$\frac{v_{\text{rad}}}{v} = \frac{d}{2 \cdot x} \quad \text{with} \quad x = \frac{C_0 \cdot m}{G} \cdot \frac{1}{k_c - 1}. \tag{8.3}$$

x [cm] is the length of the constitutional supercooling and d [cm] the diameter of a cell. Smeets [6] has carried out a detailed "examination of the conditions of formation of oxide and sulfide inclusions during the solidification of steel," presenting and evaluating the theoretical aspects. Figure 8.5 shows an example of oxysulfide excretions in the interdendritic space [6] of a solidified steel melt.

Equation (8.2) allows first to check if there was any constitutional supercooling at all in this experiment. The following data are available [6]: $v = 0.03$ cm/s, $C_0 = 0.0035\%$ (oxygen), $k_c = 0.022$ (oxygen), $m = -70$ K/% (oxygen), $D = 10^{-5}$ cm^2/s (oxygen), $G = 116$ K/cm, $d = 0.0012$ cm. It follows from Eq. (8.2) $10^6 < 10^8$, i.e. constitutional supercooling was present.

The radial growth velocity v_{rad} can be calculated from Eq. (8.3) with the above data [6]. With $x = 0.002$ cm, one obtains $v_{\text{rad}} = 0.008$ cm/s and $v_{\text{rad}}/v \leq 0.3$.

Finally, from Eq. (8.1), the enrichment of dissolved oxygen at the solidification front ($x = 0$) is calculated to be $C/C_{\text{lio}} = 45$. As a result of this high enrichment, nucleation and excretion of oxysulfides occurs in the residual interdendritic melt.

8.2 Behavior of Existing Inclusions During Solidification

If the solidification front approaches an excretion suspended in the melt, the crystal can grow around it, i.e. encompass it, or it is pushed in front of it at a small distance d [m] and does not grow around. In the second case, the inclusion reaches the solidifying residual melt and marks the crystal boundary there. Figure 8.6 shows as an example the alignment of alumina particles in cellular solidified copper [7].

The mechanical equilibrium of the repulsive force K_σ [N] (Van der Waals) acting between the particle and the crystal and the viscous force K_η [N] (Newton) of the melt

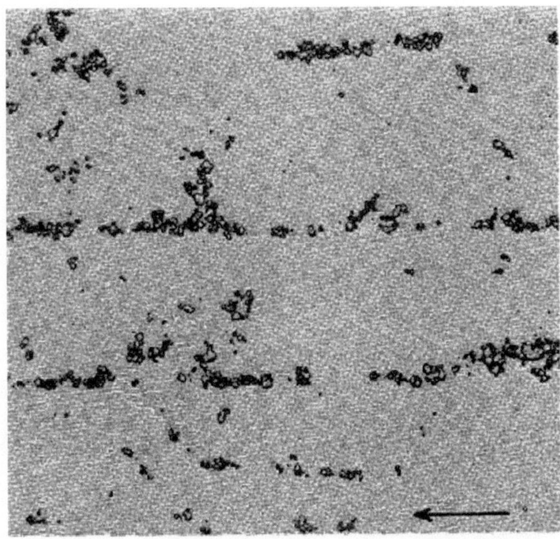

Fig. 8.6 Alumina particle arrangement ($\leq 100\ \mu$m) after cellular solidification of a copper melt. $\leftarrow$ Solidification direction (μg-experiment) [7]

flowing into the narrow gap d [cm] between the two phases serves as an explanation [8]. The following equations apply

$$K_\sigma = 2\pi R \cdot \Delta\sigma \quad \text{mit} \quad \Delta\sigma = \left(\sigma_{ps} - \sigma_{pl} - \sigma_{sl}\right) > 0 \tag{8.4}$$

and

$$K_\eta = 6\pi \cdot \eta \cdot v \cdot R^2/d. \tag{8.5}$$

Explanation: σ [J/m^2] = spec. free interfacial (boundary surface) energy (p = particle, l = melt, s = crystal), R [m] = particle radius, η [Pa s] = viscosity, v [m/s] = solidification speed. The repulsive Van der Waals force K_{VdW} and the viscosity force K_η favoring particle incorporation must be equal in the steady state. In the case of solidification of a (single-component) melt with a plane solidification front, the maximum possible solidification speed v_c [m/s] is calculated from the "critical distance" d_c [m] and taking into account the atomic distance a_o [m] in the melt to [8]

$$v_c = \frac{\Delta\sigma \cdot a_o}{12 \cdot \eta \cdot R} \ \text{[m/s]} \tag{8.6}$$

When this velocity is exceeded, one falls below the minimum stable distance d_C and the particle encompassed. The condition for the displacement is $d_C > a_o$.

If the melt consists of several components, e.g. if it is contaminated, the solidification front is curved and the model is complex, which is why reference is made to literature for precise calculation [5, 8].

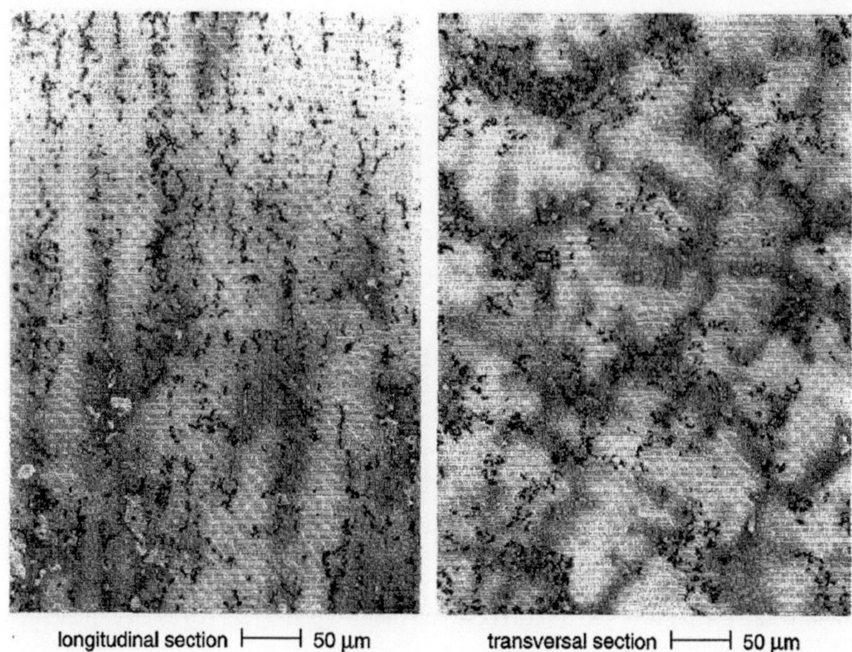

Fig. 8.7 Alumina particle arrangement (ca. 100 nm) after cellular solidification of a CMSX6-Al_2O_3 alloy (μg-experiment) [5]

Here, a simplified test is performed to determine which oxide particle size would be shifted into the intercellular microstructure region at the above-mentioned conditions [6].

The following values apply: $\Delta\sigma = (2 - 1.7 - 0.2) = 0.1$ J/m^2 (Eq. 8.4); $a_o = 2.5 \times 10^{-10}$ m; $\eta = 0.005$ Pa s; $v_C = 8 \times 10^{-5}$ m/s. Particles with diameter $2R < 10^{-5}$ m (< 10 μm) would be displaced.

Figure 8.7 shows the result of an experiment under reduced gravity showing the behavior of alumina particles (< 100 nm) in the solidifying superalloy CSMX6 [5]. Here, the displacement of the inclusions onto the sub-grain boundaries are particularly clear. The experimental findings [5] are in good agreement with the theory [7].

References

1. Chalmers, B.: Physical Metallurgy. Wiley, New York (1959)
2. Cahn, R.W.: Physical Metallurgy. North Holland Publishing Company, Amsterdam, The Netherlands (1965)
3. Tiller, W.A.: The Art and Science of Crystal Growing, Chap. 15. Wiley, New York (1962)
4. Bolling, G.F., Tiller, W.A.: Growth from the melt. J. Appl. Phys. **31**, 2040–2045 (1960)

5. Busse, P., Deuerler, F., Pötschke, J.: The stability of ODS alloy CMSX6-Al$_2$O$_3$ during melting and solidification under low gravity. J. Cryst. Growth **193**, 413–425 (1998)
6. Smeets, L.: Examination of the Formation Conditions of Oxidic and Sulfidic Inclusions Upon Solidification of Steel. Dr.-Ing. dissertation, RWTH, Aachen, Germany, Jan 1972
7. Pötschke, J., Rogge, V.: Behavior of suspended particles at the solidification front of copper. Naturwissenschaften **73**, 381–383 (1986)
8. Pötschke, J., Rogge, V.: On the behaviour of foreign particles at an advancing solid–liquid interface. J. Cryst. Growth **94**, 726–738 (1989)

Chapter 9
Electrochemical Corrosion Protection

9.1 Electrolysis in Slags

As an example of the calculation of electrolysis in molten slags, the fayalitic slag $(2FeO \cdot SiO_2)$ with a melting temperature of 1205 °C [1], studied by Sauerbrey [1] at 1400 °C, is selected. At the applied voltage of 1.5 V, a current strength of about 5 A is observed. The electric field strength is $E = 4.5$ V/cm.. The electrical conductivity of the slag is approximately $\sigma = 1 \ \Omega^{-1} \ cm^{-1}$ [2]. As a result, an electric current of density

$$j = \sigma \cdot E \ \left[A/m^2 \right] \tag{9.1}$$

flows, i.e., $j = 4.5$ A/cm^2. No polarization of the electrodes is observed. Metallic iron is deposited at the cathode, gaseous oxygen at the anode.

The reaction equations are

1. Dissociation of the (FeO) in the slag:

$$2(FeO) = 2\left(Fe^{2+}\right) + 2\left(O^{2-}\right) \tag{9.2.1}$$

2. Anode reaction:

$$2\left(O^{2-}\right) = \{O_2\} + 4e \tag{9.2.2}$$

3 Cathode reaction:

$$2\left(Fe^{2+}\right) + 4e = 2Fe_{met} \tag{9.2.3}$$

J. Pötschke, *Refractory Fundamentals in Metallurgical Practice*,
https://doi.org/10.1007/978-3-031-63709-4_9

4. Total:

$$2(\text{FeO}) = 2\text{Fe}_{\text{met}} + \{\text{O}_2\} \tag{9.2.4}$$

The calculation of the electrical voltage required for the electrolytic deposition of iron and oxygen is carried out by means of the thermodynamic calculation program Fact-Sage [3] and additionally "in classic manner" in order to make the theoretical background clear (refer to Sect. 2.1.4).

The free enthalpy of the reaction (9.2.4) is determined with Fact-Sage by writing the equation into the "Reaction Module" and calculating it at 1400 °C. One obtains $\Delta G^0 = -\,315$ kJ/mol O_2. If one sets for "P_{O_2}" $= P$ and for "Delta G (J)" $= 0$ in the "non-standard states" mode, one obtains the partial pressure of oxygen in thermodynamic equilibrium to $P_{\text{O}_2} = 1.5 \times 10^{-10}$ atm. "Classically," one also takes from the Richardson-Ellingham diagram [4] at 1400 °C, the values are $\Delta G^0 = -\,315$ kJ/mol O_2 and $P_{\text{O}_2} = 2 \times 10^{-10}$ atm (Fig. 9.1).

The activity of the (FeO) in the liquid slag ($2\text{FeO}\cdot\text{SiO}_2$) is calculated by Fact-Sage by entering 2mol FeO and 1mol SiO_2 into the "Equilib module". After selecting the "Base Phase" "FT oxide Slag A/liqu", the slag activity is calculated at 1400 °C. One finds $a_{\text{FeO}} = 0.47$ and $a_{\text{SiO}_2} = 0.5$. From the slag atlas [2], one takes from page 230 for the mole fraction $x_{\text{SiO}_2} = 0.33$ at 1350 °C the values $a_{\text{FeO}} = 0.43$ and $a_{\text{SiO}_2} = 0.7$ (Fig. 9.2). The temperature dependence of the activities is minute. Therefore, one continues to calculate at 1400 °C and $a_{\text{FeO}} = 0.45$.

Fact-Sage calculates the free reaction enthalpy by inserting the activity $a_{\text{FeO}} = 0.45$ in the "Reaction Module" after activation of the "non-standard state" for FeO. This gives at 1atm $\Delta G^0 = -\,337$ kJ/mol O_2 as result. The "classical" calculation considers the change in the free enthalpy in the standard state of the pure substances ΔG^0 due to the lower activity of the FeO dissolved in the slag:

$$\Delta G = \Delta G^0 + R \cdot T \cdot 2\ln a_{\text{FeO}} = -315{,}000 + 8.3 \cdot 1673 \cdot 2\ln 0.45 = -337 \text{ kJ} \tag{9.3}$$

Both calculations lead to the same result.

Now, the Nernst equation can be used to calculate tfhe voltage required to excrete (separate) the metallic iron and gaseous oxygen:

$$\text{EMF} = -\Delta G^0 / n \cdot z \cdot F = -(-337{,}000 \text{ J/mol})/2 \cdot 2 \cdot 96{,}500 \text{ As/mol} = 0.87 \text{ V.} \tag{9.4}$$

$F = 96{,}500$ As/mol is the Faraday constant, $n = 2$ the number of moles, $z = 2$ the ionic valence, $R = 8.3$ J/mol K the id. gas constant, $T = 1673$ K, $a = 0.45$.

Assuming pure ionic conduction of the slag, the cathodically excreted (deposited) amount of iron can be calculated from the electrochemical reaction Eq. (1.4), $(\text{Fe}^{2+}) + 2\text{e} = \text{Fe}_{\text{met}}$:

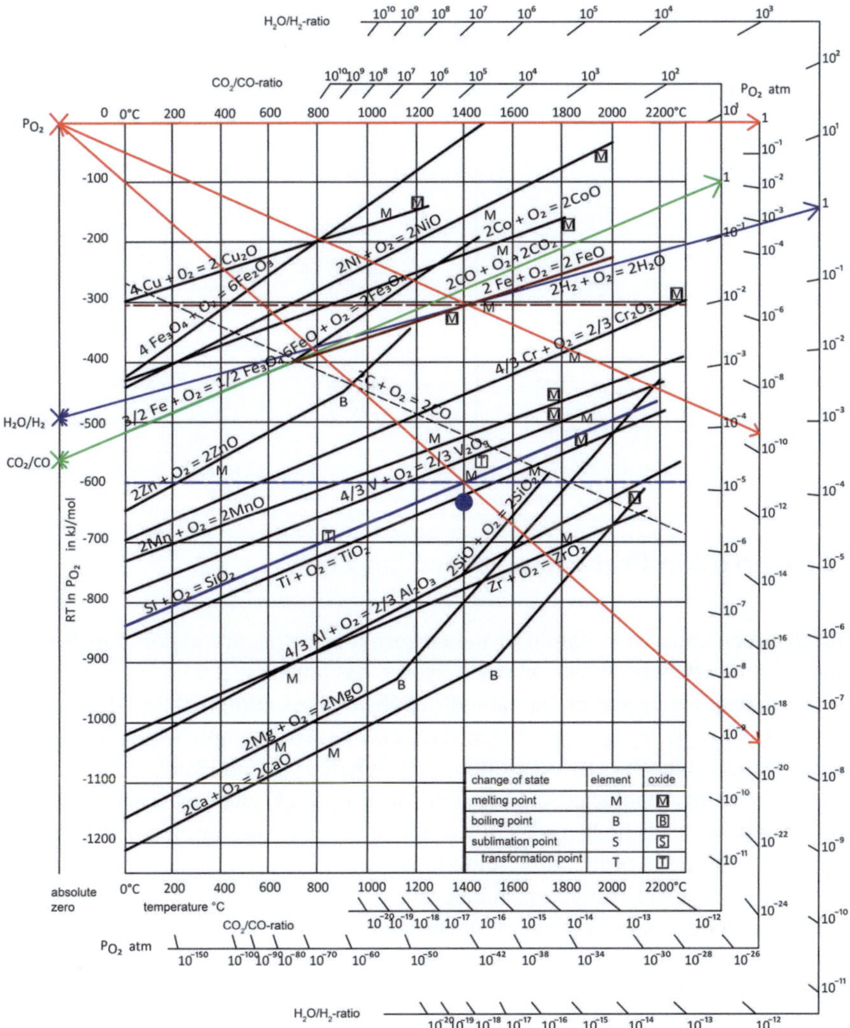

Fig. 9.1 FeO- and SiO$_2$-formation at 1400 °C illustrated in the Richardson-Ellingham diagram

$$\dot{m}_i = \frac{j \cdot M_i}{z_i \cdot F} \ [\mathrm{g/cm^2 \cdot s}]. \tag{9.5}$$

The molecular weight of iron is $M_{\mathrm{Fe}} = 56\,\mathrm{g/mol}$. Its valence is likely to be predominantly $z_{\mathrm{Fe}} = 2$. One obtains $\dot{m}_{\mathrm{Fe}} = 1.3 \times 10^{-3}\,\mathrm{g/cm^2\,s}$, which still corresponds to $4.7\,\mathrm{g/cm^2\,h}$. The anode reaction (1.3) is $\left(\mathrm{O}^{2-}\right) = 0.5\{\mathrm{O_2}\} + 2\mathrm{e}$. Using the molecular weight of oxygen, $M_{\mathrm{O_2}} = 32\,\mathrm{g/mol}$ gives us $\dot{m}_{\mathrm{O_2}} = 3.7 \times 10^{-4}\,\mathrm{g/cm^2\,s}$. In one hour, at 1400 °C (1673 K), about $6\,\mathrm{L/cm^2}$ of electrode area of gaseous oxygen are released.

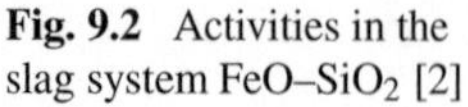

Fig. 9.2 Activities in the slag system FeO–SiO$_2$ [2]

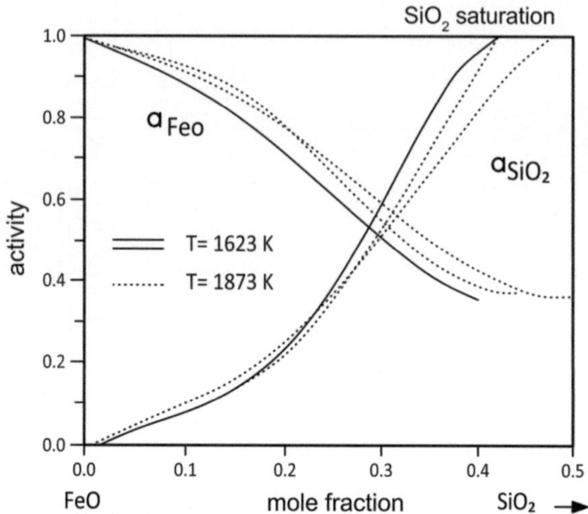

9.2   Corrosion Protection in the Slag Zone

The electrochemical retardation of the corrosion of refractory material in the slag zone is a long-cherished wish, which, however, has not yet met with any resounding success in practice. However, the cathodic deposition (excretion) of liquid aluminum at approx. 1000 °C from a cryolite slag is a well-known example which proves that such a process is technically controllable [3]. Where is the problem?

In his recently published dissertation, Sauerbrey [1] describes his laboratory experiments on the cathodic protection of refractory materials by electrical voltages, citing a large number of literature references.

Part of his results is summarized here [6]:

The electrodes are both made of magnesia. Its electrical conductivity at 1400 °C is about $\kappa = 10^{-4}$ (Ω cm)$^{-1}$ at oxygen partial pressures $P_{O_2} > 10^{-8}$ atm [7]. With decreasing oxygen partial pressure and increasing temperature, κ becomes larger (n—conductor with metal excess) [7].

The CAS slag has the composition in % by weight: 48.5% CaO; 11.5% Al$_2$O$_3$; 40% SiO$_2$ and melts at 1315 °C [1, 6]. Polarization occurs at 1400 °C with up to 10V, mainly 8V.

Figure 9.3 shows that this composition is very close to the 1400 °C isotherm, so that already a slight increase in the CaO content of the slag by just under 2% by weight leads to the excretion of 2CaO·SiO$_2$ (dicalcium silicate/C$_2$S). At a voltage of about 8 V, the measured current density is only 0.05–0.18 mA/cm^2 [1, 6]. After an experimental period of approx. 3 h, changes are observed at the cathode which differ from the "reference specimen" to which the voltage is not applied. For example, metallic silicon is excreted in addition to an accumulation of C$_2$S. In direct contact with the negative electrode, MgO-containing phases, such as merwinite (3CaO·MgO·2SiO$_2$), akermanite (2CaO·MgO·2SiO$_2$), melilite (Ca$_2$(Mg,Al)(Si$_2$O$_7$))

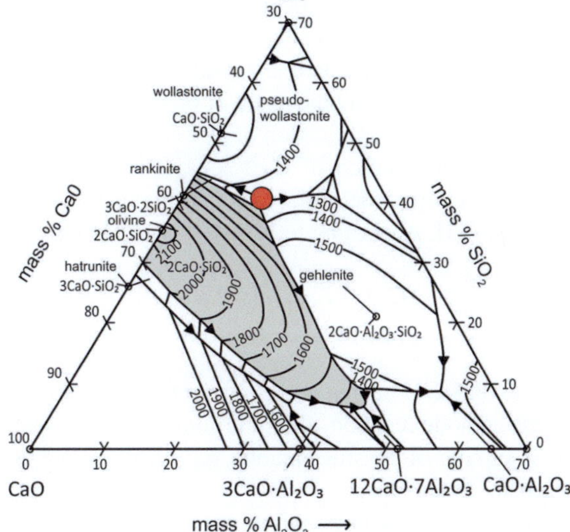

Fig. 9.3 Position of the output slag in the CAS system [1]

and monticellite ($CaO \cdot MgO \cdot SiO_2$) are formed [6]. Somewhat remotely in the slag, specifically dicalcium silicate ($2CaO \cdot SiO_2$) is formed in addition to some rankinite ($3CaO \cdot 2SiO_2$) [6]. This phase, in addition to CaO, also occurs inside the non-infiltrated pore system of the cathode, from which it is concluded that solid diffusion occurs [1].

Excretion of dicalcium silicate ($2CaO \cdot SiO_2$) occurs exclusively at the cathode and is a significant result of the experiments. The anode, on the other hand, wears out significantly, producing gaseous oxygen [1].

Three reaction equations are needed to interpret the experiments.

1. Dissociation of lime and silica in the slag:

$$2(SiO_2) + 2(CaO) = Si_2O_6^{4-} + 2Ca^{2+} \tag{9.6.1}$$

2. Anode reaction:

$$2O^{2-} = \{O_2\} + 4e \tag{9.6.2}$$

3. Cathode Reaction:

$$2Ca^{2+} + Si_2O_6^{4-} + 4e = \langle 2CaO \cdot SiO_2 \rangle + Si_{met} + 2O^{2-} \tag{9.6.3}$$

The sum is:

$$2(SiO_2) + 2(CaO) = \langle 2CaO \cdot SiO_2 \rangle + Si_{met} + \{O_2\} \tag{9.6.4}$$

In principle, the cathodic deposition of metallic silicon and the anodic deposition of gaseous oxygen are not necessary for the formation of C_2S, because the excretion of C_2S is not an electrochemical process in which the Ca^{2+} ions take up electrons from the cathode. It is solely the result of their accumulation (enrichment) at the polarized cathode. Even the application of a very low voltage leads to the thermodynamically required accumulation of Ca^{2+} ions at the cathode [7–9]. The difference in free enthalpy between the initial slag and the excretion of C2S is only $\Delta G^\circ = -2.4 \times 10^4$ J/mol [3]. According to Nernst this corresponds to the voltage

$$\text{EMF} = -\Delta G^\circ / z \cdot F = -\left(-2.4 \times 10^4 \text{ J/mol}\right)/4 \cdot 96.00 \text{ As/mol} = 0.06 \text{ V.}$$

$$(9.7)$$

The oxygen ions required to form CaO or C_2S, respectively, are provided either by the cathode reaction, which is complicated in detail (Eq. 9.6.3), or by increasing cross-linking of the silica anions, e.g., $2SiO_4^{4-} = Si_2O_7^{6-} + O^{2-}$. This increases the viscosity of the slag.

Nonetheless, the current flow observed during the excretion of the Si_{met} accelerates the formation of C_2S, since the ions move much faster in the strong electric field (8 V) than by diffusion in the weak field at 0.06V (refer to Sect. 2.1.4.1).

The free-moving Ca^{2+} ions are the main ones available for electrical conduction. The cross-linked silica anions are relatively immobile and contribute less. They even hinder the movement of the not bound cations insofar as local electroneutrality between the two types of ions must be maintained in the slag [7, 9]. Possible shares of electron conduction do not contribute to electrolysis.

Polarization phenomena at the electrodes, conduction losses in the slag and in the refractory material force the applied high voltages, which, when overcome, already excrete metallic silicon and gaseous oxygen.

From the Richardson-Ellingham diagram [4] in Fig. 9.1, the free enthalpy $\Delta G^0 = -600$ kJ ($P_{O_2} = 10^{-19} = $ atm) is taken at 1400 °C for the formation reaction of the silica. The activity of silica in the slag is calculated from Fact-Sage [3] to be $a_{(SiO_2)} = 0.06$. Thus, its free enthalpy in the slag is calculated according to the equation $Si_{met} + \{O_2\} = (SiO_2)$ to be $\Delta G^0 = -600{,}000 + RT \cdot \ln(0.06) = -639$ kJ/mol O_2 (blue dot). From the Nernst equation (9.7) $\text{EMF} = -\Delta G^0/n \cdot z \cdot F$ [V], the electrical voltage required to excrete the silicon is calculated with $n = 1$, $z = 4$, and $F = 96{,}500$ As/mol to $\text{EMF} = 639{,}000/4 \cdot 96{,}500 = 1.7$ V.

The voltage data are given with reference to a floating potential [1], because the electrical potential of the molybdenum wire, immersed in the melt and used as a reference, is not determined by any equilibrium reaction. It is convenient to use a graphite body as reference electrode, which is flushed with carbon monoxide (1atm), because then according to the reaction equation $2 <C> + \{O_2 = 2\{CO\}$, ΔG° (1400 °C) $= -517$ kJ/mol O_2 [3, 4] (refer to Fig. 9.1). Thus, the oxygen partial pressure is uniquely determined and depends only on the temperature. At 1400 °C it is 5×10^{-17} atm [3, 4] (refer to Fig. 9.1).

According to Eq. 9.6.4, one mole of C_2S is formed in parallel with the excretion of one mole of metallic silicon. Therefore, the deposited mass can be calculated:

$$\dot{m}_i = \frac{j \cdot M_i}{z_i \cdot F} \ \left[\text{g/cm}^2 \ \text{s}\right] \tag{9.5}$$

(molecular weight of C_2S, $M_{C2S} = 172$ g/mol; valence, $z_{Si^{4+}} = 4$; Faraday constant $F = 96{,}500$ As/mol; current density $j = 1.8 \times 10^{-4}$ A/cm^2 [4]). It follows $\dot{m}_{C_2S} = 8 \times 10^{-8}$ g/cm^2. Using the density of C_2S $\rho_{C2S} = 2.8$ g/cm^3 and converting 3600 s/h, the average thickness of the excreted (deposited) oxide layer is obtained to be 10^{-4} cm/h.

Experiments with graphite-bonded MgO-C refractory material at 1400 °C result in a better service life of the anode according to Sauerbrey [1], although there is already a noticeable loss of graphitic bonding matrix, which increases at higher voltage. At the cathode, a pronounced edge (hem) of metallic silicon is excreted (deposited) at only 2 V electrical voltage, which further reacts at the specimen edge to form a thin layer of SiC [1]. Thus, no overvoltage is observed with respect to the excretion of metallic silicon. The graphite matrix of the cathode is stable. The periclase inside the specimen reacts further to form melilite (Ca$_2$ (Mg,Al)(Si$_2$O$_7$)) and forsterite (2MgO·SiO$_2$) without any immediately noticeable contact with the liquid slag. It is assumed that, as a consequence of the reduction of the MgO to gaseous {Mg} and of the (SiO$_2$) to the suboxide {SiO} by the graphitic binder phase, the gas transport into the interior of the specimen takes place, where forsterite (M2S) is formed from it [1]. Only the application of 8V leads to a stable layer of dicalcium silicate in the slag, comparable to the use of MgO without graphite bonding [1].

Experiments at 1600 °C show no reduction in corrosion in either case, although the slight enrichment of C_sS is observed after only 10min [1]. The large distance of the slag composition from the 1600 °C isotherm (Fig. 9.3) causes the rapid dissolution of both electrodes despite the applied voltage.

The supposed "cathodic" protection [1] in the present case is, therefore, based only on the close proximity of the slag composition to the excretion limit of the oxides (C_2S) and not on the electrochemical reaction (refer to Fig. 9.3).

In a further series of experiments, fayalite slag (2FeO·SiO$_2$) with a melting temperature of 1205 °C [1, 6], likewise olivine (2(Mg,Fe)·SiO$_2$) [1], was used. The cathode wears less than the anode. A voltage of 1.5V is already sufficient for cathodic protection. Increasing the voltage does not affect corrosion. At this voltage a current of approx.5A flows. The current density is about 4.5 A/cm^2. The specific electrical conductivity of the slag is approx. 1 $(\Omega \ \text{cm})^{-1}$ and increases steeply with increasing FeO content [2]. The conceivable amount of electron conduction does nothing for the ion transport. The comparison of the phase analysis between the areas of the cathode and the voltage-free "reference specimen" shows no significant difference. However, slag adheres strongly to the cathode, which is explained by an increase in viscosity [1, 6]: Iron droplets are formed cathodically, which solidify at approx. 1250 °C, form a network in the liquid slag and, thus, possibly increase its viscosity [1, 6]. In addition, the excretion of metallic iron increases the silicate content (limited locally), which drastically increases the viscosity [2]. If the voltage required for the excretion of iron is also calculated according to Nernst, the result is an EMF $= 0.87$ V (see above). This value is just below the 1.5V required for electrolysis. The comparatively low overvoltage is explained by the good electrical conductivity of the slag with a high content of FeO [1].

Network formation at the anode, for example according to the reaction $2SiO_4^{4-} = Si_2O_7^{6-} + 0.5O_2 + 2e$, is also mentioned as a possibility to increase viscosity [1, 6]. Indeed, silica forms a variety of complex structures. An example with ascending electronegativity is SiO_2^0, $Si_2O_5^{2-}$, $Si_2O_6^{4-}$, $Si_2O_7^{6-}$, SiO_4^{4-}. With increasing number of the cation Si^{4+} one finds, among others, $Si_2O_7^{6-}$, $Si_3O_9^{6-}$, $Si_6O_{15}^{6-}$ ... $Si_nO_{2n+3}^{6-}$.
... $n = 2, 3, 4$... Which silicate anions are present in the slag in each case cannot be specified, which is why usually in the reactions, representative for all, the most simply ones constructed are utilized.

In summary it turns out that the cathodic increase of the corrosion resistance of refractory material by means of an electrochemical reaction is conceivable in selected cases but is not carried out here. In operational practice, however, it is likely to be difficult to meet and fulfill the necessary boundary conditions. The production conditions are in fact determined by the metallurgical requirements, which demand large equilibrium distances in order to increase the turnover. In secondary metallurgy, dolomite additions have been used for some time now to reduce the wear of the refractory material based on MgO [10]. Provided that the technical problems of polarization of the refractory material were solved, it is nevertheless unlikely to achieve a significant reduction of corrosion in the slag area by electric fields [9].

9.3 Avoidance of Clogging in the Submerged Nozzle

9.3.1 Prevention of Alumina Excretion

Clogging is the progressive reduction over time of the free cross section of a submerged nozzle when casting a steel killed with, for example, aluminum [Al]. The casting performance decreases until the casting operation is forced to stop. By using calcium and aluminum together for deoxidation, the formation of a liquid slag of approximate composition ($12CaO \cdot 7Al_2O_3$) can be accomplished, reducing the tendency to clogging (refer to Fig. 9.3).

Since this process cannot be applied to all alloys, the idea was to lower the oxygen activity on the inner side of the flow through area of the submerged nozzle to such an extent that no alumina is formed or the alumina already present goes into solution [11].

The chemical excretion reaction of deoxidation of liquid iron is as follows:

$$3[O] + 2[Al] = \langle Al_2O_3 \rangle; \quad \Delta G^\circ = +1.22 \times 10^6 - 393 \cdot T \text{ [J/FU]}. \tag{9.8}$$

If the concentrations are substituted by activities, the following applies for the deoxidation constant

$$a_{[O]}^3 \cdot a_{[Al]}^2 = K' = e^{-\frac{\Delta G^\circ}{RT}} \tag{9.9}$$

At 1800 K, i.e., just above the liquidus temperature, it follows that with $\Delta G^\circ = +5.13 \times 10^5 \, \text{J/FU}$

$$K' = e^{-\frac{5.13 \times 10^5}{8.31 \cdot 1800}} = 1.31 \times 10^{-15} \tag{9.10}$$

If 0.05% [Al] is dissolved in the steel, $8 \times 10^{-5}\%$ [O] is in equilibrium with it. In practice, however, the melt is always supersaturated with Al_2O_3, so that alumina forms on the submerged nozzle because since nucleation inhibition is lowest there (refer to Sect. 13.2.2.3).

Usually, the submerged nozzle material consists of alumina-graphite, as its porosity is low, thermal shock resistance is sufficiently high, and there is no wetting by liquid steel and casting slag. However, graphite conducts electrons. Consequently, oxygen activity can be influenced electrochemically with this material only as long as the oxide grains are in contact with each other, i.e., with much smaller graphite additions than the usual 30% by weight.

The electrolyte required is a material that conducts oxygen ions as perfectly as possible ($t_{ion} = 1$). A suitable material is zirconium dioxide, which has a disturbed anion partial lattice. By incorporating calcium ions, for example, the desired disorder is increased and the high-temperature cubic phase is stabilized [7]. This material is suitable from an electrical point of view but has poor thermal shock resistance.

Thus, if the submerged nozzle is made out of ZrO_2–CaO without the use of graphite, the oxygen potential can be influenced by setting a different electrical potential ϕ_α on the inner side through which the flow passes than on the outer side, i.e., at the mold ϕ_β, for example. This is formally:

$$\alpha |Fe_\alpha| Fe(li)_\alpha |[O]_\alpha \| ZrO_2 - CaO \| Al_2 \overset{[O]_\beta}{\overbrace{O_{3\beta} - [Al]_\beta}} |Fe(li)_\beta| Fe|_\beta \beta \tag{9.11}$$

Fe denotes two iron wires which are submerged or frozen in the melt. If their temperature is the same, there is no thermoelectric voltage between them. The following isothermal reaction chain is obtained:

$$
\begin{aligned}
2e_\alpha &= 2e_{\mathrm{Fe}\alpha} \\
2e_{\mathrm{Fe}(\alpha)} &= 2e\mathrm{Fe}(li)\alpha \\
2e\mathrm{Fe}_{(li)\alpha} + [\mathrm{O}]_\alpha &= \mathrm{O}^{2-}_{(\mathrm{ZrO_2-CaO})} \\
\mathrm{O}^{2-}_{(\mathrm{ZrO_2-CaO})} + 3/2[\mathrm{Al}]_\beta &= 1/3\mathrm{Al_2O_3}\,_\beta + 2e\mathrm{Fe}_{(li)\beta} \\
2e\mathrm{Fe}_{(li)\,\beta} &= 2e_{\mathrm{Fe}\beta} \\
2e_{\mathrm{Fe}\beta} &= 2e_\beta \\
\hline
\text{Total}: \quad 2e_\alpha + [\mathrm{O}]_\alpha + 2/3[Al]_\beta &= 1/3Al_2\mathrm{O}_{3\beta} + 2e_\beta
\end{aligned}
\tag{9.12}
$$

With the help of Nernst's equation, the difference of the electric potential $\phi_\beta - \phi_\alpha$ can be calculated (refer to Sect. 2.1.4.2):

$$
z \cdot F\left(\phi_\beta - \phi_\alpha\right) = \sum v_i \mu_i
\tag{9.13}
$$

with

$z = 2$, valency

$v_i =$ stoichiometric factors of the chemical reaction

$\mu_i = \mu_i^o + RT \cdot \ln a_i$, chem. potentials of the components (Eq. 9.12), divided into base potential μ_i^o and residual potential RT-lna_i.

$F = 9.56 \times 10^4$ J/mol $\cdot$ Volt, Faraday constant

Consequently, one obtains from (9.12) and (9.13):

$$
2F(\varphi_\beta - \varphi_\alpha) = \underbrace{1/3\mu^0_{\mathrm{Al_2O_3}} - \frac{2}{3}\mu^0_{[\mathrm{Al}]} - \mu^0_{[\mathrm{O}]}}_{1/3\Delta G^\circ(3.1.1) = -1.71\times10^5 \text{ J/FU bei 1800 K}} - RT \ln a_{[\mathrm{O}]}
\tag{9.14}
$$

i.e.,

$$
\varphi_\beta - \varphi_\alpha[V] = \frac{-1.71 \cdot 10^5}{2 \cdot 9.56 \times 10^4} - \frac{8.31 \cdot 1800}{2 \cdot 9.56 \times 10^4} \cdot \ln a_{[O]} = -0.893 - 0.078 \ln a_{[O]}
\tag{9.15}
$$

For $a_{[\mathrm{O}]} = 1.08 \times 10^{-5}$, the electrical voltage is $\phi_\beta - \phi_\alpha = 0$ V. If the oxygen potential a_O is to be lower, then $\phi_\beta - \phi_\alpha = 0$V, i.e., $\phi_\beta > \phi_\alpha$. If, for example, the mold was grounded $\left(\phi_\beta = 0\right)$, ϕ_α would have to be the negative pole.

While alumina deposition is prevented on the inner side of the submerged nozzle through which the flow passes, it excretes on the outer side, where there is supersaturation of Al_2O_3. The oxygen is "pumped" from the inside to the outside by the applied voltage against its chemical potential gradient.

If we apply the voltage $\phi_\beta - \phi_\alpha = 1$ V, the oxygen potential on the inside $\left(a_{[\mathrm{O}]}\right)_\alpha = 3 \times 10^{-11}$, i.e., Al_2O_3 should dissolve completely. It should be noted that in practical operation contact resistances and contact potentials may occur at the contact surfaces. Consequently, it is necessary to experimentally test which applied voltage is successful. If the polarity were reversed, a liquid $FeO \cdot Al_2O_3$ slag would result at $\varphi_\beta - \varphi_\alpha = -1$ V. The submerged nozzle would be purified but the purity of the steel would decrease.

9.3.2 Dissolution of the Al₂O₃ Excretion

If alumina has already been deposited on a submerged nozzle, the question arises as to the extent to which it can be dissolved electrochemically. For this purpose, we again consider an oxygen ion conductor based on doped zirconium dioxide. It transports the oxygen ions, which are formed at the melt/electrolyte phase boundary, through the electrolyte to the outside:

$$[O] + 2e = (O)^{2-}_{ZrO_2} \tag{9.16}$$

According to Faraday, the mass transported by the electric current I [A] is calculated to be

$$m_{(O^{2-})} = \frac{M_{(O^{2-})}}{z \cdot F} \cdot I \cdot t \ [g] \tag{9.17}$$

The mass of dissolved alumina (corundum) is

$$m_{Al_2O_3} = m_{(O^{2-})} \cdot \frac{M_{Al_2O_3}}{3 \cdot M_{(O^{2-})}} \ [g]. \tag{9.18}$$

If the current density j [A/m^2] is known, the layer thickness of dissolved corundum follows:

$$d_- = \frac{j \cdot t \cdot M_{Al_2O_3}}{3 \cdot z \cdot F \cdot \rho_{Al_2O_3}} \ [m] \tag{9.19}$$

At the same time, oxygen diffuses from out the melt to the phase boundary and precipitates there as Al_2O_3 if the low oxygen potential cannot be maintained electrochemically:

$$d_+ = 10^{-4} \cdot \beta \cdot \Delta c \cdot t \cdot \rho_{Fe}/\rho_{Al_2O_3} \ [m]. \tag{9.20}$$

The following numerical values are typical: $M_{Al_2O_3} = 0.1$ kg/mol, $\rho_{Al_2O_3} = 4 \times 10^3$ kg/m^3, $z = 2$, $F = 96{,}500$ A s/mol, $\beta = 3 \times 10^{-4}$ m/s (mass transition coefficient of oxygen dissolved in liquid iron), $\Delta c = 10^{-3}\%$ by weight (oxygen dissolved in liquid iron), $\rho_{Fe} = 7 \times 10^3$ kg/m^3, $\rho_{el(ZrO_2)} = 10^3$ Ω m/7/, $E = 100$ V/m and $j_{(ZrO_2)} = E/\rho_{el(ZrO_2)} = 0.1$ A/m^2.

With these numerical values, in one hour (3600 s) the decomposition of the already excreted alumina layer $d_- = 0.02$ μm is calculated, and the layer to be formed to $d_+ = 0.2$ μ. Consequently, electrochemical dissolution of the alumina excretion formed is not to be expected in practice [9].

9.4   Determination of Oxygen Content in Molten Steel

9.4.1   Experimental Setup

In steel mill practice, the dissolved oxygen content of the melts is determined electrochemically on a daily basis. It depends on the composition of the alloy, the type of deoxidation and the reactions with the refractory material. At the same, the total oxygen content is determined by combustion analysis in the graphite crucible. The difference is a direct measure of the purity of the melt, in which oxides may be present in emulsified or suspended form. Figure 9.4 shows this correlation in a laboratory experiment, the deoxidation with manganese: 20–30 ppm of the oxygen is bounded to oxides, the purity of the oxide is poor [12].

For the targeted examination of the reactions of refractory materials with steel melts, the use of electrochemical oxygen determination is a very effective tool because the lining consists for the most part of oxides. Quantitative conclusions about the thermodynamic and kinetic processes can be drawn from the change in the oxygen content of the melt [13, 14].

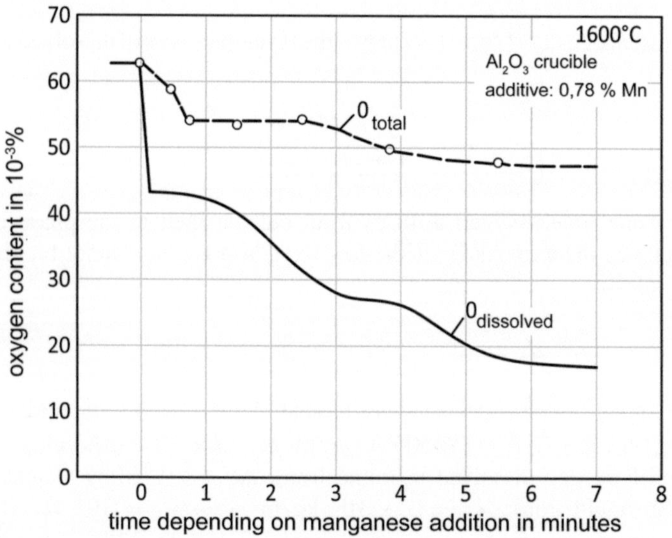

Fig. 9.4   Time Progression of the deoxidation of liquid iron with manganese [12]

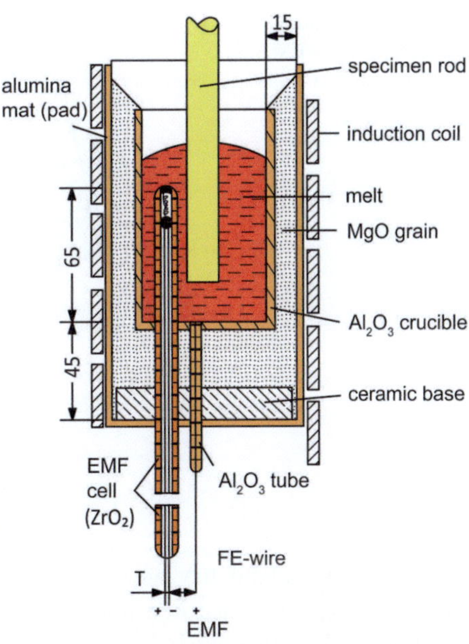

Fig. 9.5 Test setup for examination metallurgical reactions in the induction crucible [15, 16]

Figure 9.5 shows a typical test setup. The refractory material to be tested is immersed as a rod in the melt. It should be noted that it reacts not only with the melt but also with the crucible. Compatibilities can be investigated in this simple way [15]. The change in buoyancy of the immersed body additionally allows us to make some conclusions about the kinetics of dissolution [13, 15].

Whereas in the steel mill commercial probes are used to measure the oxygen content, which can no longer be used after only 30 s, service lives of several hours are required in the laboratory. The construction and setup of such a measuring cell requires experience and is, therefore, described briefly.

Partially stabilized zirconium dioxide is used as the semiconductor for the oxygen ions. Stabilization as a cubic fluoride lattice is achieved by substitution of approx. 10% by weight calcium, magnesium or yttrium. In addition, this increases the number of vacancies in the partial lattice of oxygen ions [17, 18]. Preferably, tubes closed on one side with an outer diameter of 8 mm and 1mm wall thickness are used. The thermocouple used is Pt–PtRh, type B (PtRh10/PtRh30) with a wire thickness of 0.5 mm. The welding bead must have a diameter of at least 1.5 mm, as this is the only way to ensure that the measurement is stable over time. The high-alloy (mechanically stiffer, negative) leg is used as the discharge for the electrochemical potential of the cell because a voltage of 2 mV, largely independent of the temperature, prevails between it and the iron discharge (positive) frozen into the crucible base, which does not noticeably influence the measurement and is taken into account upon completing the evaluation [13].

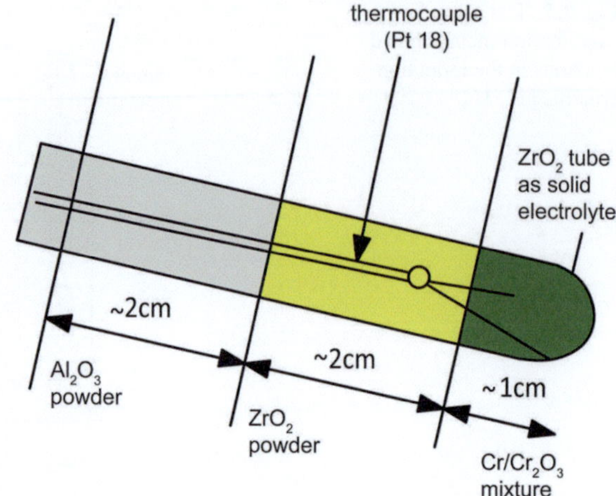

Fig. 9.6 Schematic illustration of the EMF measuring cell [14]

Figure 9.6 explains the internal structure of the cell [13, 14]: The wires of the thermocouple are guided in a thin tube out of sintered corundum with two capillary holes. The PtRh leg is led to the bottom of the tube and twisted there. It is surrounded by a densely rammed mixture out of very fine chromium oxide (Cr_2O_3) and the usually somewhat coarser chromium metal in a weight ratio of 1:10. The thermal bead must not be in contact with the chromium, otherwise the temperature measurement will change over time. The bead is rammed all the way around very densely with the powder of the same type from the zirconium dioxide tube. This is followed by ramming of alumina powder. The open end of the probe is sealed with high temperature ceramic adhesive. The probe is glued into the crucible bottom and protrudes approx. 6 cm into the crucible. The steel block used as the melting body has a corresponding hole. The probe is slowly heated together with the block.

In operation, the cell resistance is approx. 10^3 Ω, so that the input resistance of the connected measuring device must be greater than 10^6 Ω to achieve sufficient accuracy of the measurement. The evaluation of the measured voltages described below is conducted automatically with a connected computer.

9.4.2   Evaluation

Omitting the external copper wire inlets and outlets, the electrochemical concentration chain shown above is as follows:

$$\alpha\,|\,\langle Fe\rangle_\alpha\,|\,Fe(li)_\alpha - [O]_\alpha\,\|\,\langle ZrO_2\rangle\,\|\,\{\langle Cr_2O_3\rangle - \langle Cr\rangle\}_\beta\,|\,\langle PtRh\rangle\,|_\beta\,\beta \qquad (9.21)$$

$$
\begin{aligned}
2e_\alpha &= 2e_{Fe\alpha} \\
2e_{Fe\alpha} &= 2eFe_{(li)\alpha} \\
2eFe_{(li)\alpha} + [O] &= O^{2-}_{\langle ZrO_2 \rangle} \\
O^{2-}_{\langle ZrO_2 \rangle} + 3/2 \langle Cr \rangle &= 1/3 \langle CrO_{23} \rangle + 2e_{\langle PtRh \rangle \beta} \\
\underline{2e_{\langle PtRh \rangle_\beta}} &= \underline{2e_\beta} \\
\text{Total}: 2e_\alpha + [O] + 2/3 \langle Cr \rangle &= 1/3 \langle Cr_2O_3 \rangle + 2e_\beta
\end{aligned}
\tag{9.22}
$$

Using Nernst's Eq. (9.23), the difference in electric potential $\varphi_\beta - \varphi_\alpha$ can be calculated from this reaction:

$$
z \cdot F \left(\varphi_\beta - \varphi_\alpha \right) = \sum v_i \mu_i
\tag{9.23}
$$

with

$z = 2$, valency

$v_i =$ stoichiometric factors of the chemical reaction

$F = 96487\,\text{J/mol V}$, Faraday constant

$\mu_i = \mu_i^O + RT \cdot \ln a_i$, chem. potentials of the components (Eq. 9.22), divided into base potential μ_i^o and residual potential $RT \cdot \ln a_i$ (refer to Sect. 2.1.3.3).

From Eqs. (9.22) and (9.23), **one** obtains

$$
z \cdot F \left(\varphi_\beta - \varphi_\alpha \right)
$$
$$
= \underbrace{1/3 \mu^0_{\langle Cr_2O_3 \rangle} - \frac{2}{3} \mu^0_{\langle Cr \rangle} - 0.5 \mu^0_{\{O_2\}}}_{\Delta G^0_{Cr_2O_3}} - \underbrace{\mu^0_{[O]} - 0.5 \mu^0_{\{O_2\}}}_{\Delta G^0_{[O]}} - RT \ln a_{[O]}
\tag{9.24}
$$

with

$$
\Delta G^0_{Cr_2O_3} = -372{,}362.5 + 84.34 \cdot T \; [\text{J/0.5 mol}\{O_2\}]
\tag{9.25}
$$

and

$$
\Delta G^0_{[O]} = -137.118 + 7.79 \cdot T \left[\text{J/mol}[O]_{(1\%)} \right].
\tag{9.26}
$$

The measured cell voltage $(\varphi_\beta - \varphi_\alpha)$ is identical with the so-called electromotive force, EMF. If the measured EMF is expressed in [mV] and the temperature in [°C], the activity of the oxygen dissolved in the liquid iron is calculated as follows.

$$
a_{[O]} = 10^4 \cdot \exp \left(\frac{-23{,}222 \cdot \text{EMK} - 28{,}308.6}{T + 273} + 9.22 \right) [\text{ppm}]
\tag{9.27}
$$

In the concentration chain (9.21), a thermoelectric voltage arises between the iron discharge in the melt and the PtRh discharge of the cell, which is 2 mV independent of the temperature [13]. However, the correction of the measured EMF value by subtracting 2 mV is not always necessary. Indeed, at the measured EMF $= 0$ V, without correction we obtain $a_{[O]} = 27.5$ ppm and with the correction $a_{[O]} = 28.3$ ppm. This deviation is within the measurement accuracy. With respect to its oxygen content, the solution is highly diluted, i.e., $a_{[O]} = [O] = 28 \times 10^{-3}$ ppm.

If the iron is not pure, the thermodynamic interaction of the iron companions with the dissolved oxygen must be taken into account. The following applies

$$\ln a_{[O]} = \ln f_{[O]} + \ln[O]. \tag{9.28}$$

According to Wagner [4], the activity coefficient $f_{[O]}$ is linked to the so-called interaction coefficient $e^i_{[O]}$ and the concentrations of the dissolved components $[i]$ (refer to Sect. 2.1.3.3):

$$\ln f_{[O]} = \ln f^0_{[O]} + e^{[O]}_{[O]} \cdot [O] + \sum e^{[i]}_{[O]} \cdot [i], \tag{9.29}$$

with $\ln f^0_{[O]} = 0$ and $i = $ C, Mn, Si, Al and so on.

The dissolved oxygen concentration in [ppm] is thus calculated from Eq. (9.27) to $[O]_0 = 28 \times 10^{-3}$, and then $\ln f^0_{[O]}$ (9.29) is subtracted from it. The concentrations of the accompanying elements $[i]$ are analyzed or estimated, the effect parameters are available in tabular form (refer to Table 9.1) [19].

Table 9.1 Interaction coefficients of various iron companions related to dissolved oxygen (SGTE database, 1996)

Element	e (for conc. in % by weight)
Al	− 3.895
C	− 0.4536
Co	0.007
Cr	− 0.04
Mn	− 0.0211
Mo	0.0035
N	0.0564
Nb	− 0.1391
Ni	0.0473
O	− 0.2002
P	0.07
S	− 0.1328
Si	− 0.1308
Ti	− 0.5
V	− 0.3
W	0.009
Zr	− 0.3

Example The iron contains 0.3% by weight carbon, 2% by weight manganese and 0.1% by weight silicon dissolved. At 1600 °C the value EMF = 0 mV is measured. From Eq. (9.29) and Table 9.1, $[O]_0 = 0.0028\%$ by weight is used to calculate the following for the activity coefficient $\ln f_{[O]} = (-0.2002 \cdot 0.0028) + (-0.4536 \cdot 0.3) + (-0.0211 \cdot 2) + (-0.1308 \cdot 0.1) = -0.19$.

From Eq. (9.18), it follows: $\ln[O] = \ln a_{[O]} - \ln f_{[O]} = (-5.88) - (-0.19) = -5.69$, which corresponds to the oxygen content $[O] = 34 \times 10^{-3}$ by weight = 34ppm. Consequently, the correction is + 6 ppm of oxygen. The influence of the interaction becomes increasingly noticeable with increasing concentrations, which is particularly important to consider if evaluating deoxidation.

The half-cell used as a reference potential must supply a stable voltage. For measurements in steel melts, a mixture of Cr/Cr_2O_3 is used for this purpose. In liquid copper, Ni/NiO is often used [16–18]. Together with the half-cell used for the measurement $(2[O]/\{O_2\})$, a voltage should be present which is as close as possible to 0 mV. This minimizes the shares of electron conduction $(t_{ion} = 1)$ [7, 18]. In liquid copper, the EMF is [mV] and T [°C]:

$$a_{[O]_{Cu}} = 10^4 \cdot \exp\left(\frac{-23.222 \cdot \text{EMK} - 20{,}681}{T + 273} + 10.74\right) [\text{ppm}] \qquad (9.30)$$

At EMF = 0mV, oxygen is dissolved in the copper at 1100 °C $a_{[O]} = 133$ ppm.

For the determination of oxygen activity in gases, air with 0.21% by volume oxygen is often used as a reference potential. Then, the following applies

$$\text{EMK}[V] = -\frac{R \cdot T[K]}{4 \cdot F} \cdot \ln \frac{P''_{O_2}}{P'_{O_2}} = 2.15 \times 10^{-5} \cdot T \cdot \ln \frac{P''_{O_2}}{P'_{O_2}}, \qquad (9.31)$$

where $P'_{O_2} = 0.21$ atm forms the reference potential. In practice, measurements are often made at 750 °C because the EMF then changes by approximately 50 mV with each power of ten of the oxygen activity in the gas.

9.5 Capillary Electric Effects

9.5.1 *Molten Metal/Slag Interface (Lippmann Effect)*

In 1873, Lippmann [20] described the influence of an electrical DC voltage on the interfacial energy between an aqueous electrolyte and mercury. At the common phase boundary of both liquids an electrical double layer is formed, which has the effect of a capacitor of the capacity C $[A \cdot s/V \cdot m^2]$ As derived in Sect. 2.1.4.3, the interfacial tension σ changes parabolically with the applied electrical voltage U as follows

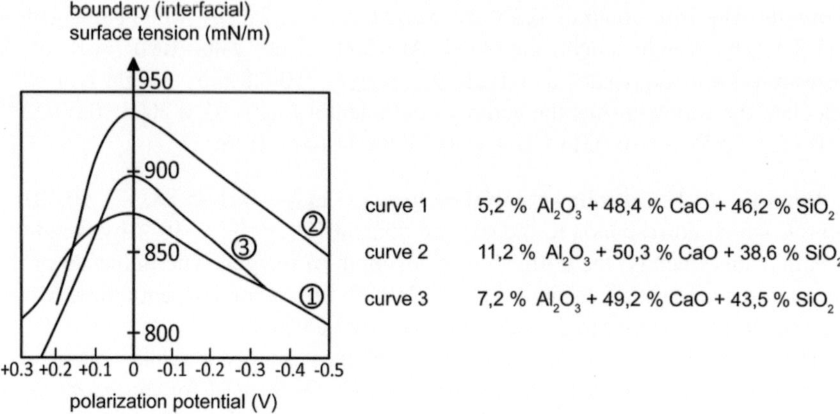

Fig. 9.7 Electrocapillary curves cast iron/oxide slags at approx. 1450 °C [21]

$$\sigma - \sigma_0 = -\frac{C}{2}U^2 \left[\frac{N}{m} = \frac{V \cdot A \cdot s}{m^2} \right] \tag{9.32}$$

In practice, the diffuse double layer and the specific adsorption of ions on the metal surface play a not to be neglected role, which is why the parabolic curve is only observed approximately. In addition, the applied voltage U requires a comparative potential, i.e., a relative value which need not be zero at the maximum of the Lippmann curve $\sigma(U)$.

In principle, these considerations also apply to the liquid steel/slag system because steel is a pure electron conductor and slag in most cases an ion conductor. Measurements [21] confirm this.

For example, Fig. 9.7 shows three electrocapillary curves of cast iron (Fe—2.6% C) over against three similar slags in the system Al_2O_3–CaO–SiO_2 at 1450–1480 °C [21].

The asymmetry of the curves is clearly visible. The maxima σ_0 are slightly shifted compared to $U = 0$ V.

The contact (wetting) angle θ can be calculated if a value typical for cast iron, e.g., $\sigma_M = 1290$ mN/m constant, is used for the surface tension of the liquid metal [22]. A typical average value of $\sigma_S = 510$ mN/m [2] is also used for slags. The equation provided by Th. Young connects the quantities with each other (M = metal, S = slag):

$$\sigma_M = \sigma_{MS} + \sigma_S \cdot \cos\theta \ [N/m] \tag{9.33}$$

In this way, the interfacial tensions between slag and iron (Fig. 9.7) are converted into wetting angles. It can be seen that with increasing polarization voltage $\pm$ U, the wetting angle decreases because σ_{MS} decreases.

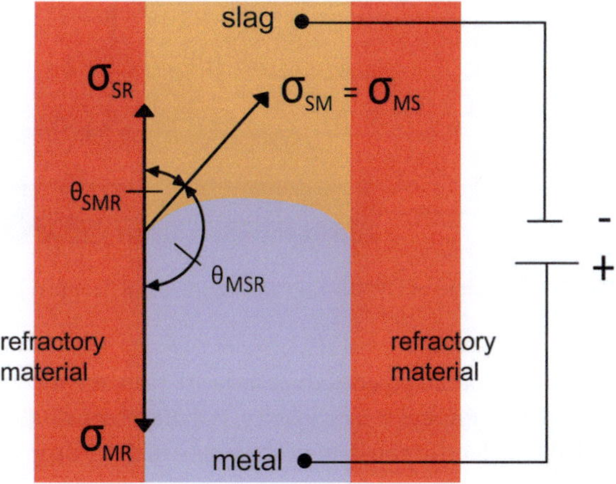

Fig. 9.8 Schematic illustration of the measurement of a capillary curve in a pore filled by metal and slag

The wetting angle between cast iron and slag is calculated as $\theta < 90°$ in the example. With applied stress, it decreases further. Thus, the wetting is enhanced further.

If one asks to what extent the **Lippmann effect** can influence the corrosion of refractory material, the "experimental setup" shown in Fig. 9.8 must be considered. It is a prerequisite for the change in the wetting angle of the liquid phases with the pore wall, which is supposed to be an insulator and the resulting capillary movement. This technical prerequisite is probably hard to accomplish in practical operation so that no influence of an externally applied electrical voltage on the corrosion can be expected.

9.5.2 Electrical Voltage—Assisted Wetting

The wetting angle θ [°] of an electrically conductive liquid drop lying on a dielectric can be changed by applying a voltage U [V] [23] (Fig. 9.9):

$$\cos\theta(U) = \cos\theta(U=0) + \frac{\varepsilon_r \cdot \varepsilon_0}{2 \cdot \sigma_{l_i}} \cdot \frac{U^2}{d} [-] \tag{9.34}$$

Equation (9.34) is derived from the Lippmann-Helmholtz equation (Eq. 9.32), taking into account Young's Eq. (9.33), where the capacity is $C = \varepsilon \cdot A/d$ [20]. The recharge of capacitor C occurs at the instant of application of an external voltage. Equation (9.34) states that the sign of the applied voltage has no influence, i.e., alternating voltage can be used since the voltage is in the form of a square. Only the thickness d of the supporting pad matters because its capacity is much lower than

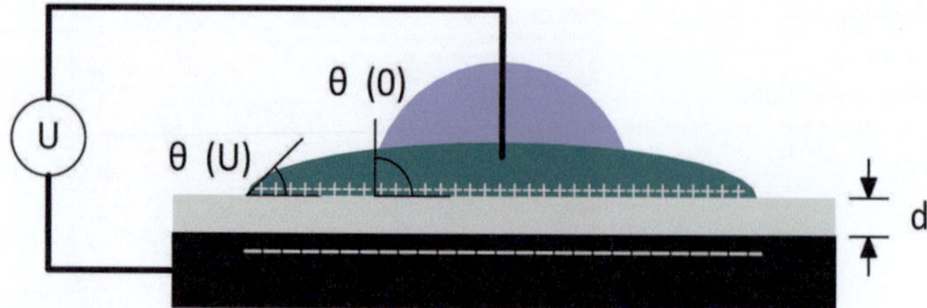

Fig. 9.9 Principle of EWOD (electrowetting on dielectrics) [23]

that of the double layer of thickness 10^{-10} m and the reciprocals of the capacities add up to the reciprocal total capacity. If a dielectric drop lies on a conductive supporting pad, the dielectric constant of the liquid and the drop radius are to be put in Eq. (9.34) [24].

The dielectric constant $\varepsilon = \varepsilon_r \cdot \varepsilon_0$ of the refractory material is of the order 10^{-10} A·s/V·m. The surface tensions of slag and iron are $\sigma_S \approx 0.5$ N/m and $\sigma_{Fe} \approx 1.5$ N/m $= 1.5$ V·A·s/m^2. For example, 10 V is used as the external electrical voltage U. The thickness d of the dielectric supporting pad then turns out to be the decisive parameter. It must be extremely thin ($< 10^{-8}$ m $= 0.01$ μm) in order to achieve an effect. This is a circumstance that does not exist in refractory engineering applications, where wall thicknesses $d \geq 10^{-3}$ m are common. Furthermore, a dielectric drop on a conductive supporting pad must be very small to achieve an effect. These conditions do not exist in practice. Consequently, no influence of an applied electrical voltage on corrosion is to be expected [25].

In a recent paper, Monaghan et al. [26] describes the increase in the corrosion rate of dense MgO measured in the laboratory. For this purpose, an electrical voltage is applied at 1450 °C between the immersed MgO rod and the liquid slag (in % by weight: 46% CaO, 46% SiO$_2$, 3% CaO, 4% Fe$_2$O$_3$, 1% FeO). As justification, they state that the DC voltage of $- 0.45$ V changes the surface tension of the liquid slag, which increases the Marangoni convection at the refractory material/slag/air three-phase line. This increases the mass transfer in time. In a second slag (50% CaO, 10% Al$_2$O$_3$, 40% SiO$_2$), this effect is not observed.

The individual results reported [26] are not sufficient for a conclusive assessment. It is not reported whether the applied electrical voltage increases or decreases the surface tension of the slag. Among other things, the velocity of the slag is assigned to the change in surface tension σ with voltage U, i.e.:u prop$\cdot \frac{\partial \sigma}{\partial U} \cdot \frac{\partial U}{\partial x}$. Consequently, the voltage U must change along the slag/air phase boundary in order to influence the velocity u. In addition to the applied voltage, the design of the experimental setup likewise influences the result.

References

1. Sauerbrey, R.K.: Corrosion behavior of refractory materials under the influence of electric fields. Dr.-Ing. Dissertation, Montan-Universität Leoben, Austria (2009)
2. Slag Atlas, VDEh, Verlag Stahleisen GmbH (1995)
3. Hack, K.: FACTSage. Thermfact & GTT Technologies (2007). www.factsage.com
4. Frohberg, M.G.: Thermodynamics for Materials Engineers and Metallurgists, 2nd edn. Deutscher Verlag für Grundstoffindustrie, Leipzig (1994)
5. Pawlek, F.: Metallurgy. Walter de Gruyter, Berlin (1983)
6. Sauerbrey, R.K., Mori, G., Majenovic, Ch., Harmuth, H.: Application of electric voltages for improved protection of MgO refractory materials. Refractories Worldforum **2**(1), 105–110 (2010)
7. Fischer, W.A., Janke, D.: Metallurgical Electrochemistry. Verlag Stahleisen, Düsseldorf (1975)
8. Müller, R.H.: Zeta-Potential of Particle Load in Lab Practice, Wissenschaftliche Verlagsgesellschaft m.b.H., Stuttgart, Band. 37 (1996)
9. Pötschke, J.: The influence of electrical forces on the corrosion of refractory materials. Refractories Manual 10–26 (2008)
10. Reisinger, P., Presslinger, H., Hiebler, H., Zednicek, W.: MgO-solubility in steel mill slags. BHM - Berg- und Hüttenmännische Monatshefte **144** (Nr. 5), 196–203 (1999)
11. Kendall, M.: Patent Specification DE 101 32575 C1: "Refractory Drain", Heraeus Electro-Nite, Patent Department, 4 July 2002
12. von Bogdandy, L., Förster, E., Klapdar, W., Richter, H.: Application of deoxidation kinetics on manufacturing of semi-killed steels. Stahl Eisen **13**(89), 704–109 (1969)
13. Markert, J.: Gravimetrical and electrochemical examination of the reaction between silicic acid and corundum in deoxidized iron melts. Dr.-Ing. Dissertation, RWTH-Aachen, Germany (1977)
14. Ollig, M.: Reactions of carbon-containing MgO-materials with liquid iron Eisen. Dr.-Ing. Dissertation at RWTH-Aachen, Germany (2000)
15. Hauck, F., Markert, J., Pötschke, J.: The reaction between silicic and alumina in carbon-deoxidized iron melt. Archiv Eisenhüttenwesen **5**(50), 189–193 (1979)
16. Maier, R., Pötschke, J.: Reaction of technical silicon carbide with oxygen-containing copper melts. Metall **11**(32), 1109–1111 (1978)
17. Middendorf, H.-W., Pötschke, J., Frohberg, M.G.: Problems and experience regarding application of solid electrolyte in galvanic cells. Metall **6**(24), 617–624 (1970)
18. Hohlfeld, J., Janke, D.: Layer ceramic sensors for supervision of metallurgical melting and refining processes. Veitsch-Radex Rundschau **1**, 40–56 (1999)
19. Schenck, H., Steinmetz, E.: Effective Parameters of Accompanying Elements of Liquid Iron Solutions and Their Interrelationships. Verlag Stahleisen m.b.H. Düsseldorf, Germany (1966)
20. Pogg. Ann. **149**, 546 (1873)
21. Slag Atlas, Verlag Stahleisen, 509 ff (1995)
22. Pohl, D., Scheil, E.: Surface tension on cast iron. Gießerei **43**(26), 833–883 (1956)
23. Mönch, W., Krogmann, F., Zappe, H.: Variable focal length through liquid micro-lenses. Photonik **4**, 44–46 (2005)
24. Shapiro, B., Moon, H., Garrell, R., Kim, C.J.: Equilibrium behavior of sessile drops under surface tension, applied external fields and material variations. J. Appl. Phys. **93**, 5794–5811 (2003)
25. Pötschke, J.: Does electrowetting influence slag infiltration? In: Refractories for metallurgy, 51st International Colloquium on Refractories, Aachen, Germany, 144–146 (2008)
26. Monaghan, B.J., Nightingale, S.A., Dong, Q., Funcik, M.: The effects of an applied voltage on the corrosion characteristics of dense MgO. Engineering **2**, 496–501 (2010)

Chapter 10
Structural Damage

The exceedance of strength of a refractory material, due to excessive stress from the outside, is not discussed here, but that due to internal stresses. Causes are as follows:

1. New phase formation (crystallization pressure).
2. Phase transformation (transformation pressure).
3. Thermal expansion.
4. Moisture in the refractory material.

10.1 New Phase Formation (Crystallization Pressure)

Refractories can decompose as a result of the formation of a new phase with an increase in volume. Known examples are as follows:

- Decomposition in CO-atmosphere by graphite precipitation.
- Decomposition of doloma as a result of moisture.
- Decomposition of mullite by formation of nepheline.
- Decomposition due to uncontrolled formation of spinel or mullite.

The mechanism is explained by means of some examples (refer to Sect. 2.1.3.2):

The thermodynamic stability of a phase is described by its free enthalpy of formation (refer to Sect. 2.1.3):

$$\Delta G = \Delta G^0 + RT \ln K \ [\text{J/FU}]. \tag{10.1}$$

(FU = formula conversion). The chemical reaction is, e.g.

$$A + 2B = AB_2, \quad T = \text{const.} \tag{10.2}$$

J. Pötschke, *Refractory Fundamentals in Metallurgical Practice*,
https://doi.org/10.1007/978-3-031-63709-4_10

Its equilibrium constant is

$$K_{AB_2} = \frac{a_{AB2}}{a_A \cdot a_B^2}.$$ (10.3)

For pure substances, $a_{AB_2} = a_A = a_B = 1$: For a gas A, $P_A = 1$ bar would be the standard state. ΔG^0 is calculated from table values, e.g. Knacke et al. [1] or by fact-sage [2] according to Eq. (10.2) from the pure substances to

$$\Delta G^0 = G_{AB_2}^0 - \left(G_A^0 + 2G_B^0\right).$$ (10.4)

At equilibrium, $\Delta G = 0$ (Eq. 10.1). If the existing activities or partial pressures present are no more 1, the free enthalpy of formation ΔG^0 is calculated from the equation.

$$\Delta G^0 = - RT \cdot \ln K \ [J/FU].$$ (10.5)

The equilibrium constant K is calculated from Eq. 10.3. Consequently, K, $G\Delta^0$, a_i, and p_i are equivalent quantities for describing equilibrium.

If the pure, solid phase AB_2 comes to excretion, its activity is in equilibrium $a_{AB_2} = 1$. If, however, excretion is impaired, e.g. by delayed nucleation or an environment which mechanically hinders growth, at least one of the components A or/and B may accumulate in the mother phase until the excretion of AB_2 is forced as a result of its so-called supersaturation. Then, the activity exceeds the value 1 and possibly considerably: $a_{AB_2} \gg 1$.

In the case of growth impairment, e.g. by the crystal material surrounding the nucleus, the increasing activity a_{AB_2} leads to often substantial crystallization pressure π, which the growing phase AB_2 exerts on its surroundings. This pressure resembles the (suppressed) mechanical work $(\pi - 1) \cdot (V_{AB_2} - V_A - 2V_B)$ and is added to the free enthalpy of formation ΔG° to calculate the equilibrium constant K_{AB_2} of the supersaturated state [3]:

$$\Delta G^\circ + (\pi - 1) \cdot \left(V_{AB_2} - V_A - 2V_B\right) \cdot R_1/R_2 + R_1 T \ln K_{AB_2} = 0$$ (10.6)

For crystallization pressure $\pi = 1$ atm, equilibrium is obtained without supersaturation of AB_2, i.e. for $a_{AB_2} = 1$. $R_1 = 8.31$ J/mol K and $R_2 = 84.8$ atm·cm^3/mol·K are the ideal gas constants. The activity of supersaturated excretion, e.g. a_{AB_2} is calculated using ΔG° and considering the modified equilibrium constant $K_{AB_2} > 1$ and Eqs. (10.3) and (10.5): $\Delta G^0 = -RT \cdot \ln(K_{AB_2})$.

From Eq. (10.6), it follows directly for the crystallization pressure π to:

$$\pi - 1 = \frac{-\Delta G^o - R_1 \cdot T \cdot \ln K_{AB_2}}{(V_{AB_2} - V_A - 2V_B) \cdot R_1/R_2} \ [atm].$$ (10.7)

Some of the following examples are calculated using Fact Sage 5.5 [2] in the reaction module: After entering the data as "non-standard states," the "Table Reaction" is executed at T under the boundary condition $\Delta G = 0$ and $P = p$ at $a = 1$. The crystallization pressure P is obtained.

Example $MgO + Al_2O_3 = MgAl_2O_4$, $T = 1400\,°C$, $\Delta V = 2.9\ cm^3/mol$, $P = 119 \times 10^3$ atm.

10.1.1 CO-Decomposition at 500 °C (CO Bursting)

Refractory material can be destroyed, e.g. at $500\,°C$ ($773\ K$) in a CO-containing atmosphere, by precipitation of solid, crystalline carbon (graphite) in the microstructure (Fig. 10.1). Parallel to this, the excretion of metallic iron is observed, which acts as a catalyst [3]. (refer to Sect. 3.1.1.1).

From literature, at $500\,°C$ and a pressure of 1 atm, for the Boudouard reaction [1, 2]

$$2CO = C + CO_2,\ \Delta G°(773\ K) = -35522\ kJ/mol. \tag{10.8}$$

Fig. 10.1 Decomposition of refractory material as a result of graphite excretion, personal communication Krause and FGF (2013)

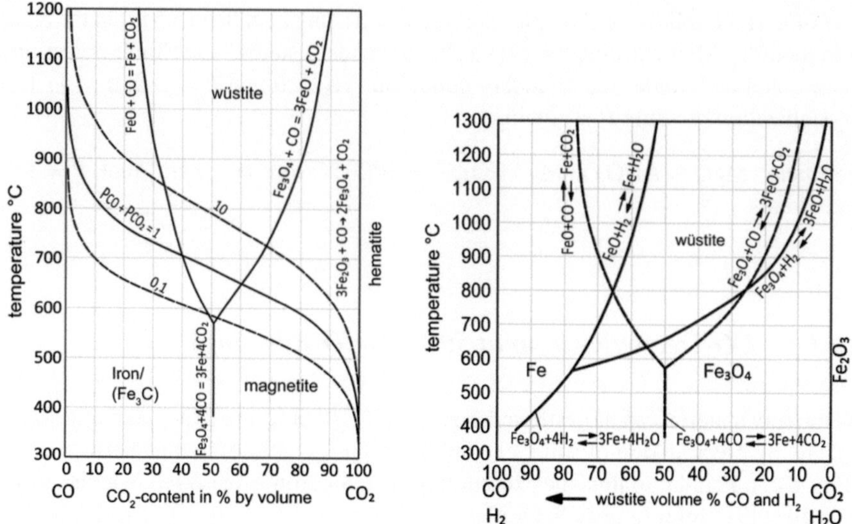

Fig. 10.2 Reduction of iron oxides in CO_2/CO– and H_2O/H_2-mixtures [4]

The Baur-Glaessner diagram (Fig. 10.2) [4] shows that magnetite (Fe_3O_4) can be immediately reduced to metallic iron (Fe) at low total pressure, where the volume shares of CO and CO_2 are just equal in size. Then, the equilibrium constant of the Reaction (10.8) for ac = 1:

$$K_c = \frac{P_{CO_2}}{P_{CO}^2} = 2.$$

(10.9)

The activity of the supersaturated carbon is calculated with

$$\Delta G° = -R_1 T \cdot \ln(K_C\, ac)$$

(10.10)

at 500 °C and 1 atm to

$$-35522 = -8.31 - 773 - \ln(2 \cdot a_c), \text{ i.e. } a_c = 126.$$

Consequently, at 500 °C and a total pressure of 1 atm, with sufficiently high carbon activity, magnetite (Fe_3O_4) is immediately reduced to iron, which can react further to form carbide. Above 705 °C, graphite at 1 atm total pressure directly reduces wüstite (FeO) to iron (refer to Figs. 10.2 and 10.12). The oxygen partial pressure is about 10^{-21} atm. As the temperature increases, the CO_2/CO ratio decreases, whereas H_2O/H_2 increases. The reducing effect of hydrogen increases compared to that of CO and predominates above 800 °C.

The molar volume of the graphite is $V = 5.3$ cm³/mol. With $R_1/R_2 = 0.1$ [J/atm · cm³] it follows from Eq. (10.7).

$$(\pi - 1) = \frac{35522 - 8.31 \cdot 773 \cdot \ln 2}{5.3 \cdot 0.1} = 58 \times 10^3 \, \text{atm} \qquad (10.7.1)$$

This crystallization pressure of the graphite destroys the structure of the refractory material.

Remark   Although the reduction potential (μ°_{O2}) of a CO_2/CO mixture below 800 °C is greater than that of a H_2O/H_2 mixture of the same composition (compare Fig. 10.2, right-hand side and 10.13), the addition of a small amount of hydrogen (approx. 3% by volume) to the CO gas accelerates carbon excretion noticeably [5]. Since the transport of CO to the reaction site determines the speed for carbon excretion [3], the effect of hydrogen must be catalytic. Therefore, J. Pötschke recommended to ISO/ TC33 WG 18 on February 2, 1996, corresponding experiments with the addition of 3% hydrogen. These were carried out at the Gesteinshütteninstitut of RWTH Aachen, Germany, by Dietrichs [6] on fireclay bricks of the grade AO with the result that the same damage is obtained in a much shorter time (approx. 1/5). Tests at Corus/ Ijmuiden and at the Luoyang Institute of Refractory Research, China, confirm this result. Nevertheless, the ISO standard has not yet been changed because of concerns about comparability with earlier results.

Investigations by Krause and Pötschke [7] show that the graphite occurs in two growth forms, as a "bucky onion" and as a "nano tube." In the first case, the graphite envelops the iron core, which acts as a catalyst, and exerts only a negligible effect on the microstructure. As a "nano tube," on the other hand, especially in large numbers, it drives the structure apart and the refractory material disintegrates. This is due to the enormous strength of this single-crystal form of growth. As Fig. 10.3 shows, an iron droplet sits at the tip of each crystal. Graphite condenses in it from the CO atmosphere, which is then incorporated into the crystal lattice on the bottom side. This is referred to as the VLS (vapor–liquid–solid) mechanism.

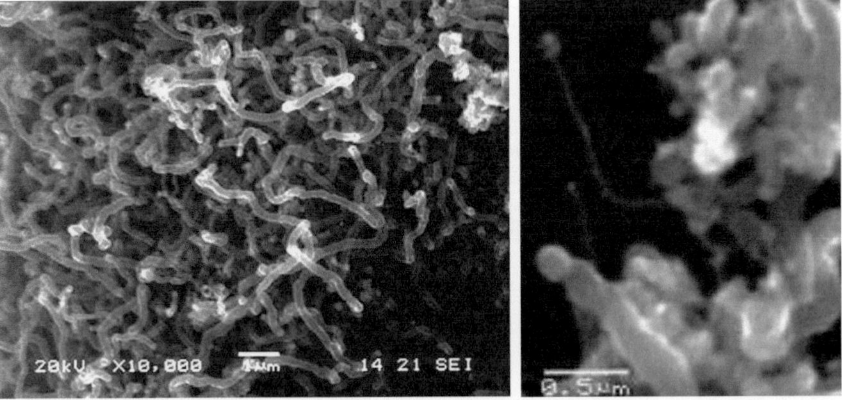

Fig. 10.3   Graphite excretion as "nano tubes" in refractory material: left side [13], right side [7]

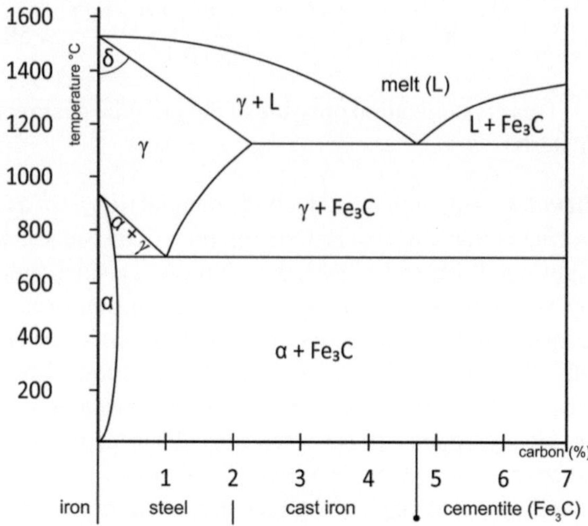

Fig. 10.4 Iron-carbon diagram (Wikipedia)

A rough calculation shows that already above 300 °C liquid iron droplets are stable. Dissolved carbon (4.3% C) drops the melting temperature of liquid iron from 1539 °C to 1147 °C at the eutectic point (Fig. 10.4). Thereby, the activity of the dissolved carbon is just one.

The solidus and liquidus lines of a phase diagram are calculated from the equation of Clausius—Clapeyron (refer to Sect. 2.1.3, Eq. 2.1.17)

$$\ln \frac{a_{li}}{a_{\text{sol}}} = -\frac{\mu_{li}^o - \mu_{\text{sol}}^o}{RT} = \frac{\Delta H_m^o}{R} \cdot \left(\frac{1}{T_m} - \frac{1}{T} \right). \tag{10.11}$$

For the metastable system Fe − (Fe₃C), the enthalpy of fusion at 1.800 K can be estimated to $\Delta H_m = 38.7$ kJ/mol [2]. With $R = 8.31$ J/K mol, it follows for the melting temperature of the eutectic iron melt being strongly supersaturated in dissolved carbon or [Fe₃C] at the activity $a_C = 126$

$$\frac{1}{T} = \frac{1}{T_m} - \frac{\ln \frac{a_{li}}{a_{\text{sol}}}}{\frac{\Delta H_m^o}{R}} = \frac{1}{1447} - \frac{\ln \frac{1}{126}}{\frac{38,700}{8.31}} = \frac{1}{579} [\frac{1}{K}] \tag{10.12}$$

The melting point of the iron, which is very heavily supersaturated with carbon, is reduced to $T = 306$ °C. Needle growth is observed above 335 °C in the heating microscope [7].

10.1.2 Decomposition of Refractory Castable as a Result of Graphite Excretion

Clay-bonded ramming mixes and refractory castables containing organic plasticizers may warp or even be destroyed by the excretion of graphite during heating-up. Black excretions are observed just below the surface or in the core. This defect is particularly observed during improper drying.

Prerequisites are as follows:

- Organic shares
- High temperature
- Dense sintered surface
- Lack of atmospheric oxygen.

Figure 10.5 shows two examples.

Graphite is formed as a result of the decomposition of hydrocarbons (HC) to graphite and hydrogen under exclusion of air. If air were present, CO_2 and H_2O could form instead.

The decomposition ("cracking") of the hydrocarbons takes place gradually, so that an atmosphere of very different composition is produced in addition to the graphite. Aside from hydrogen, methane (CH_4) is the lightest component of a series, hydrogen. Methane decomposes increasingly with increasing temperature. Figure 10.6 compares the thermochemical stability of methane and ethane. With increasing molecular weight, increasing temperature and decreasing pressure, the stability of the hydrocarbons decreases.

Fig. 10.5 Left: ULCC based on "fused silica"; right: ramming mix (approx. 1300 °C), personal communication Krebs (2012)

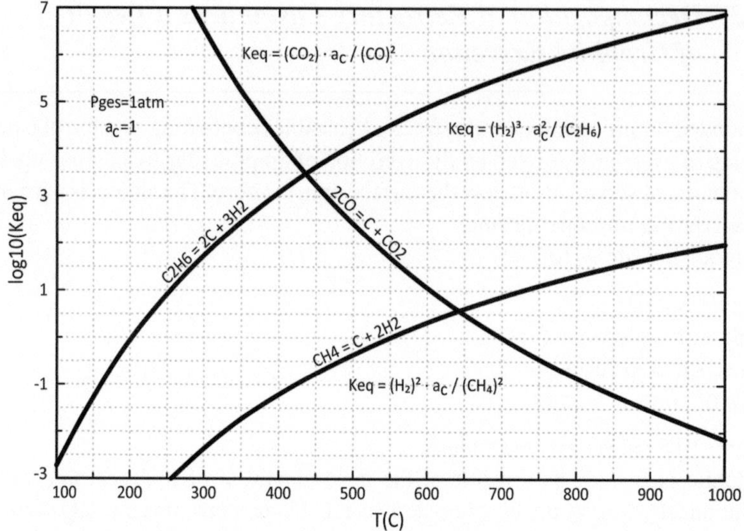

Fig. 10.6 Thermochemical stability of components C_2H_6, CH_4, and CO (FactSage)

In addition, the above-mentioned decay of carbon monoxide (CO) according to the Boudouard is shown. The tendency to decomposition decreases with increasing temperature and decreasing pressure.

The excretion of graphite weakens the structure. One example is the reaction [2]

$$\{CH_4\} = \langle C \rangle + 2\{H_2\}. \Delta G^o (1000K, 1\,\text{atm}) = -19395\,\text{J/FU} \tag{10.13}$$

$$K_{eq} = \frac{P_{H_2}^2 \cdot a_C}{P_{CH_4}} = \exp\left(-\frac{\Delta G^o}{R_1 \cdot T}\right) = \exp\left(-\frac{19395}{8.31 \times 1000}\right) = 10.3. \tag{10.14}$$

In equilibrium

$$\Delta G = \Delta G^o + R_1 T \cdot \ln K_{eq} + (1 - \pi) \cdot \Delta V_C = 0. \tag{10.15}$$

The equilibrium constant is chosen to be $K_{eq} = P_{H_2}^2 / P_{CH_4} = 1$. The supersaturation of the graphite is thus given by Eq. (10.14) to be $a_C = 10.3$. The molar volume of the graphite is $\Delta V = 5.3$ cm^3/mol and the gas constants are $R_1 = 8.31$ J/mol·K $\Delta G = \Delta G^o + R_1 T \cdot \ln K_{eq} + (1 - \pi) \cdot \Delta V_C = 0$ K and $R_2 = 84.8$ atm cm^3/mol · K. The crystallization pressure of graphite at 727 °C is therefore calculated from Eq. 10.7 to be

$$(\pi - 1) = \frac{-\Delta G^o - R_1 T \cdot \ln K_{eq}}{\Delta V \cdot R_1 / R_2} = \frac{+19395 - 8.31 \times 1000 \cdot \ln 1}{5.3 \cdot 0.1} = 36 \times 10^3\,\text{atm}. \tag{10.16}$$

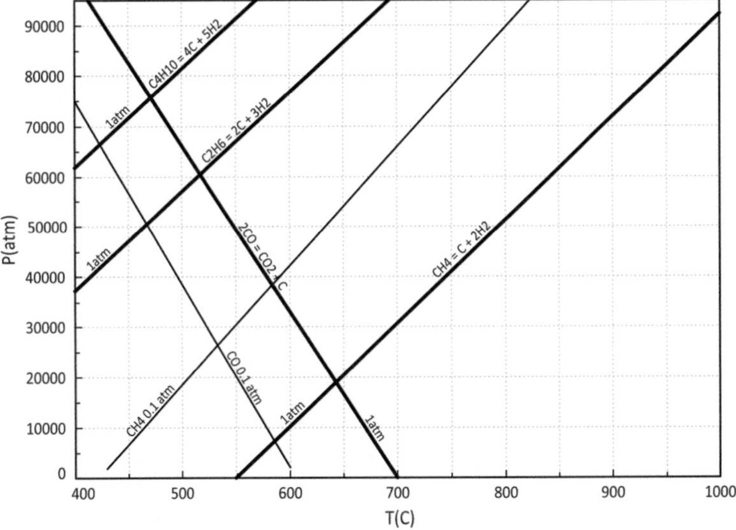

Fig. 10.7 Crystallization pressure of graphite formed from different gas components /FactSage/

The structure cannot withstand this pressure.

Figure 10.7 compares the crystallization pressures of graphite formed from different gas components as a function of pressure and temperature.

The crystallization pressure of the carbon formed from CO decomposition decreases with increasing temperature and decreasing pressure. Hydrocarbons behave in the opposite way. The crystallization pressure increases with rising temperature. The higher the molecular weight and the lower the gas pressure, the stronger the crystallization pressure. The decomposition of CO at 550 °C and of methane at 800 °C with excretion of graphite occurs at 50,000 atm each (Fig. 10.7).

10.1.3   Decomposition of Doloma (27 °C)

Doloma consists of periclase (MgO) and lime (CaO). Both components do not form a compound with each other, but solidify as eutectic. They hydrate in an atmosphere containing steam (water vapor). Both reactions are treated separately.

10.1.3.1   Formation of Brucite (Mg(OH)$_2$) at 300 K

$$MgO + H_2O = Mg(OH)_2; \Delta G° (300\,K,\ 1atm) = -35567\ J/mol. \quad (10.17)$$

ΔG° is calculated for the standard state, i.e. $a_{\mathrm{MgO}} = a_{\mathrm{Mg(OH)_2}} = 1$ and $P_{\mathrm{H_2O}} = 1$ atm from table values [1] or using FactSage [2].

At equilibrium ($\Delta G = 0$, Eq. (10.5)), the equilibrium constant is calculated from

$$K_0 = \exp\left(\frac{-\Delta G^\circ}{RT}\right) = \exp\left(\frac{+35567}{8.31 \cdot 300}\right) = 1.57 \times 10^6. \tag{10.18}$$

with

$$K_0 = \frac{1}{P^0_{\mathrm{H_2O}}} \tag{10.19}$$

It follows for the equilibrium partial pressure of steam

$$P^0_{\mathrm{H_2O}} = 6.37 \times 10^{-7}\ \mathrm{atm.}$$

The highest partial pressure available at 300 K is present above the aqueous solution and is (FactSage) $P = 0.0354$ atm. The corresponding equilibrium constant is $K = 28.2$. The activity of the supersaturated brucite is $a_B = P/P_0 = 0.0354/6.37 \times 10^{-7} = 56 \times 10^3$.

The change in volume is

$$(V_{\mathrm{Mg(OH)_2}} - V_{\mathrm{MgO}}) = 24.5 - 11.2 = 13.3\ \mathrm{cm^3/mol.}$$

Placed in Eq. (10.7), it follows with $R_1/R_2 = 0.1$ [J/atm $\cdot$ cm^3] as the crystallization pressure π:

$$\pi - 1 = \frac{R_1 T \cdot \ln P/P_O}{\left(V_{\mathrm{Mg(OH)_2}} - V_{\mathrm{MgO}}\right)R_1/R_2}$$

$$= \frac{8.31 \cdot 300 \cdot \ln\left(0.0354/6.37 \times 10^{-7}\right)}{(24.5 - 11.2) \cdot 0.1} = 20 \times 10^3\ \mathrm{atm.} \tag{10.20}$$

The formation of brucite depends on the temperature and vapor pressure (Fig. 10.8). At room temperature up to 267 °C, it is stable above a partial pressure of 5×10^{-7} to 1 atm.

With the assumption of diffusion-controlled growth, the speed of formation can be calculated. The diffusion flow of the steam in the pores is [8].

$$J_{St} = \frac{D_{St}}{\mu \cdot R \cdot T} \cdot \left(\frac{\theta_R - \theta_{\mathrm{H_2O}}}{\theta_R}\right) \cdot \frac{\Delta P}{L/2} \tag{10.21}$$

and its diffusion coefficient [8]

$$D_{St} = 3.3 \times 10^{-5} \cdot \left(\frac{T}{273}\right)^{1.8}\ \left[\mathrm{m^2/s}\right]. \tag{10.22}$$

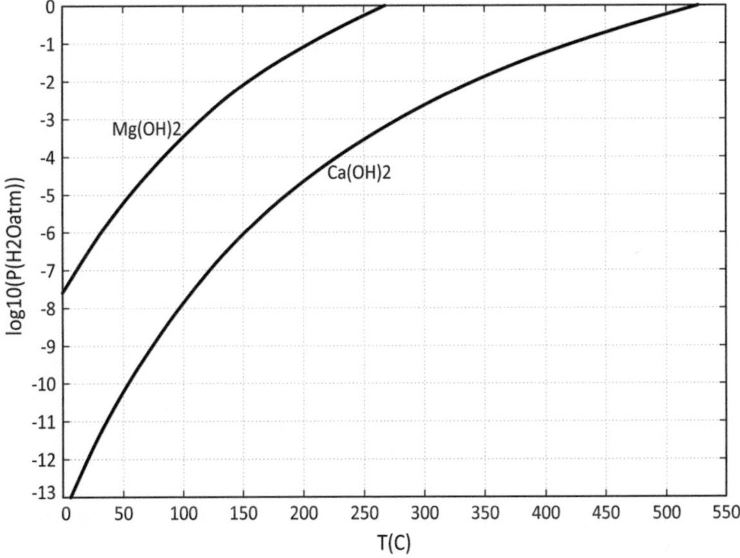

Fig. 10.8 Thermochemical stability of brucite and portlandite (Fact Sage)

The resistance coefficient of mass transport in the pores is $\mu = 10$ [8]. The specific gas constant for steam is $R = 460$ J/kg $\cdot$ K [8]. The humidity would be $\theta_{H_2O} = 5\%$ and the apparent porosity $\theta_R = 20\%$. The steam pressure of water at 100 °C is 10^5 N/m^2 and replaces ΔP. The diffusion coefficient at 100 °C is $D_{St} = 5.8 \times 10^{-5}$ m^2/s. Let the thickness of the castable be $L = 0.2$ m, its area $F = 1$ m^2. The transformation occurs from both sides towards the center. Then, the total, two-sided diffusion flux is $J_{st} = 2.5 \times 10^{-5}$ kg/s $\cdot$ m^2. Since one mole of water forms one mole of brucite, it follows that

$$J_{Br} = 3600\frac{M_{Br}}{M_{H_2O}} \cdot J_{St} = 3600\frac{58}{18} \cdot 2.5 \times 10^{-5}$$
$$= 0.3\,\text{kg/h} \cdot \text{m}^2. \tag{10.23}$$

The weight of the plate (slab) is $G = \rho_R \cdot L = 2700 \cdot 0.2 = 540$ kg/m^2. It is in the time $t = G/J_{Br} = 1.840$ h converted to brucite. At room temperature, the saturation steam pressure is only 0.035×10^5 N/m^2 and the conversion time $t = 57 \times 10^3$ h.

10.1.3.2 Formation of Portlandite (Ca(OH)$_2$) at 300 K [2]

$$CaO + H_2O = Ca(OH)_2; \quad \Delta G°(300\,K, 1\,atm) = -66235\,\text{J/mol} \tag{10.24}$$

$$P_{H_2O}^0 = \frac{1}{K_0} = \exp\left(\frac{-66235}{8.31 \cdot 300}\right) = 2.9 \times 10^{-12} \text{atm.} \qquad (10.25)$$

From FactSage, the partial pressure of steam over the aqueous solution at 300 K is calculated to be $p = 0.0354$ atm. According to Eq. (10.7), with $\Delta V = 17.4$ cm^3 / mol, the crystallization pressure turns out to be

$$\pi - 1 = \frac{8.31 \cdot 300 \cdot \ln\left(0.0354/2.93 \times 10^{-12}\right)}{(34.1 - 16.7) \cdot 0.1} = 33 \times 10^3 \text{ atm.} \qquad (10.26)$$

Since the affinity of steam for CaO is twice as great as that for MgO, the formation of portlandite is likely to be the main factor contributing to the decomposition of doloma.

The kinetics of the formation of portlandite is calculated as is for brucite.

The thermochemical stability of both oxides depends on the partial pressure of the steam and the temperature. Figure 10.8 shows the correlation.

In a steam atmosphere at 1 atm, brucite loses its stability above 270 °C and portlandite above 550 °C. Brucite formed at room temperature ($P_{H_2O} = 0.035$ atm) loses its stability above 180 °C and portlandite above 380 °C. The original oxide is formed once again with elimination (separation) of water.

Doloma must be stored at room temperature under exclusion of humidity. At service temperatures above 600 °C, doloma is thermochemically stable.

10.1.4   Decomposition of Spinel-Forming Refractory Material (1400 °C)

Very dense refractory material, which forms spinel, can disintegrate as volume expansion occurs [2]:

$$\Delta V = \left(V_{Sp} - V_{MgO} - V_{Al_2O_3}\right) = (39.7 - 11.3 - 25.5) = +2.9 \text{ cm}^3/\text{mol.}$$

$$MgO + Al_2O_3 = MgO \cdot Al_2O_3; \ \Delta G^\circ (1.673 \text{ K}) = -34.581 \text{ J/mol.} \qquad (10.27)$$

Spinel is either in equilibrium with Al_2O_3 (corundum) or with MgO (periclase). Stoichiometric spinel (72% by weight Al_2O_3) is expected. Marginal solubilities are not taken into account. In general, $K = 1$ if there is a pure phase in equilibrium with the supersaturated phase, here MgO or Al_2O_3. The supersaturation of the spinel is

$$\Delta G_{Sp}^\circ = -RT \ln a_{1Sp}, \text{ i.e.} - 34581 = -8.31 - 1673 \cdot \ln\left(a_{Sp}\right), \text{ i.e. } a_{Sp} = 12. \qquad (10.28)$$

It follows from Eq. (10.7):

$$(\pi - 1) = \frac{-\Delta G^0_{Sp}}{\Delta V \cdot R_1/R_2} = \frac{34581}{2.9 \cdot 0.1} = 119 \times 10^3 \text{ atm.} \tag{10.29}$$

This max. crystallization pressure can destroy the microstructure.

10.1.5 Decomposition of Refractory Material as a Result of Mullite Formation (1400 °C) [2]

$$3Al_2O_3 + 2SiO_2 = 3Al_2O_3 \cdot 2SiO_2, \; \Delta G^{\circ}_M (1.673 \text{ K}) = -25962 \text{ J/mol} \tag{10.30}$$

where $\Delta V = \left(V_M - 3V_{Al_2O_3} - 2V_{SiO_2}\right) = (135 - 3 \cdot 25.5 - 2 \cdot 25.7) = +7.0 \text{ cm}^3/$mol.

Mullite is in equilibrium with either SiO_2 or Al_2O_3. Ignoring edge solubilities, their free enthalpy at 1400 °C $\Delta G^{\circ} = 0$. According to Eq. 10.10 $\Delta G^{\circ}_M = - R_1 T \cdot \ln(a_M)$, the supersaturation of mullite is $a_M = 6.5$. From Eq. (10.7), it follows for the crystallization pressure of mullite

$$(\pi - 1) = \frac{-\Delta G^0_M}{\Delta V \cdot R_1/R_2} = \frac{25962}{7.0 \cdot 0.1} = 37 \times 10^3. \tag{10.31}$$

10.1.6 Decomposition of Mullite as a Result of the Formation of Nepheline (1400 °C) [2]

$$Na_2O + Al_2O_3 + 2SiO_2 = Na_2O \cdot Al_2O_3 \cdot 2SiO_2,$$
$$\Delta G^{\circ}_N (1.673 \text{K}) = - 339000 \text{ J/mol}. \tag{10.32}$$

The refractory lining consists of mullite with 75% Al_2O_3. Taking Eq. 10.7 both stoichiometric mullite and nepheline are calculated. $a_N = 24.4$.

$$\Delta V = \left(V_N - V_{Al_2O_3} - 2V_{SiO_2} - V_{Na_2O}\right)$$
$$= (108.4 - 25.5 - 2 \times 25.7 - 27.3) = +4.2 \text{ cm}^3/\text{mol}$$
$$\pi - 1 = \frac{-\Delta G^0_N}{\Delta V \cdot R_1/R_2} = \frac{339000}{4.2 \cdot 0.1} = 545 \times 10^3 \text{ atm} \tag{10.33}$$

An example is the corrosion of flat glass tanks.

10.1.7 Decomposition of Mullite as a Result of the Formation of Leucite (1400 °C) [2]

$$K_2O \ + \ Al_2O_3 + \ 2SiO_2 = \ K_2O \cdot Al_2O_3 \cdot 2SiO_2;$$

$$\Delta G_L^\circ(1.673) = -425,000\,\text{J/mol}.$$
$$\Delta V = \left(V_L - V_{Al_2O_3} - 2V_{SiO_2} - V_{K_2O}\right)$$
$$= (126.5 - 25.5 - 2 \times 25.7 - 40.5) = +9.1\,\text{cm}^3/\text{mol}. \tag{10.34}$$

It follows from Eq. (10.7):

$$\pi - 1 = \frac{-\Delta G_L^0}{\Delta V \cdot R_1/R_2} = \frac{425000}{9.1 \cdot 0.1} = 467 \times 10^3\,\text{atm} \tag{10.35}$$

10.1.8 Decomposition of Refractory Castable as a Result of the Formation of β-Corundum (1400 °C) [2]

$$Na_2O + 9Al_2O_3 = Na_2O \cdot 9Al_2O_3;\ \Delta G°K(1.673\,K)$$
$$= -182,200\,\text{J/mol}. \tag{10.36}$$

The volume change ΔV is

$$Na_2O \cdot 9Al_2O_3 - (Na_2O + 9Al_2O_3) = 297$$
$$- (27.3 + 9 \cdot 25.5) = 40.2\,\text{cm}^3/\text{mol}$$

It follows from Eq. (10.7) to get the crystallization pressure:

$$\pi - 1 = \frac{-\Delta G_K^0}{\Delta V \cdot R_1/R_2} = \frac{182200}{40.2 \cdot 0.1} = 45 \times 10^3\,\text{atm} \tag{10.37}$$

* In practice, $Na_2O \cdot 9Al_2O_3$ (FactSage) and not $Na_2O \cdot 11Al_2O_3$ is present.

10.1.9 *Oxidation of SiC by Steam (1100 °C)*

According to ASTM C 863–00, the resistance of SiC is tested in a steam atmosphere at 1100 °C for 500 h, among other conditions:

$$SiC + 3H_2O = SiO_2 + 3H_2 + CO; \quad \Delta G°(1.373\ K, 1\ atm) = -323.116\ J/FU \tag{10.38}$$

The equilibrium constant is ($a_{Cristobalite}$ and $a_{SiC} = 1$):

$$K = \frac{a_{SiO_2} \cdot P_{H_2}^3 \cdot P_{CO}}{a_{SiC} \cdot P_{H_2O}^3} = \exp\left(\frac{323116}{8.31 \cdot 1373}\right) = 1.95 \times 10^{12}. \tag{10.39}$$

If the partial pressures are each 1 atm, then $a_{SiO_2} = 1.95 \times 10^{12}$.

However, the "off-gas" would contain about 0.75 atm H_2, 0.25 atm CO and only traces of H_2O (3.8×10^{-5} atm) if the reaction proceeds stoichiometrically. The activities of SiC and SiO_2 (cristobalite) are considered as 1.

However, once the test is carried out in practice, the ratios are just the opposite, i.e. the contents of H_2 and CO are small compared to the steam used in excess.

Setting $P_{H_2O} = 0.9$ atm, $P_{H_2} = 0.075$ atm, and $P_{CO} = 0.025$ atm, then $a_{SiO2} = 1.3 \times 10^{17}$, this means cristobalite is strongly supersaturated. (The formation of tridymite is kinetically impaired.) SiO_2 is formed with the increase in volume $\Delta V = V_{SiO_2} - V_{SiC} = 25.7 - 12.5 = 13.2$ cm^3/mol. Consequently, according to Eq. 10.7, the crystallization pressure during cristobalite precipitation at 1373 K is at least

$$(\pi - 1) = \frac{\Delta G_C^0}{\Delta V \cdot R_1/R_2} = \frac{+323116}{13.2 \cdot 0.1} = 245 \times 10^3\ atm. \tag{10.40}$$

This high pressure deforms the SiC material and can lead to cracks.

10.1.10 *Corrosion of SiC in Molten Salt*

The corrosion resistance of SiC materials in molten salt (54% $CaSO_4$, 27% K_2SO_4, 6% Na_2SO_4, 8% NaCL, 5% KCl) is determined at approx. 900 °C. The salt consists mainly of calcium sulfate (anhydride) with proportions of alkali sulfates and alkali chlorides. The most important reaction is considered to be:

$$\langle SiC \rangle + (CaSO_4) = \langle CaSiO_3 \rangle + \{CO\}$$
$$+ \{S\}, \Delta G^0(1173\ K) = -339774\ J/mol\ Si \tag{10.41}$$

If the activities of the condensing phases are one, at a total pressure of 1 atm, the equilibrium constant of the reaction is

$$K = P_{CO} \cdot P_S = 0.5 \cdot 0.5 = 0.25.$$

The crystallization pressure of the calcium silicate is calculated as

$$(\pi - 1) = \frac{-\Delta G^0_{CS} - R_1 T \ln K}{\left(V_{CaSiO_3} - V_{SiC}\right) \cdot R_1/R_2} = \frac{339774 - 8.3 \cdot 1173 \cdot \ln 0.25}{(40.1 - 12.5) \cdot 0.1} = 1.3 \times 10^{+5}\,\text{atm} \quad (10.42)$$

$R_1 = 8.3$ J/mol $\cdot K$ and $R_2 = 84.8$ atm $\cdot$ ccm/mol $\cdot K$ are the ideal gas constants. This high crystallization pressure is capable of stretching and destroying the microstructure as a result of the excretion of calcium silicate.

Table 10.1 summarizes the results obtained so far. It becomes obvious that in all cases very high pressures are generated in the microstructure, which can lead to its mechanical failure. In practice, the proven method to avoid damage is to give the phase transformations enough time by slow heating-up so that the slow relaxation processes in the microstructure can proceed for a sufficiently long time.

10.1.11 Formation of Spinel in Aluminum Furnaces

Aluminum furnaces usually receive a lining of purest tabular alumina with addition of $BaSO_4$ as protection against infiltration. $BaSO_4$ reduces the wettability and thus no tendency towards infiltration. However, aluminum melts containing dissolved magnesium frequently wet the lining and creep up the wall. So-called "corundum bulbs/nodules" are formed, which are firmly attached to the wall. They can lead to its damage. They contain spinel ($MgOAl_2O_3$). The question arises as to how this reaction takes place.

Figure 10.9 shows the activity curves of the components involved in the formation of spinel in contact with the lining of an aluminum furnace at 1100 °C. The magnesium content of the melt is calculated to be increased up to %. In this process, the magnesium content of the melt is increased mathematically up to 10% by weight. The activity of the magnesium dissolved in the melt is in equilibrium with the atmosphere and, therefore, the same in both phases. In the concentration range observed, alumina (Al_2O_3) and periclase (MgO) are not stable, since their activities are less than one. Only the activity of spinel ($MgO \cdot Al_2O_3$) is equal to one. Spinel is therefore formed in the concentration range observed directly in contact with both the melt and the gas phase as well as the lining and not via the intermediate step of a formation of periclase.

The high crystallization pressure (117×10^3 atm.) of spinel (refer to Sect. 10.1.4) can destroy the microstructure of the lining if magnesium penetrates into pores or crevices of the microstructure.

Table 10.1 Compilation of thermodynamically calculated crystallization pressures in the microstructure of different oxides as a result of the formation of new phases

Product	Reaction	ΔV (cm^3 mol)	T (°C)	$-\Delta G^0$ (J/mol)	$\Pi - 1$ (atm $\times$ 10^{-3})
Graphite	$2CO = CO_2 + C$	5.3	500	31,100	58
Graphite	$CH_4 = 2H_2 + C$	5.3	700	16,450	31
Brucite	$MgO + H_2O = Mg(OH)_2$	13.3	RT	27,200	20
Portlandite	$CaO + H_2O = Ca(OH)_2$	16.3	RT	57,800	33
Spinel	$MgO + Al_2O_3 = MgAl_2O_4$	2.9	1400	34,581	119
Mullite	$2SiO_2 + 3Al_2O_3 = Si_2Al_6O_{13}$	7.0	1400	25,962	37
Nepheline	$2SiO_2 + Al_2O_3 + Na_2O = Na_2Al_2\,Si_2O_8$	4.2	1400	339,000	545
Leucite	$2SiO_2 + Al_2O_3 + K_2O = K_2\,Al_2\,Si_2O_8$	9.1	1400	425,000	467
β-Corundum	$Na_2\,O + 9\,Al_2O_3 = Na_2Al_{18}O_{28}$	40.2	1400	182,200	45
Cristobalite	$SiC + 3H_2\,O = SiO_2 + 3H_2 + CO$	13.2	1100	323,116	245
Calcium silicate	$SiC + CaSO_4 = CaSiO_3 + CO + S$	27.6	900	339,774	128

10.2 Phase Transformation

10.2.1 Conversion of Silica

The crystal structure of silica (SiO_2) changes with temperature and pressure: At 1 atm pressure, find the following transformation temperatures in Fig. 10.10: At 573 °C, α-quartz transforms to β-quartz with a volume increase of $\Delta V = 3.9\%$. At about 780 °C, β-quartz transforms to β-tridymite, $\Delta V = 14.3\%$, and this transforms to β-cristobalite at about 1.470 °C, where $\Delta V = 15.4\%$. Figure 10.10 additionally shows the transformation temperatures of the most important phases at higher pressures, calculated with FactSage [2]. (refer also to Sect. 10.3).

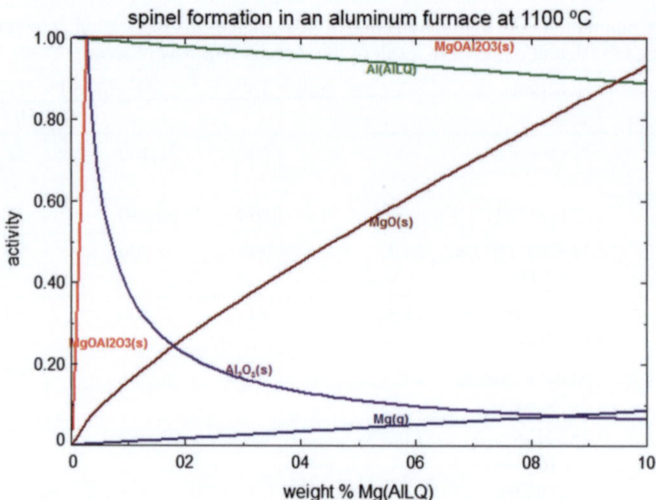

Fig. 10.9 Activity curves (progressions) of the components involved in the formation of spinel in contact with the lining in an aluminum furnace at 1100 °C (FactSage)

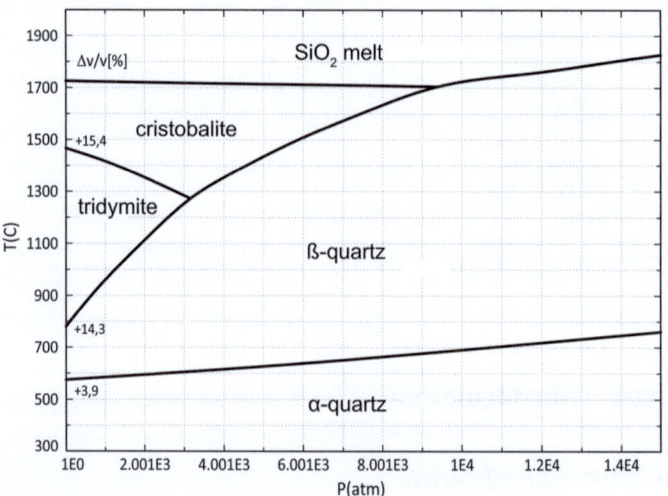

Fig. 10.10 Phases of silica as a function of pressure and temperature [2]

10.2.2 Decomposition of Dicalcium Silicate (2CaO · SiO₂) (620–500 °C)

Dicalcium silicate slags decompose at slow cooling below 500 °C, since a volume increase of 12% occurs during the transformation of β_H—Ca_2SiO_4 into γ—Ca_2SiO_4 [9]. At the (theoretical) transformation temperature (620 °C), the free enthalpy of both phases is equal, i.e. $\Delta G = 0$. Upon further cooling without excretion of the γ-phase, the β_H—phase becomes increasingly supersaturated with γ. Because of the supercooling to 500 °C $\Delta G = G_{C2S(\beta)}$ (773 K) $- G_{C2S(\gamma)}$ (893 K) $= -31{,}752$ J/mol. The activity a_{C2S} is calculated at 500 °C to be $a_{C2S} = 140$ (FactSage). The molar volume of Ca_2SiO_4 is $V_{C2S} = 52$ cm^3/mol and the volume jump during the conversion $\Delta V = +6.25$ cm^3/mol. Thus, from Eq. (10.7), the crystallization pressure turns out to be

$$\pi - 1 = \frac{-\Delta G^0}{\Delta V \cdot R_1/R_2} = \frac{31752}{6.25 \times 0.1} = 51 \times 10^3 \, \text{atm} \tag{10.43}$$

If the temperature drops below 500 °C, the supersaturation increases and with it the crystallization pressure.

10.3 Expansion Cracks

It is important to also state that not only the thermodynamically based crystallization pressure but also the thermally induced volume change by itself is a reason for the slow heating of the quartz required in practice. The pressure arising in the microstructure depends on the relative volume change $\Delta V/V$ and the compressibility κ [1/atm] [10]:

$$\Delta P = -\Delta V/(V \cdot \kappa)[\text{atm}]. \tag{10.44}$$

For the transformation of quartz (573 °C), for example, $-\Delta V/V \approx 0.04$ and $\kappa \approx -2 \times 10^{-6}$ atm^{-1}. The pressure change is therefore approx. 2×10^4 atm. This corresponds to a crystallization pressure in its magnitude (Table 10.1) and is likely to be decisive for the risk of cracking. Cristobalite transforms between 225 and 270 °C. Here, $-\Delta V/V \approx 0.03$, i.e. the resulting pressure increase is about 1.5×10^4 atm. This temperature range must be traversed very slowly to avoid thermal stresses applying temperature equalization.

By definition, the relative volume change $\Delta V/V$ can be expressed by the cubic thermal expansion coefficient β [1/K] and the temperature difference ΔT [10]: $\Delta V/V = \beta \cdot \Delta T$. With Eq. 3.1, it follows:

$$\Delta P = -(\beta/\kappa) \cdot \Delta T \,[\text{atm}]. \tag{10.45}$$

The compressibility κ [1/atm] depends on the modulus of elasticity E [GPa] of the material and its transverse contraction number ν [−] (Poisson's ratio) [10]:

$$\kappa = -\frac{3(1-2\nu)}{E}[1/\text{atm}].\tag{10.46}$$

For refractory materials, $\nu \approx 0.2$ [11].

$\beta \approx 7 \times 10^{-4}\,\text{K}^{-1}$ is the thermal expansion coefficient of cristobalite in this temperature range. Thus, with $\kappa \approx -2 \times 10^{-6}\,\text{atm}^{-1}$ and $\Delta T = 45\,°\text{C}$, the resulting pressure increase would be about 1.6×10^4 atm upon rapid heating (Eq. 10.45). Table 10.2 summarizes the specific pressures calculated for the quartz transformations and the resulting modifications.

It should be emphasized that this thermomechanical effect always plays a role if the process is accompanied by a volume change. It can even be a greater danger than the thermodynamic crystallization pressure.

For example, (Eq. 10.45) yields for the thermal expansion of corundum

$$\Delta P = -\left(3.85 - 10^{-6}/ - 3.85 \cdot 10^{-7}\right) \cdot \Delta T = 10 \cdot \Delta T \,[\text{atm}],\tag{10.47.1}$$

as for

$$\text{periclase } \Delta P = -\left(4.2 \cdot 10^{-6}/ - 8 \cdot 10^{-7}\right) \cdot \Delta T = 5 \cdot \Delta T \,[\text{atm}],\tag{10.47.2}$$

For partially stabilized zirconia

$$\Delta P = -\left(2.4 \cdot 10^{-6}/ - 6 \cdot 10^{-7}\right) \cdot \Delta T = 4 \cdot \Delta T \,[\text{atm}]\tag{10.47.3}$$

and for silicon carbide

$$\Delta P = -\left(2 \cdot 10^{-6}/ - 4 \cdot 10^{-7}\right) \cdot \Delta T = 5 \cdot \Delta T \,[\text{atm}],\tag{10.47.4}$$

to mention just a few examples.

The material values used in Table 10.3 were compiled by Eschner [11].

Table 10.2 Calculated transformation pressures of SiO_2

Transformation	ΔV [%]	T [°C]	ΔP [atm]
α-quartz → β-quartz	3.9	573	$\approx 4 \times 10^4$
β-quartz → β-tridymite	14.3	870	$\approx 14 \times 10^4$
β-tridymite → β-cristobalite	15.4	1470	$\approx 15 \times 10^4$
α-cristobalite → β-cristobalite	3.3	225–270	$\approx 3 \times 10^4$
α-tridymite → β-tridymite	0.6	117	$\approx 0.5 \times 10^4$

Table 10.3 Calculated pressure change of different oxides as a result of a temperature change

Compound	$\beta \cdot 10^{+6}$ (1/K)	$-K \cdot 10^{+7}$ (1/atm)	$\Delta P/\Delta T$ (atm/K)
Corundum	3.85	3.85	10
Periclase	4.2	8.0	5
ZrO_2 (partial)	2.4	6.0	4
SiC	2.0	4.0	5

10.4 Moisture in Refractory Material

10.4.1 Steam Explosion

If moist refractory material and molten material, e.g. liquid slag, come into contact with each other at 1.500 °C, the water can evaporate abruptly and rupture the lining. One mole of water, i.e. 18 g have a volume of 18 cm^3 at room temperature. If the water evaporates, it has a volume of 22.4 L. If it is heated further to 1.500 °C (1800 K), the volume (1 atm) is created.

$$V = 22.4 \cdot \frac{m_{H_2O}}{M_{H_2O}} \cdot \frac{T}{\vartheta} = 22.4 \cdot \frac{18}{18} \cdot \frac{1800}{300} = 134\,\text{L}. \tag{10.48}$$

The initial volume has increased 7500 times.

The brick shown in Fig. 10.11 had a weight of $m = 100$ kg and contained moisture. It was installed at a height of $h = 2$ m in the lining and was heated up too quickly. At an estimated temperature of 200 °C, the steam explosion occurred causing it to fly approximately $x = 2.5$ m into the furnace vessel and then hit the bottom. What amount of water develops this energy? The conservation of energy is

$$P \cdot V = \frac{m}{2} \cdot u_x^2 \tag{10.49}$$

With u [m/s] being the horizontal velocity of the brick. Its vertical velocity is

$$u_h = \sqrt{2g \cdot h}. \tag{10.50}$$

Both times $t_x = x/u_x$ and $t_h = h/u_h$ are equal, thus follows

$$P \cdot V = \frac{m \cdot g \cdot x^2}{h} [\text{Pa} \cdot \text{m}^3]. \tag{10.51}$$

The steam pressure results in a sufficient approximation from the equation

$$P[\text{Pa}] = 5 \times 10^{10} \cdot \exp(-\frac{2257}{0.460 \cdot (T_{°C} + 273)}), \tag{10.52}$$

Fig. 10.11 Refractory brick after a steam explosion (Lenzner 2012)

Fig. 10.12 Steam pressure curve of water /FactSage/

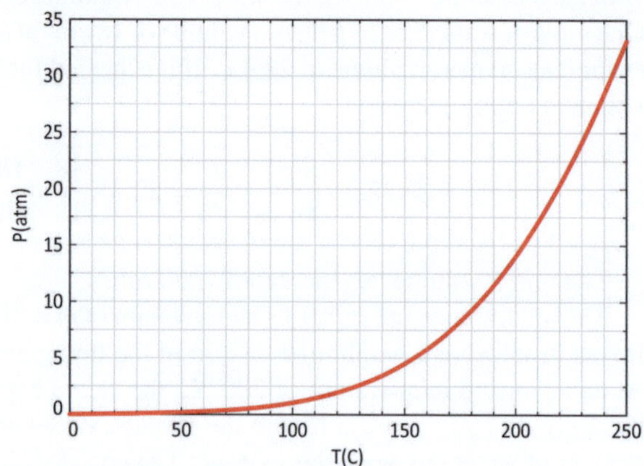

where $R_D = 0.46$ [kJ/kg $\cdot$ K] is the specific gas constant of steam and $\Delta H_D = 2257$ [kJ/kg] is its enthalpy of vaporization [2]. At 200 °C, $P = 1.5 \times 10^6$ Pa is obtained, which is about 15 atm (refer to Fig. 10.12).

From Eq. (4.1.4), the steam volume at 15 atm is calculated to be

$$V = \frac{100\,\text{kg} \cdot 9.81\,\text{m/s}^2 \cdot (2.5\,\text{m})^2}{1.5 \cdot 10^6\,\text{kg/m} \cdot \text{s}^2 \cdot 2\,\text{m}} = 2 \times 10^{-3}\,\text{m}^3 = 2\,\text{L}. \tag{10.53}$$

At normal pressure (1 atm) this is 30 L.

Using Eq. (10.48), the mass of the required water is given by

$$m_{\text{H}_2\text{O}} = \frac{V \cdot M_{\text{H}_2\text{O}} \cdot T_{KR}}{22.4 \cdot T_K} = \frac{30\,\text{L} \cdot 18\,\text{g/mol} \cdot 300\,\text{K}}{22.4\,\text{L/mol} \cdot 473\,\text{K}} = 15\,\text{g}. \tag{10.54}$$

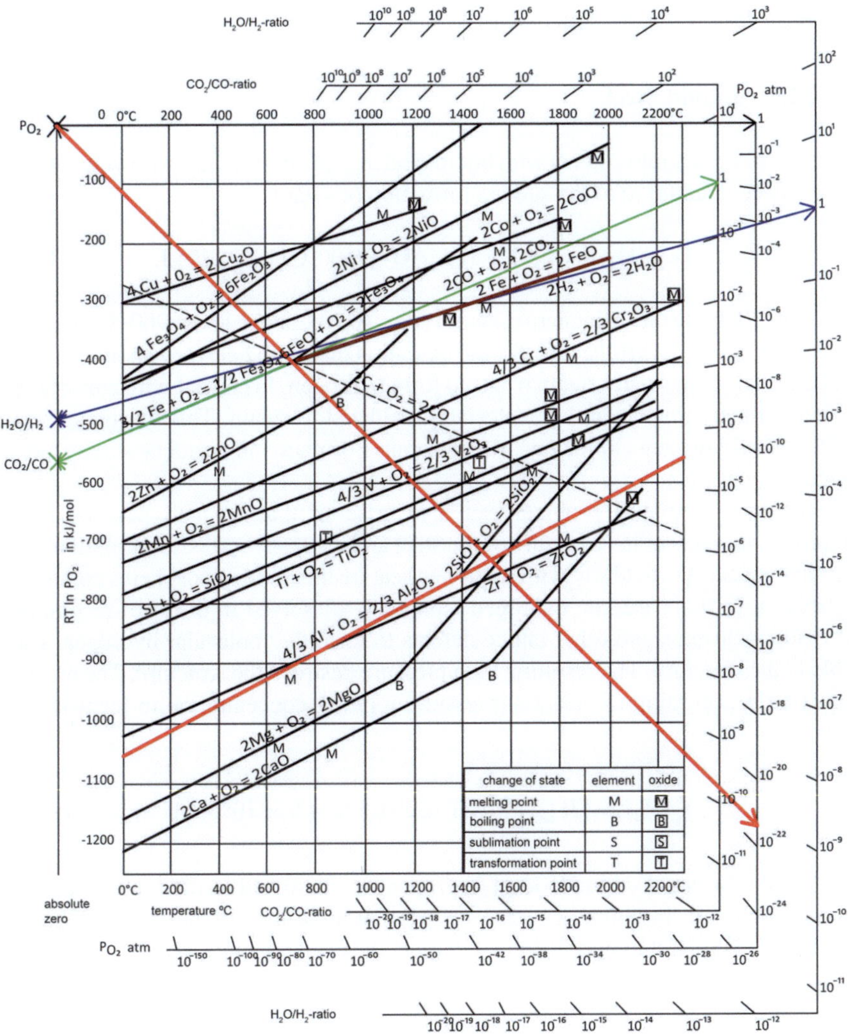

Fig. 10.13 Richardson-Ellingham diagram with H_2O/H_2 and CO_2/CO—equilibria [4]

During heating-up, the formation of steam begins on the hot side of the brick and pushes the still liquid water into the cold area. There it collects in cavities or pores and, once a limit temperature is exceeded, abruptly forms high-energy steam. This results in a steam explosion.

10.4.2 Hydrogen Explosion

10.4.2.1 Liquid Steel

If moisture comes into contact with liquid steel, e.g. at 1600 °C, the following reaction takes place in addition to the sudden formation of steam

$$Fe + H_2O = FeO + H_2; \quad \Delta G^\circ(1600°C) = -220\,kJ. \tag{10.55}$$

In the Richardson-Ellingham diagram (Fig. 10.13), above about 900 °C, one finds the ratio $H_2O/H_2 \approx 1$ (blue line) for the oxidation of iron to FeO (brown line). Consequently, half of the water steam reacts to form hydrogen. In contact with atmospheric oxygen, the hydrogen can be explosively oxidized to steam. This reaction, known as the "oxyhydrogen explosion," was the cause of many an accident at the smelter plant.

However, a quality problem can also arise in that a small portion of the hydrogen remains in the structure of the steel and causes so-called flocculation. This problem occurs particularly with improper heat treatment in atmospheres containing hydrogen. In this case, the hydrogen atomically dissolved in the steel recombines at oxide inclusions and other lattice defects to form the molecular hydrogen insoluble in the structure. The resulting high pressure destroys the structure. The reaction equation is according to Sieverts (pressure in [atm], concentration in [ccm/100 g]) [2]:

$$\frac{1}{2}\{H_2\} = [H]_{\alpha-Fe}; \quad \Delta G^0(300\,K) \simeq +5 \times 10^3\,J/FU \tag{10.56}$$

At equilibrium, X_S [cm^3/100 g] hydrogen are dissolved in the melt at a partial pressure of P_S. In the lattice, X_G [cm^3/100 g] are soluble at P_G. If the content dissolved in the melt is forced to dissolve in the lattice, the pressure increases:

$$P_G = \frac{P_S \cdot X_S}{X_G} \tag{10.57}$$

In liquid iron, about 25 cm^3 H$_2$/100 g Fe is dissolved at 1 atm, and only about 0.1 cm^3 H$_2$/100 g Fe [12] at room temperature. If, by rapid cooling 25 cm^3 H$_2$/100 g Fe are in forced solution at room temperature, this would correspond to the pressure

$$P_G = \frac{1\,atm \cdot 25\,ccm/100\,gFe}{0.1\,ccm/100\,gFe} = 250\,atm. \tag{10.58}$$

At temperatures above approx. 200 °C, atomically dissolved hydrogen diffuses relatively rapidly in iron, but the share remaining at room temperature is problematical if it recombines to form the molecule and builds up pressure according to Eq. (10.56).

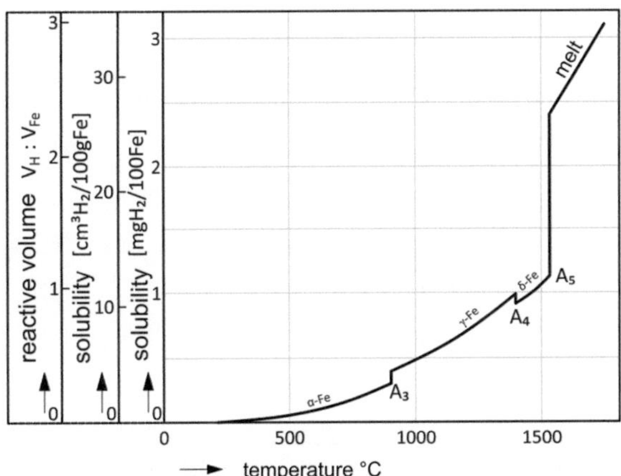

Fig. 10.14 Solubility of hydrogen in iron [12]

Looking at Fig. 10.14, it can be seen for practical operations that an increase in gas pressure of more than 15 atm is still possible during cooling of the solidified steel. This pressure destroys the microstructure.

10.4.2.2   Liquid Aluminum

Basically, any metal, whose free enthalpy of formation of its oxide is more negative than the formation of water, reduces the latter to hydrogen (refer to Fig. 10.13, blue line). For example, chromium metal reduces steam to form chromium oxide and hydrogen. This is also applies for aluminum:

$$3H_2O + 2\,Al = Al_2O_3 + 3H_2;\ \Delta G^0(700\,°C) = -790\,kJ. \tag{10.59}$$

Water vapor is, thus, quantitatively reduced to hydrogen. The following effects can be observed in practice (refer to Fig. 10.15).

The melt starts to bubble because at about 800 °C only 2 cm^3/kg Al hydrogen is soluble.

During solidification, the melt "boils" strongly, because only 0.04 cm^3/kg Al hydrogen is soluble in the lattice.

The solidified metal contains a large number of pores, an effect that has been utilized on various occasions in the laboratory to produce metal foam.

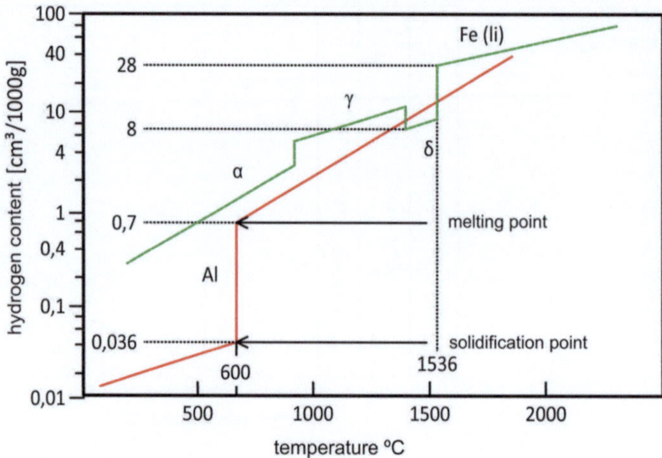

Fig. 10.15 Solubility of hydrogen in aluminum /Trimet Comp. (2005)

An explosion of oxyhydrogen gas occurs in the air Al_2O_3 skins can form and render the microstructure unusable.

10.4.3 Moisture Increases Infiltration

If refractory material still contains residual moisture when metals, e.g. steel, are melted down, this can lead to catastrophic operational accidents in the plant (refer to Sect. 10.4.2).

When steam comes into contact with the melt, it reacts with it chemically to form iron II oxide $(FeO)_x$ and hydrogen:

$$Fe + H_2O = (FeO)_x + H_2, \Delta G^0 = -22749 + 11.7 \cdot T \text{(J/mol)}. \tag{10.60}$$

The iron II oxide $(FeO)_x$ formed dissolves immediately in the total melt because the oxygen content there is only approx. $10^{-3}\%$ by weight. It imposes on the melt the highest oxygen content dissolved in the iron directly at the phase boundary. It corresponds to the equilibrium of the reaction (10.73). These values were calculated using FactSage in the temperature range from 1550 to 2000 °C and shown as a blue line in Fig. 10.16. The relationship is

$$[O]_{Fe} = 10^{-20} \cdot T_c^{6.04} \text{[wt\%]}.$$

Oxygen dissolved in iron is very capillary-active, i.e. it considerably reduces the surface tension of the melt. The surface tension of conventional steel melts is approx. 1600 mN/m. As a result of the contact of the steam with the molten steel, it drops to

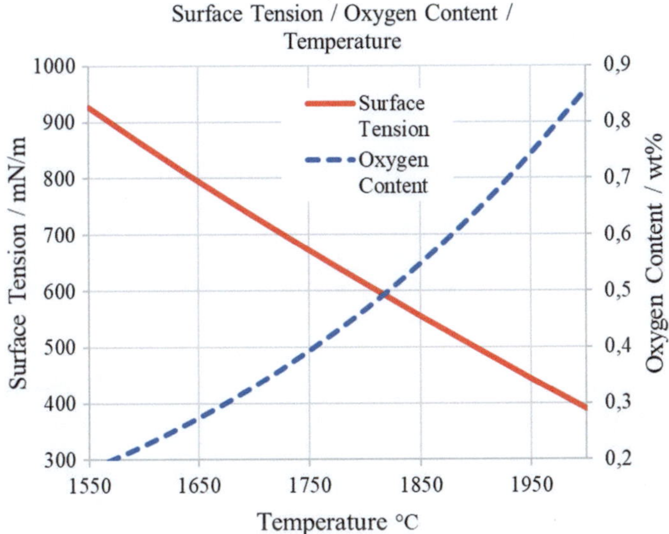

Fig. 10.16 Surface tension of liquid iron as a function of the maximum dissolved oxygen content at temperature T

below 1000 mN/m, i.e. it is almost halved. The surface tension curve shown as a red line in Fig. 10.16 is calculated from various measurements by many authors (refer to Eq. 7.15):

$$\sigma_{\underline{O},T_2}, = (584 - 464 \cdot \log \underline{O}) - 0.5(T_2 - 1550) \tag{10.61}$$

As a result of the reduced surface tension, the infiltration of the molten steel into the pores and cracks of the lining is fostered, because the wetting angle θ is also reduced considerably. It drops from 120° to about 105°. If the diameter of the pores infiltrated by the melt is calculated according to the known Eq. 5.2.2:

$$d = -\frac{4\sigma \cdot \cos \theta}{P_a + \rho_{Fe} \cdot g \cdot h_{Fe}} \tag{10.62}$$

and leaves the external conditions, such as atmospheric pressure P_a and ferrostatic pressure $\rho_{Fe} \cdot g \cdot h_{Fe}$, unchanged. The pore diameter decreases in a ratio of 1:4, i.e. now the infiltration of the molten steel already takes place into pores which have only 25% of the initial diameter.

If the iron II oxide (FeO)x formed in contact with the melt does not dissolve again immediately, but remains stable as a thin layer of slag because steam is resupplied from the pores quickly enough, the (FeO)x slag impregnates the lining. The wetting angle θ of the slag is less than 90°, i.e. the brick is wetted by it. The penetration depth of the slag into the pores is calculated from the equation given by Hagen-Poiseuille and the capillary pressure (refer to Eq. 5.11):

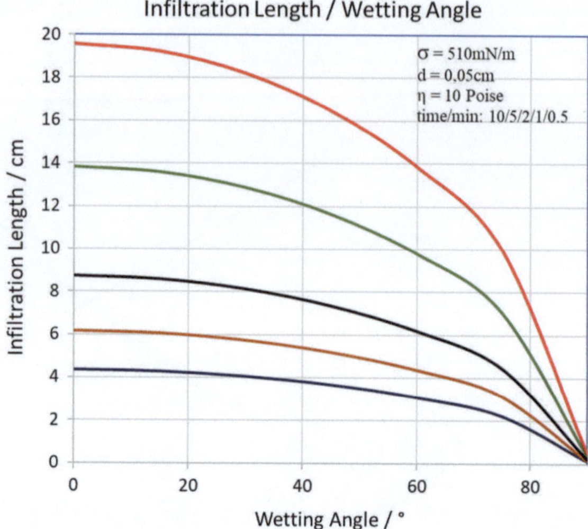

Fig. 10.17 Infiltration depth of a $(FeO)_x$ slag into the refractory material as a function of the wetting angle and time

$$l = \left(\frac{\sigma \cdot \cos\theta \cdot d}{4\eta} \cdot t \right)^{0.5}. \tag{10.63}$$

Using the typical values for the surface tension of a $(FeO)_x$ slag $\sigma = 510$ mN/m [2], the slag viscosity $\eta = 10$ g/cm · s and the pore diameter $d = 0.05$ cm, the infiltration depths l (cm) shown in Fig. 10.17 are calculated as a function of the wetting angle $0 < \theta < 90°$. At a wetting angle of $40°$, the slag penetrates 12 cm deep into the refractory material in 5 min. The temperature decline below the solidification temperature on its own stops further slag penetration. The iron II oxide penetrating into the pores dissolves the microstructure and leads to its deterioration.

10.4.4 Formation of Ice During Evacuation

The drying of the still moist refractory material in an induction furnace is occasionally accelerated by evacuation. Caution is required here, as otherwise ice will form which destroys the microstructure of the lining.

If water is supercooled by only 1 °C, i.e. to 272 K, thus $\Delta G^0 = -63$ J/mol [2]. During solidification the volume of water increases by $\Delta V = 1.6$ ccm/mol, resulting in a crystallization pressure of 40 atm:

$$(\pi - 1) = \frac{-\Delta G^o}{\Delta V \cdot R_1 / R_2} = \frac{63}{1.6 \cdot 0.1} = 39 \text{ atm.} \tag{10.64}$$

Comparably destructive is the volume expansion of the ice cooling down further $(-\Delta T)$. Its compressibility is

$$\kappa = -\frac{\Delta V}{V \cdot \Delta P} \simeq -10^{-5} \, \text{atm}^{-1}. \tag{10.65}$$

The cubic expansion coefficient is (10.8)

$$\beta = \frac{\Delta V}{V \cdot \Delta T} \simeq 1.5 \times 10^{-4} \, \text{K}^{-1}. \tag{10.66}$$

Together it then follows

$$\frac{\Delta P}{\Delta T} = -\frac{\beta}{\kappa} = \frac{1.5 \times 10^{-4}}{10^{-5}} = 15 \, \text{atm/K}. \tag{10.67}$$

If ice cools by only 10 °C, it already exerts a pressure of 150 atm on its surroundings. This was used in quarrying operations in the past and occasionally causes damage to refractory material stored improperly.

Figure 10.18 shows the phase equilibrium of water as a function of pressure and temperature. At the triple point (0.6 kPa/0.01 °C), the Gibbs phase law gives $F = K + 2\text{-}P = 0$ with $K = 1$ (water) and $P = 3$ (sol, li, g). If the water solidifies, its evaporation slows down significantly. This is another reason to avoid the formation of ice.

Without a proven program for drying, the lining should be heated to no more than 110 °C to safely avoid steam explosion. How much water heated in this way can evaporate adiabatically before ice formation starts? The energy balance is

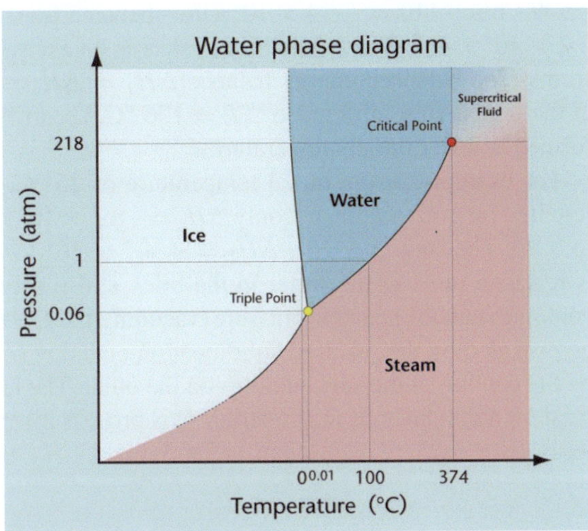

Fig. 10.18 Phase equilibrium of water (triple point), Internet: Angel Qu (3.5.2021), Eye, Students Newspaper (nai)

$$n \cdot \Delta H_V = (1 - n) \cdot C_P \cdot \Delta T \tag{10.68}$$

n mass share steam
$1-n$ mass share water
ΔH_V 2.257 kJ/kg, vaporization enthalpy
C_P 4.3 kJ/Kg x K, specific heat
ΔT Temperature change at adiabatic evaporation.

Consequently,

$$n = \frac{1}{\frac{\Delta H_V}{C_P \cdot \Delta T} + 1} = 0.17 \tag{10.69}$$

Ice is formed if more than 17% of the hot water evaporates adiabatically. Heat must therefore be transported in to prevent ice formation upon continued evaporation.

However, it must be taken into account that the evaporating water cools down the entire refractory material. Focus is on 1 m^2 of thickness $d_R = 0.2$ m. Let the pure (true) density of the refractory material be $\rho_R = 2.835$ t/m^3. With the porosity filled by water $\varepsilon = 0.05$, the total mass turns out to be:

$$\bar{m} = d_R\big(\rho_R(1 - \varepsilon) + \rho_{H_2O} \cdot \varepsilon\big) = 0.2(2835 \cdot 0.95$$
$$+ 1000 \cdot 0.05) = 549 \, \text{kg/m}^2. \tag{10.70}$$

The water content is $m_{H_2O} = 549 \cdot 0.05 = 27.4$ kg. It changes the specific heat of the system only slightly to $C_P = 1.06$ kJ/kg $\cdot$ K. For the complete evaporation of the water, $\Delta H_\uparrow = \Delta H_V \cdot m_{H_2O} = 2257 \cdot 27.4 = 6.18 \times 10^4$ kJ/m^2 must be added to the system. By cooling from 110 °C to 0 °C $\Delta H_\downarrow = -m \cdot C_p \cdot \Delta T = -549 \cdot 1.06 \cdot 110 = -6.4 \times 10^4$ kJ/m^2 become free. The sum is $\Delta H_\uparrow + \Delta H_\downarrow = -2.2 \times 10^3$ kJ/m^2. The cooling thus releases an excess of heat, which is why no ice forms. The balanced energy balance $\Delta H_\uparrow = \Delta H_\downarrow$ is just fulfilled in our example at the temperature difference $\Delta T = 106$ °C, i.e. just slightly below 110 °C. Ice is formed at lower initial temperatures.

For example, at the initial temperature of 25 °C, $\Delta H_\uparrow = 6.18 \times 10^4$ kJ/m^2 is required for evaporation, but only $\Delta H_\downarrow = -1.5 \times 10^4$ kJ/m^2 is released when cooling to 0 °C. The sum is $\Delta H_\uparrow + \Delta H_\downarrow = +4.7 \times 10^4$ kJ/m^2. Consequently, once 0 °C is reached, most of the water in the brick still remains liquid. If evaporation now continues into the negative pressure (vacuum), the necessary heat of evaporation must be provided by the released solidification enthalpy of 6 kJ/kg on the one hand and by the cooling of the surroundings on the other. The ice formed near the surface can blast off a thin layer of the material. This process is repeated, resulting in significant wear.

With linearly decreasing temperature, the partial pressure of the water drops exponentially. Accordingly, the evaporation rate decreases (e.g. at 25 °C $\dot{m} = 5.7 \times 10^{-4} \mathrm{kg/m_R^2 \cdot s}$). For this reason, the temperature gradient flattens, causing the icy zone to extend further into the interior of the material. The usual temperature drop from the surface to the interior of the material amplifies this effect.

If the evaporation takes place into a "vacuum," with the temperature-dependent steam pressure (10.14)

$$P[\mathrm{Pa} \equiv \mathrm{Kg/ms^2}] = 5 \times 10^{10} \cdot \exp\left(-\frac{2257}{0.460 \cdot (T_{°C} + 273)}\right) \qquad (10.71)$$

the evaporation rate will become (10.15,10.16)

$$\dot{m} = F_{\mathrm{eff}} \cdot 4.4 \times 10^{-3} \cdot \sigma \cdot P \cdot \sqrt{\frac{M}{T}} \; [\mathrm{kg/s}] \qquad (10.72)$$

The prerequisite is that the water does not boil. (Using the appropriate values, this equation also describes the evaporation rate of liquid metals).

The water evaporates from the porous surface of the lining into the reduced pressure space ("vacuum"). 5 kg/100 kg of mixing liquid is used, which is 5% or an evaporation area of $F = 0.05 \, \mathrm{m^2/m_R^2}$.

The molecular weight is $M = 18$ kg/kmol. The evaporation coefficient σ of water varies between 3×10^{-4} and 3×10^{-2} (10.17). We choose $\sigma = 3 \times 10^{-3}$.

With these values and Eqs. (10.67) and (10.68), we obtain the following for the evaporation rate of water

$$\dot{m} = \frac{B}{\sqrt{T_K} \cdot \exp(\frac{A}{T_K})}, \; \mathrm{mit} \; A = 4907 \, \mathrm{K} \; \mathrm{und} \; B = 1.4 \times 10^5 \, \mathrm{kg \cdot K^{1/2}/s}. \qquad (10.73)$$

The progression is shown by the orange line in Fig. 10.19.

To answer the question of how much time elapses before ice formation begins, it is convenient to form the integral mean value of the evaporation rate $\dot{m}_{imv}$ for fixed specified temperature intervals ΔT. From Eq. (10.69), by integration over the temperature difference ΔT, the relation follows directly:

$$\dot{m}_{imv} = \frac{60}{T_u - T_l} \int_{T_l}^{T_u} \frac{1.4 \times 10^5}{\sqrt{T} \cdot \exp(\frac{4907}{T})} dT \quad (\mathrm{kg/m_R^2 \cdot min}). \qquad (10.74)$$

The graphical illustration is shown in Fig. 10.19 as a red line. For the entire temperature range from 383 to 273 K, for example, the value pair $\overline{T} = 343K$ $(70°C)$ *und* $\dot{m}_{imv} = 0.28 kg/m_R^2 \cdot$ min is calculated (refer to Fig. 10.17, red line at $T = 110$ °C, or orange line at 70 °C).

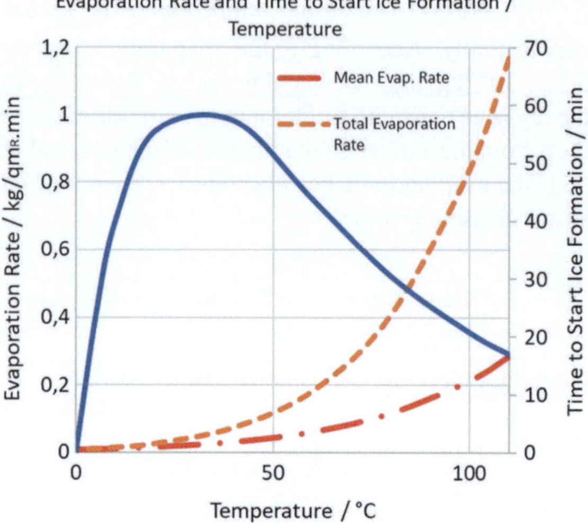

Fig. 10.19 Evaporation rate of water ($F = 0.05\,\mathrm{m^2/m_R^2}, \sigma = 3 \times 10^{-3}$) and time progression of beginning icing

To evaporate all the water at $\overline{T} = 70\,°\mathrm{C}$ in our example per $1\mathrm{sqm_R}$ of the lining the representative time required would be required time

$$t = m_{\mathrm{H_2O}}/\dot{m}_{imv} = 27.4/0.28 = 98 \text{ min} \tag{10.75}$$

In fact, however, ice is already formed if only $n \simeq 0.16 \cdot \Delta T$ share of the water has evaporated. In this example, this is 17.3%. The time available for evaporation until the formation of ice begins is, therefore, calculated from the equation

$$t_n = m_{\mathrm{H_2O}} \cdot (0.16 \cdot \Delta T)/(100 \cdot \dot{m}_{imv})(\text{min}). \tag{10.76}$$

The blue line in Fig. 10.19 shows the total time progression of ice formation t_n (min), calculated taking into account the amount of water actually evaporated until ice formation (Fig. 3.2, red line).

The lower the initial temperature, the longer it takes for ice to be excreted before the temperature falls below 30 °C. Then, the time available for evaporation drops very quickly to zero.

For practical purposes, the example tells us not to start a vacuum-assisted drying process below 30 °C, to carefully include it in the process schedule, and to closely monitor the process.

To clarify whether the heat flux $\dot{q} = \lambda \cdot \Delta T/\Delta x$ in the refractory material is sufficient to ensure this rate of evaporation, it is compared with the heat of evaporation $\Delta H_\uparrow$. Due to evaporation, the temperature gradient at the surface is positive. At 70 °C, according to Fig. 10.17, $\dot{m} = 0.28\,\mathrm{kg/m_R^2} \cdot$ min. The average thermal conductivity is $\lambda = 120\,\mathrm{KJ/min} \cdot \mathrm{m} \cdot \mathrm{K}$ [21]. The negative pressure in the boiler is kept constant. Then, the (required) gradient is

$$\frac{\Delta T}{\Delta x} = \frac{\dot{m} \cdot \Delta H_V}{\lambda_R} = \frac{0.28 \cdot 2257}{120} = +5.3\,\text{K/m} \tag{10.77}$$

The heat flux is $\dot{q} = 636\,\text{kJ/m}^2 \cdot \text{min}$ and the quotient.

$$\frac{\Delta H_\uparrow}{\dot{q}} = \frac{6.18 \times 10^4}{636} = 98\ \text{min} \tag{10.78}$$

The comparison with the result of Eq. (10.71) shows that the heat flux is sufficient to carry the heat of evaporation to the surface.

10.4.5 Formation of Thenards Blue (CoAl$_2$O$_4$)

It is observed that in corundum additions the spinel CoAl$_2$O$_4$ (Thénards blue) is formed during melting of alloys containing cobalt (Fig. 10.20). This can lead to high crystallization pressures and spalling, refer to Sect. 10.2.

Valve steel 1.4971 (X 12 Cr Co Ni 21 20) with 20% by weight cobalt serves as an example here.

Without the presence of water steam, only corundum (Al$_2$O$_3$) is present in the lining, no co-spinel. It is true that steam considerably increases the activity of the cobalt dissolved in the iron: at 1600 °C, the activity of the cobalt in the melt increases from 0.64 to 0.98, even if the refractory material contains only a few percent water, e.g. 5%. But, the spinel is not formed according to the reaction (10.1).

Fig. 10.20 Formation of Thénards blue in the lining of a 4t—vacuum induction furnace (Saveway 2018)

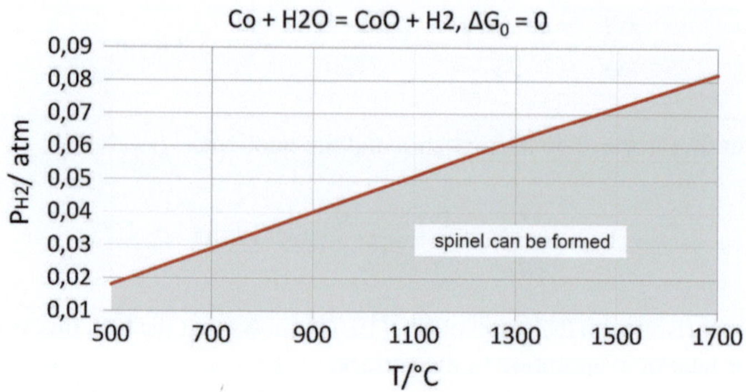

Fig. 10.21 Stability limit of the hydrogen partial pressure for the formation of CoO

$$Al_2O_3 + Co + H_2O = CoAl_2O_4 + H_2 \quad \Delta G^0_{1600°C} = +1.1 \times 10^6 \, J/mol, \quad (10.79)$$

as long as the atmosphere consists almost completely of hydrogen (92%). The Richardson-Ellingham diagram (refer to Fig. 10.13) shows that cobalt oxide (CoO) is less stable than H_2O and is, therefore, reduced by hydrogen to form H_2O. However, only when cobalt oxide exists, is it possible to form spinel with the lining (10.1):

$$CoO + Al_2O_3 = CoAl_2O_4, \Delta G^0 = -3.4 \times 10^4 + 4.36 \cdot T \quad (J/mol) \quad (10.80)$$

However, especially in areas with high humidity, where the content of steam is high, cobalt oxide can be formed if the hydrogen formed is sufficiently diluted by the gases and steam present in the atmosphere and, consequently, the free enthalpy of formation of CoO is negative ($\Delta G^o \leq 0$) is (10.2):

$$Co + H_2O = CoO + H_2, \Delta G^o \leq 0. \quad (10.81)$$

Figure 10.21 shows the calculated partial pressure of hydrogen as a function of the temperature at which this condition is fulfilled (FactSage). As soon as the calculated hydrogen partial pressure falls below the indicated value, cobalt oxide and, subsequently, cobalt spinel are formed: $P_{H_2} = 5 \times 10^{-5} \cdot T_C - 0.083 \, [atm]$.

In addition, it cannot be ruled out that steel splashes, which oxidize if the crucible cools in air and forms cobalt oxide, react with the corundum lining to form blue Co-spinel already below the temperature of liquid steel.

References

1. Knacke, O., Kubaschewski, O., Hesselmann, K.: Thermochemical Properties of Inorganic Substances I/II, 2nd edn. Springer-Verlag, Verlag Stahleisen (1991)
2. Hack, U.M.K.: FACTSage, Thermfact & GTT Technologies (2007). www.factsage.com
3. Baukloh, R., Knacke, O., Löscher, W.: The growth pressure of carbon. Archiv. Eisenhüttenwesen. **27** FEB (1956) 95–99
4. Eisenhütte.: 5th Edition 94 (1961)
5. Walker, P.L., Rakszawski, J.F., Imperial, G.R.: Carbon formation from carbon monoxide-hydrogen mixtures over iron catalysts. J. Phys. Chem. **63**, 140–149 (1959)
6. Dietrichs, P.: Test Report GA-No. 2309, GHI-RWTH, Aachen, Germany, 20th Nov 1996
7. Krause, O., Pötschke, J.: Resistance of Refractory Products to Carbon Monoxide… In: UNITECR'07, 10th Biennial Worldwide Congress on Refractories, pp 242–244. Dresden, Germany (2007)
8. Grunewald, J.: Lecture: Construction Physics. TU-Dresden, Germany (2009)
9. Geiseler, J., Bau, R., Heinke, R.: Slags in the manufacturing of high-grade steels. In Koch, K., Janke, D. (eds) Slags in Metallurgy" Verlag Stahleisen mbH, pp. 221–234, Düsseldorf, Germany (1984)
10. Meschede, D.: Gerthsen Physics. Springer 23rd Printing (2005)
11. Eschner, A.: VDI- Heat Atlas (Material Values for Refractory Materials)/Google 1–9 Dec (2006)
12. Schumann, H.: Metallography; VEB—Verlag Leipzig, Germany (1962)
13. Hocquet, S., Mastroianni, I., Tirlocq, J., Lardot, V., Cambier, F.: Resistance to carbon monoxide of refractory materials: towards a reliable test. In: 56th International Colloquium on Refractories. Aachen, Germany, Sept 80–83 (2013)
14. Routschka, G., Wuthnow, H.: Practice Manual Refractory Materials, 5th Edition. Vulkan Verlag, Essen, Germany (2011)
15. Pötschke, J.: A novel model for drying refractory castables. Refract. Worldforum **2**, 99–106 (2010)
16. Winkler, O.: Theory and practice of vacuum melting. Metall. Rev. **5**(17), 1–33 (1960)
17. Wutz, M.: Theory and Practice of Vacuum Technology. Vieweg, Braunschweig, Germany (1965)
18. Grassmann, P.: Basics of Chemical Technology, Salle und Sauerländer (1983)

Chapter 11
Drying of Refractory Castable

11.1 Introduction

The drying of ceramic materials containing water is a complex process. For example, the drying of bricks differs considerably from that of a refractory castable. The brick mass contains about 15% by weight water, an Ultra-Low Cement Castable (ULCC) about 5% by weight. Bricks are prone to shrinkage cracks, while ULCC is prone to explosive spalling if not dried properly.

In both cases, a programmed sequence is preferred for drying. Bricks are heated slowly up to approx. 110 °C, whereas refractory material is heated up to more than 600 °C. Calculation models facilitate the determination of the temperature/time program. For the brick industry, Junge and Telljohann [1], Tretau [2] have developed a numerical calculation model. For the drying of refractory castables, this was done by Großwendt [3].

Based on industrial experience, a simple analytical model for drying refractory castables is developed.

11.2 Basic Information

11.2.1 Refractory Castables

Chemically speaking, refractory castables consist of Al_2O_3, CaO and SiO_2, but may also contain ZrO_2, MgO, SiC and BaO. The focus of this consideration is on Ultra-Low Cement Castables (ULCC), which, in addition to Al_2O_3, contain only small amounts of CaO as calcium aluminate and SiO_2 to improve hydraulic bonding. The main components are spinel/corundum and andalusite, respectively. The grain size distribution of the fine bond matrix is often bimodal, with about 50% by weight of

© The Author(s), under exclusive license to Springer Nature Switzerland AG 2024
J. Pötschke, *Refractory Fundamentals in Metallurgical Practice*,
https://doi.org/10.1007/978-3-031-63709-4_11

the grains having 10^{-2}–1 μm Ø and 50% by weight having 1–100 μm Ø. The coarse grain ranges up to about 6 mm in diameter.

The dry mass has a bulk density of approx. 2700 kg/m^3. Approx. 5% by weight water is added to it, so that after a mixing time of approx. 5 min and subsequent compaction, a fresh castable with a bulk density of 2835 kg/m^3 is produced.

After its processing, the castable rests for $\geq$ 24 h, during which hydraulic binding phases are formed. These are mainly gibbsite ($Al(OH)_3$), monocalcium aluminate hydrate ($CaO \cdot Al_2O_3 \cdot 10H_2O$) and, at slightly elevated temperature, tricalcium aluminate hydrate ($3CaO \cdot Al_2O_3 \cdot 6H_2O$) [4]. These hydrates clog fine pore channels in particular, resulting in a noticeable obstruction of moisture conduction in the microstructure. Usually, the distribution of open pores is determined after a preliminary drying at 110 °C, since it is assumed that all hydrates have formed after the long setting time. A typical pore size distribution of an ULCC is shown in Fig. 11.1, according to which 90% of the open pores are smaller than 0.1 μm. The largest open pores are 3–5 μm.

The hydraulic binder phase decomposes with splitting off of the water at temperatures above 150 °C. A distinction is made between the three drying phases:

- Adsorbed water evaporates ($\geq$ 110 °C).
- Calcium aluminate hydrates with low hydrate content (CAH10, C2AH8, C3AH6) decompose; the released water boils (< 250 °C). There is evidence that further hydrates can be formed up to 200 °C (Krebs, R., priv. communication).
- The remaining hydrates, especially gibbsite ($Al(OH)_3$), decompose ($\leq$ 400 °C). The smallest, insignificant residues in gussets should hold up to approx. 600 °C [5].

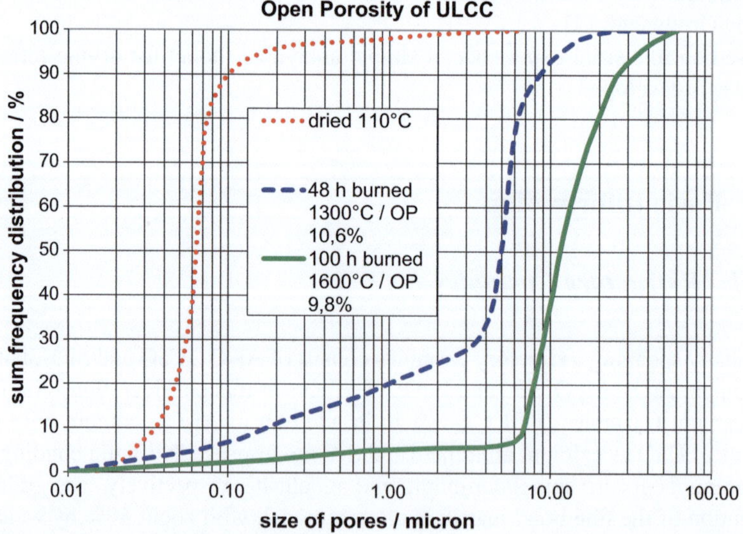

Fig. 11.1 Pore size distribution of an ULCC

Curing should take place at elevated temperature (< 100 °C) and, thus, as completely as possible to avoid the formation of stable hydrates at higher temperatures (up to 200 °C). Although these contain less water, they decompose only at higher temperatures. This could lead to a vapor explosion (Krebs, R., priv. communication).

About 50% of the added water is bound as hydrate, the rest is free in the structure and binds the grains by van der Waals forces.

11.2.2 Mass Transfer

The mass transfer differs in the three drying phases:

11.2.2.1 First Drying Phase

In the first drying phase, water flows from the interior to the surface under the influence of capillary force and evaporates there into the heat-carrying air. This evaporation occurs only over the free surface of the castable. The mass flow density is [1–3]:

$$\dot{m} = \frac{\beta}{R_D \cdot T}(P_S(T) - P_L)\left[\frac{\text{kg}}{\text{m}^2\,\text{s}}\right] \tag{11.1}$$

ß [m/s] is the mass transfer coefficient, $R_D = 0.46 \left[\frac{\text{kJ}}{\text{kg K}}\right]$ the special gas constant for vapor, $P_S(T)$ [Pa] is the saturation vapor pressure of water at the outer surface of the castable, and P_L [Pa] is the vapor pressure in the ambient air. The saturation vapor pressure follows the relationship [6].

$$P_S(T) = 5 \times 10^{10} \cdot \exp\left(-\frac{2257}{0.46 \cdot T}\right) \text{ [Pa]}. \tag{11.2}$$

At $\vartheta = 100$ °C ($T = 373$ K), $P_s = 10^5$ Pa results. At temperatures > 250 °C, values get calculated somewhat too high, since the critical point of the water ($\vartheta_S = 373$ °C, $P_s = 217$ bar) is approached. (More exact approximations, which are not needed here, can be found in [7, 8]).

The mass transfer coefficient depends largely on the flow velocity u of the drying gas and can, therefore, be influenced by the flow

$$\beta = \sqrt{\frac{4D \cdot u}{\pi \cdot z}} \text{ [m/s]}, \tag{11.3}$$

where D [m²/s] is the diffusion coefficient of vapor in the drying gas and z [m] is a characteristic length.

The diffusion coefficient of vapor follows the relationship [10].

$$D_D(T) = 2.3 \times 10^{-5} \left(\frac{T}{273} \right)^{1.8} \ [\text{m}^2/\text{s}] \tag{11.4}$$

and results for 90 °C (363 K) to D_D (363) $= 3.8 \times 10^{-5}$ m²/s.

Assuming that there is a turbulent gas flow in the drying chamber, then $\beta \approx$ is 0.01 m/s and $\Delta P_S \approx 0.9 \times 10^5$ Pa, so that from Eq. (11.1) the high mass flux density $\dot{m} = 0.005$ kg/$(\text{m}^2$ s) results.

We will see that, in accordance with practical experience, it is not the mass transfer in the flow boundary layer that determines the drying rate of the refractory castable, but much more the temperature control, i.e. the heating through of the castable, that is, most decisive here.

Shrinkage, which can lead to crack formation during drying of the brick, should be briefly mentioned. It is calculated from the ratio of the bulk density of the dry mass to the wet mass ρ_{dr}/ρ_{we}:

$$S_l = \frac{100}{3} \left(\frac{\rho_{we}}{\rho_{dr}} - 1 \right) \ [\%]. \tag{11.5}$$

With $\rho_{dr} = 2700$ kg/m³ and $\rho_{we} = 2835$ kg/m³ few get linear shrinkage $S_L = 1.6\%$. This value is generally not dangerous for refractory castables compared to brick drying ($\approx 6\%$).

For drying of refractory castables, the first drying stage is usually passed through in a short time due to the rapid increase in temperature to over 100 °C and is, therefore, of little importance in operational practice. It is completed as soon as the capillary transport of liquid water to the surface is no longer possible over a large area and the drying level retreats into the interior.

Pores clogged by hydrate play a major role here because capillary pressure is inversely proportional to pore diameter:

$$P_\sigma = \frac{4\sigma_w \cdot \cos\theta}{d} \ [\text{Pa}]. \tag{11.6}$$

Small pores, therefore, suck the water out of the large ones and transport it to the surface. The more the fine pore channels are clogged, the sooner this effect comes to a standstill.

Upon putting the capillary pressure Eq. (11.6) into the Hagen-Poiseuille equation [6] for laminar pipe flow, the following applies for the mass flow density:

$$\dot{m} = \frac{\rho_w}{16} \cdot \frac{\sigma_w \cdot \cos\theta \cdot d}{\eta \cdot L \cdot \mu} \ \left[\frac{\text{kg}}{\text{m}^2 \ \text{s}} \right] \tag{11.7}$$

$\dot{m}$ declines linearly with decreasing capillary radius d [m]. For the mean pore diameter of the refractory castable shown in Fig. 16.1, $d = 0.06$ μm$(6 \times 10^{-8}$ m) one gets, with tension of the water, $\sigma_w = 0.07$ N/m, its contact angle with the capillary wall θ $= 0°$ (complete wetting), dynamic viscosity of the water, $\eta = 3 \times 10^{-4}$ Pa s and its density $\rho_w = 1000$ kg/m^3 as well as the thickness of the castable layer $L = 0.2$ m, the value $\dot{m} = 0.00044$ kg/m^2 s and for the smallest diameter $d = 0.02$ μm the value $\dot{m} = 0.00015$ kg/m^2 is obtained. Here, $\mu = 10$ is the diffusion resistance factor [1, 11] which corrects the mass transport actually taking place in the microstructure compared to that in free space: The transport is reduced by a factor of 10. This empirically determined value corresponds approximately to the labyrinth factor used elsewhere.

11.2.2.2 Second Drying Phase

In second drying phase, the drying level retreats into the castable. The free surface is dry. Evaporation takes place inside at the drying level. This condition is reached during the drying of a refractory castable after only a short time, as the temperature at the surface rises rapidly to over 100 °C. The transport of the vapor in the already dried layer L_K occurs (initially) via diffusion [1, 2]:

$$\dot{m} = -\frac{D_D}{\mu \cdot L_K} \cdot \frac{P}{R_D \cdot T} \cdot \ln \frac{P - P_{D0}}{P - P_{DK}} \left[\frac{kg}{m^2\ s} \right] \tag{11.8}$$

D_D is calculated according to Eq. (11.4); $\mu = 10$; L_K [m] is the thickness of the already dried layer. P [Pa] is the total pressure of the drying gas, e.g. $P = 10^5$ Pa (1 bar). The driving partial pressure gradient is the mean logarithmic difference between the vapor partial pressure at the drying level P_{DK} and that at the blank (green) surface P_{DO} [1, 2]. For example, if the temperature at the drying level is 60 °C (333 K) and at the free surface 90 °C (363 K), Eq. (11.2) gives the partial pressures $P_{DK} = 0.2 \times 10^5$ Pa and $P_{DO} = 0.67 \times 10^5$ Pa. The average diffusion coefficient for 75 °C (348 K) is D_D (348 K) $= 3.6 \times 10^{-5}$ m^2/s. With $L_K = 0.05$ m it follows that at 75 °C $\dot{m} = 4 \times 10^{-5}$ kg/m^2 s. Because of the rising outside temperature, the water in the material > 110 °C begins to boil [11] and a flow of vapor occurs in the capillaries. Applying the equation of Hagen–Poiseuille [6] to the flowing vapor, we obtain as vapor flux density

$$\dot{m}_p = \frac{d^2 \cdot \rho_D}{32 \cdot \eta_D \cdot \mu \cdot L_K} \cdot (P_S - P) \left[\frac{kg}{m^2\ s} \right]. \tag{11.9}$$

With $d = 3 \times 10^{-6}$ m, $\rho_D = 0.6$ kg/m^3 $\eta_D = 1.2 \times 10^{-5}$ Pa s, $\mu = 10$, $L_K = 0.05$ m at $\vartheta = 150$ °C (423 K) $P_S = 4.6 \times 10^5$ Pa, it follows that $\dot{m} = 0.01$ kg/m^2 s.

This result is an estimate for two reasons:

- The equation of Hagen-Poiseuille is valid for a laminar flow. However, if the water is very superheated, a turbulent flow is observed [11].
- Especially in very fine pore channels the mean free path length of the water molecules is greater than the pore diameter. There, mass transport takes place by Knudsen diffusion [6], i.e. possibly more rapidly than with molecular diffusion or convection [3]. This effect, which is to be expected for capillary diameters $d < 0.1$ μm (1 bar), therefore has no influence on the calculation of a possible impairment of drying of refractory castables, but favors mass transport.

The sudden spalling of refractory castable parts in the lining, as a result of the spontaneous formation of a vapor bubble, is dangerous (refer to Sect. 10.4.1). Temperatures above 200 °C [11, 12] are particularly dangerous. It is observed that the vapor bubbles form in the area where liquid water is still present, which is considerably overheated, i.e. close behind the drying level.

This is supported by the capillary pressure. Because of the complete wetting of the capillary wall by the water (wetting angle $\theta = 0°$), the meniscus of the water surface in a pore is to be calculated negatively. With the numbers used for Eq. (11.7), Eq. (11.6) would hypothetically give $P_\sigma = -4.6 \times 10^5$ Pa. This is equivalent to the fact that compared to the boiling temperature of an extended water surface of 100 °C (1 atm), a boiling temperature is now required which corresponds to the overpressure of $+4.6$ atm, i.e. $\vartheta = 149$ °C. (Nucleation impairments are not taken into account here).

This consideration explains the experimentally confirmed fact that water can certainly be overheated—compared to 100 °C—well over a hundred degrees [11, 12] without boiling. Consequently, the task is to control the heating rate in such a way that the temperature at the drying level allows a uniform and rapid removal of the vapor, so that the liquid water behind it does not evaporate explosively. The fundamental difficulty is that the temperature of the drying level and its location are practically unknown.

11.2.2.3   Third Drying Phase

In the third drying phase, all liquid water has already been expelled. Residual moisture in very narrow spandrels and hydrates is removed as the temperature continues to rise. This is less than 5% of the initial water content [3], and there is usually no longer any risk of spalling due to vapor explosion. However, if further hydrates with a high decomposition temperature have formed during heating and these split off a sufficiently large quantity of water, which evaporates abruptly, an unexpected vapor explosion can still occur at this stage (Krebs, R., personal communication).

11.2.3 Heat Transfer

For drying, it is necessary to provide the overall system with the amount of heat

$$Q = q_1 + q_2 + q_3:$$

Heating of water: $q_1 = h_0 \cdot \rho_w \cdot C_{pw} \cdot \Delta\vartheta \; [\text{kJ/m}]^3 \qquad (11.10)$

Evaporation of water: $q_2 = h_0 \cdot \rho_w \cdot \Delta H_w \; [\text{kJ/m}]^3 \qquad (11.11)$

Heating of castable: $q_3 = \rho_{\text{Ref}} \cdot C_{p\text{Ref}} \cdot \Delta\vartheta \; [\text{kJ/m}^3]. \qquad (11.12)$

h_0 is the initial water content, e.g. 0.05 [kg/kg] equals 5%.

The heat quantity Q is supplied by means of the heat flux

$$\dot{Q} = \alpha(\vartheta_L - \vartheta_0) = \frac{\lambda_{FF}}{L_K}(\vartheta_0 - \vartheta_K) \left[\frac{W}{m^2} \right] \qquad (11.13)$$

The quotient $\dot{Q}/Q$ describes the evaporating mass flux density of water

$$\dot{m}_\vartheta = \rho_w \cdot \frac{\dot{Q}}{Q} \left[\frac{\text{kg}}{\text{m}^2 \, \text{s}} \right] \qquad (11.14)$$

From this, the drying rate is given by [13, 14]:

$$\dot{m}_\vartheta$$

$$= \frac{\rho_w \cdot \lambda_{FF}(\vartheta_0 - \vartheta_K)}{L_K[(\vartheta_K - \vartheta_a) \cdot (\rho_{FF} \cdot C_{PFF} + h_0 \cdot \rho_w \cdot C_{Pw}) + h_0 \cdot \rho_w \cdot \Delta H_w]} \left[\frac{\text{kg}}{\text{m}^2 \, \text{s}} \right]$$
$$(11.15)$$

ϑ_a is the temperature of the entire refractory body at the beginning of the process.

ϑ_0 is its rising temperature at the surface and ϑ_K that at the drying front (drying level).

An example is used to compare the calculated evaporation rates from mass transfer (Eq. 11.9) and heat transfer (Eq. 11.15). Of the total thickness $L = 0.2$ m of the castable wall, $L_K = 0.09$ m has already dried at the time of calculation. The temperature there is $\vartheta_K = 150\,°C$. The surface temperature has increased linearly in time from $\vartheta_a = 30\,°C$ to $\vartheta_0 = 210\,°C$. L_K corresponds to the location of the integral mean value of the temperature distribution T_{IMV} in the castable layer.

The **data** used are.

Thermal conductivity of refractory castable: $\lambda_{Ref} = 2.5 \times 10^{-3}$ kW/m K.
Initial temperature of the wet castable: $\vartheta_a = 30\ °C$.
Surface temperature of the castable: $\vartheta_0 = 210\ °C$.
Already dried section: $L_K = 0.09$ m (from 0.2 m).
Temperature at drying level L: $\vartheta_K = 150\ °C$.
Average heating of water and refractory castable: $(\vartheta_K - \vartheta_a) = 120\ °C$.
Temperature difference between surface and dry level: $(\vartheta_0 - \vartheta_K) = 60\ °C$.
Density of refractory castable: $\rho_{Ref} = 2700$ kg/m³.
Specific heat of concrete: $\mathbf{C_{PRef}} = 1$ kJ/kg K.
Mixing water content: $h_0 = 4.5$ kg/100 kg castable $= 0.045$.
Density of water: $\rho_W = 917$ kg/m³.
Specific heat of water: $\mathbf{C_{PW}} = 4.3$ kJ/kg K.
Heat of evaporation of water: $H\Delta_w = 2257$ kJ/kg.
Density of vapor: $\rho_D = 2.5$ kg/m³.
Viscosity of vapor: $\eta_D = 1.4 \times 10^{-5}$ Pa s.
Detour factor: $\mu = 10$.
Vapor pressure $P_S = > 5 \times 10^5$ Pa.
Air pressure $P = 10^5$ Pa.
Pore diameter: $d = 3 \times 10^{-6}$ m.

Using Eq. 11.9, the difference in vapor pressure is used to calculate the vapor flux density to be

$$\dot{m}_P = 5 \times 10^{-9} \cdot \left(Ps - 10^5\right)/L_K \left[\frac{kg}{m^2\ s}\right],$$

$$\dot{m}_p \geq 20 \times 10^{-4}/L_K \left[\frac{kg}{m^2\ s}\right] \tag{11.16}$$

The decisive factor is the temperature range $> 150\ °C$, i.e. $P_s \geq 5 \times 10^5$ Pa.
Due to the heat input, the following is calculated from Eq. 11.15.

$$\text{The drying rate to } \dot{m}_\vartheta = 3 \times 10^{-4}/L_K \left[\frac{kg}{m^2\ s}\right]. \tag{11.17}$$

As already mentioned, mass transfer is faster than heat transfer, i.e. there is no vapor jam and the risk of explosion is low if the correct heating-up curve is selected. Overheated liquid pockets, i.e. boiling delay, must be avoided.

11.2.4 Spalling

Spalling occurs when the liquid water still present beyond the drying level evaporates abruptly due to a boiling delay and the resulting pressure ruptures the material.

The nucleation of vapor bubbles is homogeneous, since the wetting angle is $\theta = 0$ (refer to Sect. 2.1.3.5). The vapor pressure required for the formation of a spherical bubble nucleus can be estimated from the tensile/tear strength of the water [15, 16]:

$$\Delta P = \frac{8\sigma}{3d} \; [Pa]. \tag{11.18}$$

For the largest pore diameter of 3 μm (3×10^{-6} m) with the surface tension of water ($\sigma \cong 0.06$ N/m at 100 °C) $\Delta P = 5 \times 10^4$ Pa (+0.5 atm) follows relative to ambient pressure. This would be the minimum gauge pressure at which vapor bubbles form. For $d = 0.3 \times 10^{-6}$ m, $\Delta P = 5$ bar already follows as the overpressure. In practice, pressure and temperature will be significantly higher, since the nucleation frequency increases exponentially with decreasing bubble radius [15].

However, the pore structure of the refractory material also acts as a "boiling stone" in that any extraneous gas present is forced into the water during its thermal expansion, thus acting as a bubble nucleus.

It becomes problematic once water from a hot area is forced into cold areas of the refractory material by the high vapor pressure prevailing there. There, in fact, the effect of the "boiling stone" is much less given, since the saturation vapor pressure of the inflowing water is higher than the ambient pressure. The result can be spontaneous vapor generation, the pressure of which ruptures the castable. The process in a microwave oven is comparable. In the boiling pot, on the other hand, the bubbles form at the hot bottom with only a slight boiling delay at approx. 110 °C.

Since the vapor explosion is a stochastic process, a lower limit value for the pressure must be sought, which, if exceeded, poses a danger [5]. For this purpose, the tensile strength σ_s of the castable in the set state, e.g. at 110 °C, can be used. In literature, one can find quite a lot of different data between 1.4 MPa [5] and 40 MPa [12], where the tensile strength was calculated to be at most 1/3 (up to 1/10) of the compression (crushing) strength ($\sigma_s \sim 1/3 \, \sigma_c$) [17]. According to the vapor pressure formula of water (Eq. 11.2), 1.5 MPa, as saturation vapor pressure, corresponds to a temperature of almost 197 °C and 6 MPa to about 269 °C (refer to Fig. 11.2). This is the temperature range in which spalling occurs. The critical temperature range is between 200 and 300 °C [12, 13]. Consequently, the task is to pass through this temperature range so slowly that the vapor is transported to the outside without any problems.

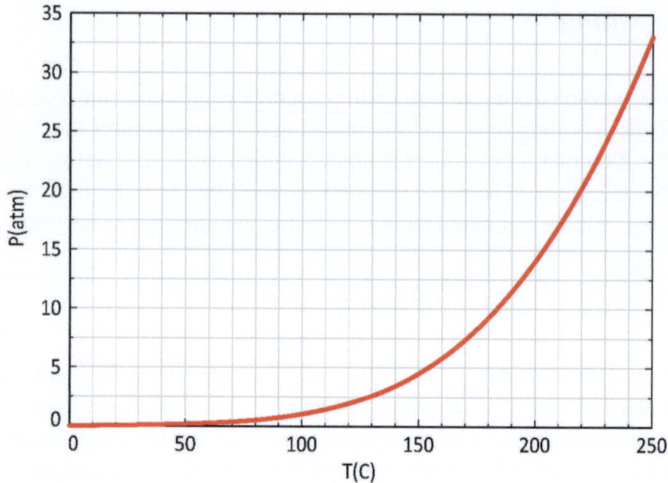

Fig. 11.2 Steam pressure of water as a function of its temperature [6]

11.3 Drying Model

In accordance with operational practice, the temperature control determines the drying process of refractory castables. For example, the following heating recommendation is given to the customers by the manufacturer of a ULCC—refractory castable (H. Dünnes, Calderys, private communication). After setting and removal of the molds, heat up at 15 K/h and allow a holding time of 1 h/0,01 m wall thickness at 150, 350 and 600 °C. The holding (dwelling) time serves to equalize the temperature.

Four assumptions are made to derive a simple model for drying refractory castables:

- It is not the mass transport but the temperature progression that determines the drying process [3, 13, 14].
- The first and second drying process periods of a refractory castables can be treated in the same way, since the first is comparatively short and no steam explosions are observed [3–5, 11–14].
- Once the water is largely removed, there is no danger for the castable (> 350 °C) during further heating-up [4, 5, 11, 12].
- The position L_{IMV} of the integral mean temperature ϑ_{IMV} formed over the entire thickness of the castable layer L, is a quantity which approximately replaces the knowledge of the exact location of the drying plane and can, therefore, be used as a supporting value for the calculation of the temperature gradient in the castable over time. Experimental results [11, 12] support this assumption insofar as the temperature gradient at this point is supposed to be so steep that there is no longer any danger of a steam explosion in the hot, already dried area. This danger

decreases very quickly in the still moist area because of the cooling. Consequently, the greatest danger exists shortly behind the drying front L_{IMV}, i.e. in the vicinity of the integral mean temperature $_{IMV}$.

The linear heating of an infinitely strong plate (slab) is described by the following temperature field [18].

$$\frac{\vartheta(L, t, k, a) - \vartheta_a}{\vartheta_0 - \vartheta_a} = \left[\left(1 + \frac{1}{2F_0}\right) \cdot \mathrm{erfc}\left(\frac{1}{2\sqrt{F_0}}\right) - \frac{1}{\sqrt{\pi \cdot F_0}} \cdot e^{-\frac{1}{4F_0}}\right] \quad (11.19)$$

$\vartheta_0 = \vartheta_a + kt$ is the surface temperature and k [K/s] is the constant heating-up rate. ϑ_a is the initial temperature, e.g. 30 °C.

$F_0 = \frac{a \cdot t}{L^2}$ denotes the Fourier number. If the thickness of the plate (slab) exceeds the value.

$L = 4 \cdot \sqrt{a \cdot t}$ it can be viewed as being infinitely thick (refer to Sect. 2.2.2.2).

In the present example ($L = 0.5$ m, $a = 9 \times 10^{-7}$ m^2/s), this limit is exceeded after $t = 5$ h.

The time progressions of the integral average temperature ϑ_{IMV} and the corresponding location L_{IMV} are shown in Fig. 11.3 as an example ($T_a = 30$ °C, $L = 0.5$ m; $k = 15$ K/h; $a = 9 \times 10^{-7}$ m^2/s). As the drying progression L_{IMV} increases, the temperature T_{IMV} rises steadily. The equations required for the derivation are taken from literature [18, 19].

The temperature curves in Fig. 11.4 have been calculated with Eq. (11.19). The industrially proposed heating-up curves ($k = 15$ K/h) with the interposed holding times (1 h/0.01 m wall thickness) are compared with the continuous heating with $k = 7.5$ K/h according to the theory offered here. It can be seen that for a given wall thickness of $L = 0.2$ m, the temperature curves are comparable and even surpassed by the temporal increase of the holding temperature. The method of targeted continuous

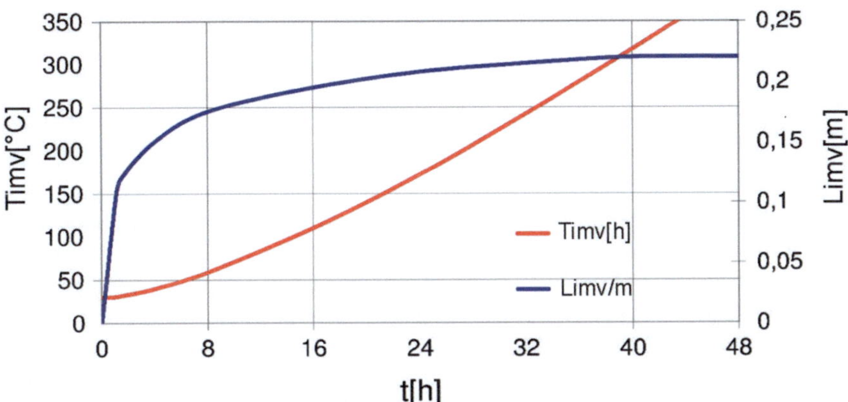

Fig. 11.3 Course of the integral mean value of the temperature ϑ_{IMV} and the location L_{IMV} as a function of time t $\left(\vartheta_a = 30\,°C;\ L = 0.5\ \text{m};\ k = 15\ \text{K/h};\ a = 9 \times 10^{-7}\ \text{m}^2/\text{s}\right)$

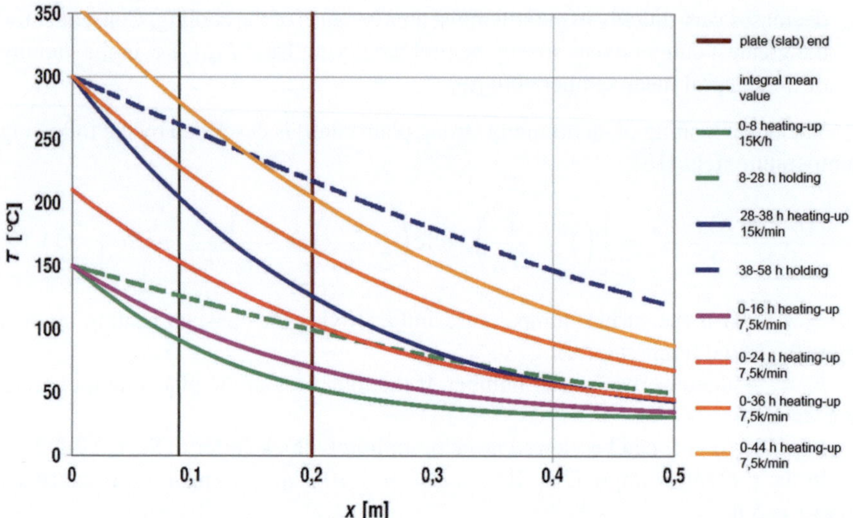

Fig. 11.4 Comparison between step by step heating-up according to a recommendation gained from practice and continuous heating-up according to the model calculated here $\left(\vartheta_o = 30\ °C;\ k = 15\ \text{K/h and } 7.5\ \text{K/h; resp.,}\ a = 9 \times 10^{-7}\ \text{m}^2/\text{s};\ L = 0.2\ \text{m};\ L_{\text{imv}} = 0.09\ \text{m}\right)$

heating-up is thus gentler and leads faster to the objective. Practical testing is still pending. In Fig. 11.4, the location corresponding approximately to the integral mean value of the temperature has also been drawn in. It is $L_{\text{IMV}} = 0.45\ L$.

The result of the derivation is a simple numerical equation for practical use [19]:

$$\frac{\vartheta_0 - \vartheta_a}{\vartheta_{IMV} - \vartheta_a} = \frac{\sqrt{L}}{3000 \cdot a^{0.25}} \cdot \frac{t}{3.7 \times 10^{-3} \cdot t - 33.5} \tag{11.20}$$

The required heating-up rate is then

$$k = \frac{\vartheta_0 - \vartheta_a}{t}\ [\text{K/s}]. \tag{11.21}$$

Let ϑ_a=30°C, ϑ_{IMV}=150°C

The characteristic parameter is the integral mean value ϑ_{IMV} of the temperature curve in the refractory castable. It provides information about the average temperature in the dry material. The heating-up rate should be selected so that the temperature curve in the castable is flat over time. If possible, the temperature should be $< 200\ °C$ until the cold end of the castable exceeds $110\ °C$. With this stipulation, k can be calculated approximately: Let $= 30\ °C$, $\vartheta_{IMV} = 150\ °C$; $L = 0.2\ \text{m}$; $t = 86{,}400\ \text{s}$ (24 h) and $a = 9 \times 10^{-7}\ \text{m}^2/\text{s}$ (thermal conductivity of castable), so that Eq. (11.20)

Fig. 11.5 Heating rate k/°C/ h and surface temperature T/C_0° as a function of the wall thickness L/m of the castable plate (slab) after 24 h at a given mean temperature $\vartheta_{IMV}/°C$, initial temperature $\vartheta_a = 30\ °C$ and thermal conductivity $a = 9 \times 10^{-7}\ m^2/s$

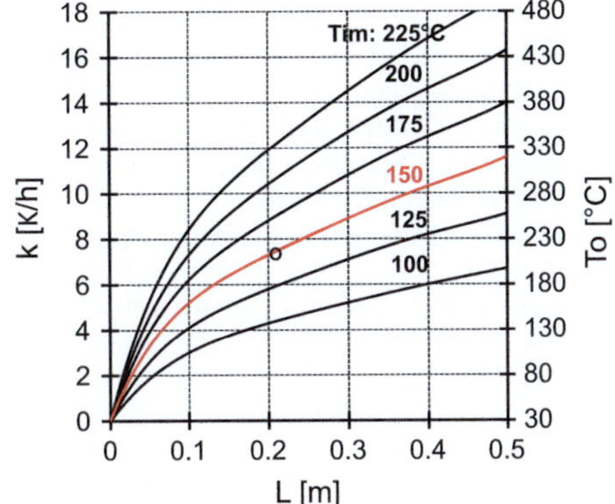

calculates $\vartheta_0 = 205\ °C$. From this, according to Eq. (11.21), the heating-up rate is $k = 0.002$ K/s (7.3 K/h). The temperature on the cold side can be read off from the curves shown in Fig. 11.4 for the total thickness L of the plate and its heating-up time t at a given heating-up rate k. For $L = 0.2$ m and $k = 7.5$ K/h, it is just 105 °C after 24 h (red line). Here, for $L_{IMV} = 0.45\ L = 0.09$ m $= 150\ °C$.

Figure 11.5 also shows the result on the red curve. The curves show the heating-up rate k [°C/h] and the surface temperature ϑ_0 [°C] after the linear heating-up time of 24 h as a function of the wall thickness L [m] of the castable plate (slab) at a given mean temperature ϑ_{IMV} [°C], initial temperature $\vartheta_a = 30\ °C$ and thermal conductivity $a = 9 \times 10^{-7}\ m^2/s$. This representation offers the possibility to decide on the boundary conditions of drying without calculating. For example, it is recommended to select a decreasing mean temperature with increasing wall thickness or to heat up more slowly. If, however, experience shows that a mean temperature of e.g. 150 °C is not a problem, heating-up can even be conducted more rapidly. After more than 24 h, the risk of a steam explosion is generally significantly lower.

Equations 11.20 and 11.21 are only valid for slow heating-up and for times > 3 h. For the calculation of laboratory tests, please refer to the corresponding publication [19].

References

1. Junge, K., Telljohann, U. , Drying of green bricks/tiles—influence of moisture conductivity on shrinkage progression and drying. ZI-Yearbook 21 ff (2005)
2. Junge, K., Tretau, A.: Energy input for drying green bricks/tiles in chamber dryers. ZI-Yearbook 25 ff (2007)
3. Großwendt, I.: Model for description of drying refractory concretes. Dr.-Ing. Dissertation, RWTH-Aachen, Germany, 9 Aug 2007
4. Akiyoshi, M.M., Cardoso, F.A., Innocenti, M.D.M., Pandolfelli, V.C.: Key properties for the optimization of refractory castable drying. Refractory Appl. News **9**(4), 14–16 (2004)

5. Innocenti, M.D.M., Miranda, M.F.S., Cardoso, F.A., Pandolfelli, V.C.: Vaporization process and pressure buildup during dewatering of dense refractory castable. J. Am. Ceramic Soc. **86**, 1500–1503 (2003)
6. Gerthsen, C., Meschede, D.: Physik. Springer, Berlin (2005)
7. Reid, R.G., Prausnitz, J.M., Poling, B.E.: The Properties of Gases and Liquids, pp. 208–209. Mc. Graw-Hill, New York
8. Junge, K.: Data on moist air drying technique. ZI-Yearbook 128–137 (1996)
9. Bird, R.B., Stewart, W.E., Lightfoot, E.N.: Transport Phenomena (1960) 541. Wiley, New York
10. Grunewald, J.: Lecture "Bauphysik." TU-Dresden, Germany (2009)
11. Abe, H.: Drying mechanism of castable refractories. J Tech Assoc Refractories Japan **23**(3), 164–170 (2003)
12. Abe, H.: The effect of permeability on the explosion of castables during drying. J. Tech. Assoc. Refractories Japan **24**(2), 115–119 (2004)
13. Block, F.R., Geber, H., Gross, C., Pernot, J., Zografou, C.: Temperature and drying supervision in manufacturing of monolithic construction parts. In: Proceedings of 37th International Colloquium on Refractories, Aachen, Germany, pp. 210–217 (1994)
14. Schmitz, W., Bank, H., Block, F.R., Geber, H., Groß, C., Manga, D., Maier, H.R., Franken, K., Metzlaff, K.: Measuring technique for supervision of heating-up and drying of the refractory lining. In: Proceedings 38th International Colloquium on Refractories, Aachen, Germany, pp. 48–56 (1995)
15. Volmer, M.: Kinetics of Phase Formation, p. 161. Steinkopf, Dresden (1939)
16. Pötschke, J.: Emergence and Behavior of Excretions During the Solidification in "Solidification of Metallic Melts", pp. 221–231. DGM (1981)
17. Routschka, G.: Refractory Materials, 2nd edn. Vulkan Verlag, Essen (2004)
18. Carlslaw, H.S., Jaeger, J.C.: Conduction of Heat in Solids, 2nd edn. Oxford University Press, Oxford (1959)
19. Pötschke, J.: A Novel Model for Drying Refractory Castables; Refractories Worldforum, pp. 99–106 (2010)

Chapter 12
Thermal Shock Resistance

12.1 Wedge Splitting Method

Figure 12.1 shows the crack progression in a refractory material subjected to a complex notch / tensile / bending stress. The strong branching of the incipient cracks, starting from a center, is clearly visible. The picture indicates how complex the measurement of fracture-mechanical properties of these materials is.

A variety of measurement methods are used to test mechanical properties of refractory materials [1]. The theoretical connections used for the evaluation are mainly derived from linear elastic fracture mechanics. However, this leads to inaccurate results because nonlinear elastic fracture mechanics better describe the properties of refractory materials and concretes (castables). A detailed description can be found, among others, in the works Harmuth, and Tschegg [2]. Characteristic quantities for the thermal shock resistance are the material properties: fracture energy G_f [J/m^2], fracture stress σ_f [GPa], modulus of elasticity E [GPa], thermal conductivity a [m^2/s] and coefficient of expansion α_T [1/K]. Added to this are dimension [m], temperature difference ΔT [K], heat transfer coefficient h [W m^{-2} K^{-1}] and time [s].

The "wedge splitting method" developed by Tschegg [3] for the determination of fracture mechanics parameters is shown schematically in Fig. 12.2.

For the wedge splitting test, a wedge out of densely sintered corundum is slowly (0.5 mm/min) pressed into the specimen (approx. $10 \times 10 \times 7$ cm^3) via a roller system, causing it to crack along a prefabricated narrow groove. Usually, the crack consequently branches off only slightly and the crack edge widening is immediately monitored visually. The application is carried out up to approx. 1500 °C. The horizontal force F_H is calculated from the vertical component F_V and the wedge angle α (10°):

$$F_H = F_V/2 \cdot \tan(\alpha/2) \; [\text{N}] \tag{12.1}$$

© The Author(s), under exclusive license to Springer Nature Switzerland AG 2024

J. Pötschke, *Refractory Fundamentals in Metallurgical Practice*,

https://doi.org/10.1007/978-3-031-63709-4_12

Fig. 12.1 Crack progression
in refractory material [23]

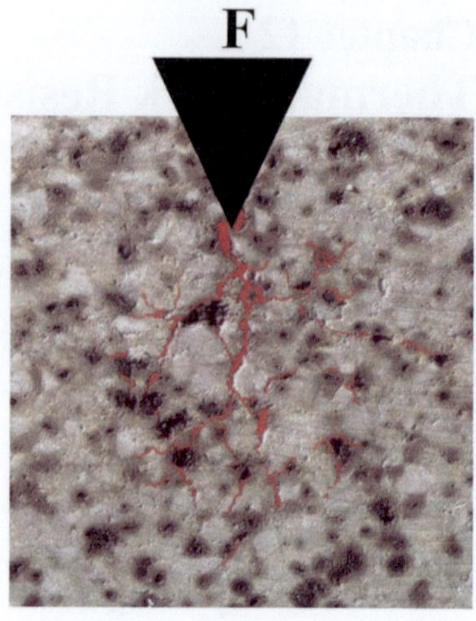

Fig. 12.2 Wedge splitting
method according to
Tschegg [3, 18]

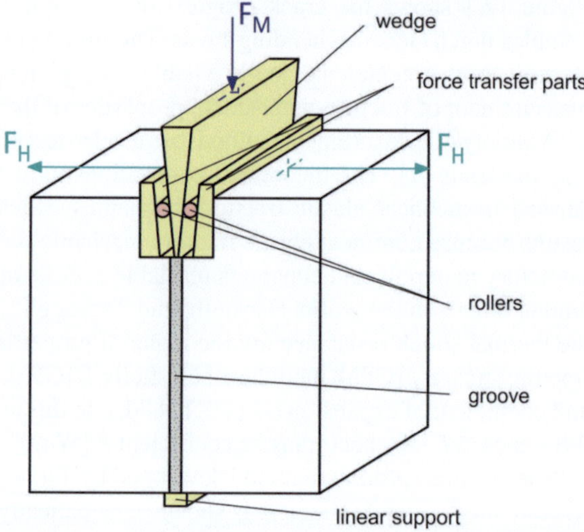

The results of the measurement are force/displacement diagrams of the form
illustrated in Fig. 12.3.

The fracture energy G_f and the adequate fracture work γ_{WOF} are calculated as the
integral under the force/displacement measurement progression:

$$G_f = \frac{1}{A} \int_0^\delta F(a) \cdot \mathrm{d}a \; \left[\mathrm{J/m^2}\right] \qquad (12.2)$$

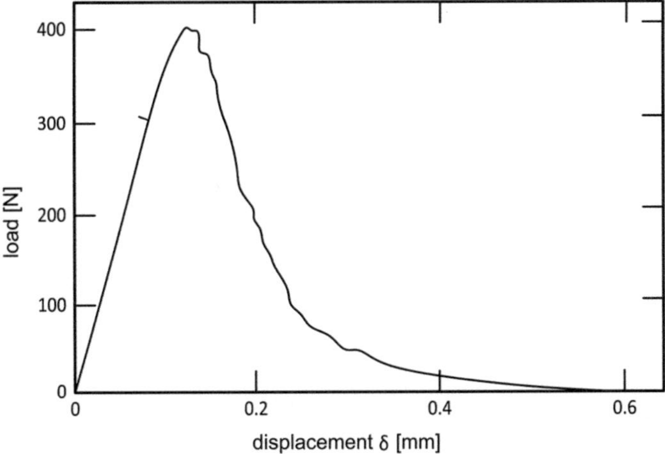

Fig. 12.3 Load/displacement diagram [18]

The work of fracture γ_{WOF} (work of fracture) refers (correctly) to the total newly created ligament area, i.e. to 2A:

$$\gamma_{WOF} = G_f/2 \ \left[\text{J/m}^2\right] \tag{12.3}$$

Due to the roughness of the fracture surface, the actual surface area is 20 to 30% larger than the geometrically measured surface area, to which, however, the evaluation refers, as agreed.

Advantages of the wedge splitting method are:

- The sample size is adapted to the coarse structure of refractory material and enables reproducible measurement results
- As a result of the acute wedge angle, crack progression is already caused by small forces, i.e. only a negligible excess energy is stored in the system. The crack progression takes place steadily.
- As a result of the acute wedge angle, the crack opens only slightly, which keeps measurement inaccuracies low. In fact, there is complex mechanical stress on the specimen [2], with the stress components due to notch effect, bending and tension: σ_{kbz}.
- The crack edge opening (load attack point) can be observed visually or with optical aids, so that the calculation of the fracture energy G_f is directly possible by integrating the measurement curve.
- The application of the method is possible up to a maximum of 1500 °C. At higher temperatures, creep and deformation of the wedge out of sintered corundum start to become noticeable. The E-modulus is determined only inaccurately.

The modulus of elasticity E [GPa] is determined more accurately in the 3-point bending test (EN993-6 and -7), since the linear elastic range is rarely well developed in refractory materials. Specimens ($25 \times 25 \times 150$ mm^3) are supported along a length of 125 mm and subjected to a continuously increasing load (0.15 MPa/s) in the middle until fracture occurs once F_{max} is reached. The fracture stress σ_f and the modulus of elasticity E are then:

$$\sigma_f = \frac{3 \cdot F_{max} \cdot l}{2 \cdot b \cdot h^2} \text{ [GPa]} \tag{12.4}$$

and

$$E = \frac{F \cdot l^3}{4 \cdot \delta \cdot b \cdot h^3} \text{ [GPa]} \tag{12.5}$$

The measured deflection is $\delta(F)$ [mm].

The rough approximation to nonlinear-elastic fracture mechanics is possible by formally applying Hook's Eq. (12.1) to the measured V_E modulus. The V_E modulus roughly considers the real deformation behavior of refractory material. To determine it, the measured strength maximum σ_{max} is linearly connected to the coordinate origin and the gradient $\sigma_{max}/\varepsilon_{max} = V_E$ is used as a substitute for the modulus of elasticity. It is always $V_E < E$ [GPa]. Experience shows that a high ratio of cold bending strength/V_E-modulus points to good thermal shock resistance.

12.2   Thermal Shock Quality Values

Determining the resistance of different materials to fracture in service as reliably as possible has always been a problem. Refractory ceramics are a particularly difficult material to estimate because their microstructure (structural constitution) is very heterogeneous. Coarse grains, fine-grained or even amorphous phases, which bind them, determine thermomechanical behavior together with the pores. In addition, there are the high thermal and chemical stresses. Figure 12.4 shows an overview of the complex stress factors related to refractory material.

Basically, two types of thermal shock quality values can be distinguished from one other:

The "**damage parameter**" describes the prevention of fracture by making crack initiation more difficult.

The "**damage tolerance parameter**" describes the prevention of fracture by making crack branching more difficult.

For easier understanding of the various thermal shock quality values, we consider Hook's basic equation of linear elastic fracture mechanics in the two-dimensional stress state:

$$\sigma = E \cdot \varepsilon / (1 - v) \text{ [GPa]} \tag{12.6}$$

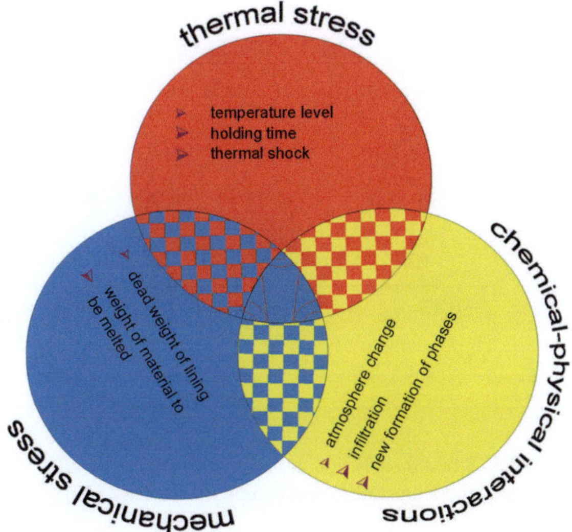

Fig. 12.4 Stress on refractory material

It combines the modulus of elasticity E [GPa] with the acting mechanical tensile stress σ [GPa] and the relative elongation (stretching).

$\varepsilon \equiv \Delta l/l_0$. $v \approx 0.2$ is the transverse contraction number named after Poisson. The coefficient of linear thermal expansion is

$$\alpha_T = \frac{1}{l_0} \cdot \frac{l - l_0}{T - T_0} = \frac{\varepsilon_{th}}{\Delta T} \; [1/K] \tag{12.7}$$

where the relative length change is set in relation to the temperature difference.

The mechanical elongation (stretching) ε and the thermal εth are equal under the same stress conditions:

$$\sigma_f = \frac{\alpha_T \cdot E}{(1 - v)} \cdot (T_m - T) \; [\text{GPa}] \tag{12.8}$$

T_m is the integral mean value with parabolic temperature profile $T(x)$

$$\frac{T_R - T(x)}{T_R - T_0} = \left(1 - \frac{x^2}{R_x^2}\right) \tag{12.9}$$

at a given time t over the cross-Section 2Rx, $(- R_x \leq x \leq + R_x)$ of the body. The mean value of the temperature $T_m = 2/3T_0 + 1/3T_R$ separates the tension region ($T_m > T$, $\sigma > 0$) from the compression region ($T_m < T$, $\sigma < 0$). For $T_0 = 0$, $T_m = 1/3T_R$ and for $T_R = 0$, $T_m = 2/3T_0$. Figure 12.5 shows the stationary stress distribution during cooling of a large, thin plate.

In a basic derivation of the thermal shock quality values, Harmuth [4] shows that for a parabolic temperature progression the following applies:

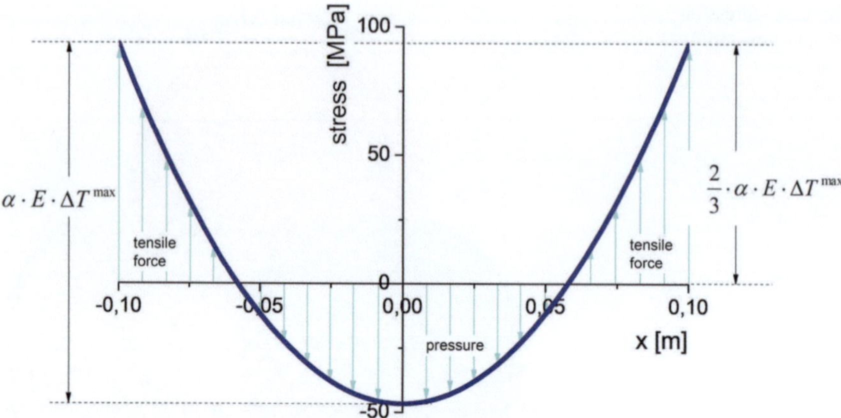

Fig. 12.5 Stress distribution at a time t in a plane-parallel plate cooling uniformly on both sides with parabolic temperature distribution (E. Brochen [20])

$$(T_m - T) = -\frac{k}{a} \cdot \left(\frac{R_x^2 - 3x^2}{6} \right) \ [\text{K}] \tag{12.10}$$

$k = \frac{\partial T}{\partial t}$ [K/s] is the constant rate of change speed of temperature throughout the whole plate and its surroundings (stationary condition) and $a\,[\text{m}^2/\text{s}]$ is the thermal conductivity of the ceramic material. Putting this in Eq. 12.8 gives as critical cooling rate

$$\left(\frac{\partial T}{\partial t} \right)_C = \frac{a \cdot \sigma_f \cdot (1 - \nu)}{\alpha_T \cdot E} \cdot \left[\frac{6}{R_x^2 - 3x^2} \right] \left(\frac{\text{K}}{\text{s}} \right) \tag{12.11}$$

$2R_x$ [m] is the thickness of the plate and $- R_x \leq x \leq + R_x$ is the running coordinate. Upon cooling $(\partial T/\partial t < 0)$, the tensile stress develops at the surface $x = \pm R_x$, i.e. the square bracket of Eq. 12.11 is $\left[-3/R_x^2 \right]$.

Upon heating $(\partial T/\partial t > 0)$, tensile stress occurs in the core $x = 0$, i.e. in Eq. 12.11 the following is valid $\left[+6/R_x^2 \right]$. Consequently, heating may occur at twice the rate of cooling. The geometry flows into the result quadratically.

From Eq. 12.10, the maximum temperature difference is calculated as the difference between the temperatures responsible for the maximum stresses under tension $(x = \pm R_x)$ and compression $(x = 0)$ to be $(T_{Rx} - T_0) = R_x^2 - k/2a$ (refer to Fig. 12.5).

Kingery [5] developed the first model for estimating the thermal shock resistance of brittle ceramic materials. According to Eq. (12.8), his damage parameter (thermal shock quality value I) is $R \equiv (T_m - T)_C$

$$R = \frac{\sigma_f}{\alpha_T \cdot E} \cdot (1 - \nu) \ [K] \tag{12.12}$$

R stands for the just barely tolerable temperature difference occurring abruptly at the surface of the body ($T_m - T_C$). The greater the damage parameter R, the higher the resistance to thermal stress, i.e. the better the thermal shock resistance of the material should be σ_f [GPa] is the fracture stress at the initial temperature.

An improved approximation for more difficult heat transfer, i.e. low Biot number $Bi = h \cdot L/\lambda$ (refer to Sect. 2.2.6.2, Eq. 2.2.66), is the equation derived by Büssem [6]

$$R' = \frac{\sigma_f \cdot \lambda}{\alpha_T \cdot E} \cdot (1 - v) = R \cdot \lambda \ [\text{W/m}] \tag{12.13}$$

which additionally considers the thermal conductivity λ [W/m K] of the ceramic material (thermal shock quality value II). h [W m^{-2} K^{-1}] is the heat transfer coefficient.

For rapid temperature change, i.e. at large Biot number $Bi = h \cdot L/\lambda$, (Eq. 2.2.66) Büssem [6] derived the damage parameter R'', which takes into account the thermal conductivity a [m^2/s] (thermal shock quality value III):

$$R'' = \frac{\sigma_f \cdot a}{\alpha_T \cdot E} \cdot (1 - v) = R \cdot a \left[\text{K m}^2/\text{s}\right] \tag{12.14}$$

The damage parameters R, R' and R'' describe the resistance of the intact material to its failure by crack formation. In addition, the rarely used parameter

$$R''' = \frac{E}{\sigma_f^2 \cdot (1 - v)} \ [1/\text{Pa}] \tag{12.15}$$

must be mentioned. This already indicates the so-called kinetic crack progression (refer to (Eq. 12.16) and is inversely proportional to the elastic energy stored during crack initiation [4].

The introduction of nonlinear fracture mechanics by considering the energy balance of the fracture process is owed to Griffith [7] and Irwine [8]. The following two damage tolerance parameters were derived by Hasselman [9]. (refer to Sect. 2. 3.2.2):

$$R'''' = \frac{G_f \cdot E}{2 \cdot \sigma_f^2 \cdot (1 - v)} \ [\text{m}] \tag{12.16}$$

and

$$R_{st} = \sqrt{\frac{G_f}{2 \cdot \alpha_T^2 \cdot E \cdot (1 - v)}} \left[\text{K m}^{1/2}\right] \tag{12.17}$$

G_f [J/m^2] is the fracture energy, σ_f [MPa] is the initial strength, E [GPa] the modulus of elasticity and α_T [K^{-1}] the coefficient of linear thermal expansion.

As R'''' increases, the formation of incipient cracks becomes more difficult and, as R_{St} increases, their growth becomes more difficult. Both damage tolerance parameters R'''' and R_{st} describe the resistance to initiation of cracks and their branching (spreading) in the already pre-damaged material. They are, therefore, indicative for refractory material containing many pores.

Comparing the parameter for kinetic crack progression Eq. (12.16) with that for stationary crack growth Eq. (12.17), the following is apparent [3, 10].

The modulus of elasticity is in the numerator in the kinetic cracking process and in the denominator in the stationary case. The optimization of a refractory material is, therefore, problematic. The modulus of elasticity should (theoretically) be large (R'''') or small (R_{St}), respectively, which is not possible at the same time. Therefore, it depends on the required stress, while the refractory material is in service. It determines the selection.

During rapid cooling, tensile stresses prevail in the colder surface compared to the core and ***many*** cracks may develop, which jointly reduce the stresses and thus increase thermal shock resistance. Consequently, the elastic energy stored in the body may be so low that the influence of the catastrophic crack progression R'''' is minute compared to the stationary crack propagation R_{St}. As cooling increases, the stresses continue to rise, causing the cracks to become progressively longer that is stationary. To make the stationary crack progression R_{St} more difficult, E must be small and G_f must be large (Eq. 12.17).

However, if this concept does not work out, catastrophic fracture can occur. This is often observed with dense, fine-grained ceramics. Their E-modulus is usually high, so that the elastic energy stored in the structure is large.

If, on the other hand, the temperature is increased very rapid, the still cold core is still subjected to tensile stress. In that area, however, often only ***a few*** cracks develop, on which the entire stored elastic energy is concentrated and which can, therefore, grow kinetically (Eq. 12.16). This tears the body apart and represents an incalculable risk. Stationary crack progression then plays a minor role. E and G_f should be chosen large even at small σ_f to make kinetic crack growth more difficult.

Heating-up is less dangerous than cooling down (refer to Eq. 12.11). Depending on the heat transfer (Biot number), the permissible temperature difference during heating-up may be two to three times as large as during cooling [4] (refer to Sect. 12.3). During heating-up, the tensile stress acting in the core, prop. ($T_m - T$), is only about one third of the existing total stress, prop. ($T_R - T_0$). Upon cooling down, on the other hand, it is two thirds on the outside surface (refer to Fig. 12.5). As the dimension increases, the risk of component failure increases quadratically (Eq. 12.11) (Figs. 12.6 and 12.7).

In order to achieve the greatest possible structural flexibility, a low modulus of elasticity (refer MgO-C in Fig. 12.8) often proves effective in steel mill practice with very high fracture stress. This would suggest choosing R_{St} as the damage tolerance parameter [10–13]. In fact, however, the so-called characteristic length L_C, which is regarded as a measure of the structure flexibility, in other words the residual resistance force after preceding mechanical stress. It is derived from R'''' [2, 14], often proves its worth:

Fig. 12.6 Typical crack pattern of the slide plate after the first use (service) [23]

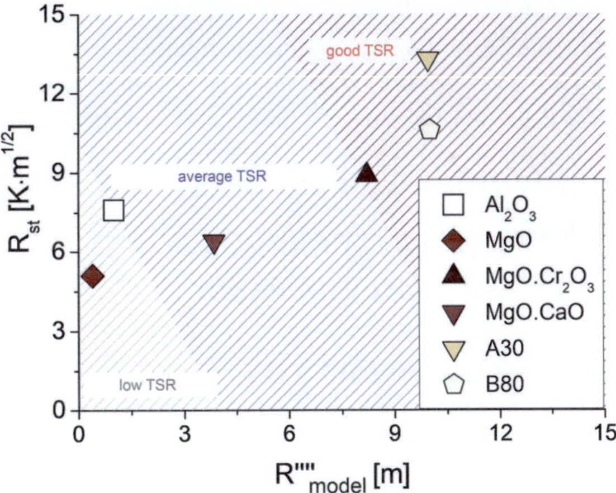

Fig. 12.7 Thermal shock resistance behavior of known refractory materials [20, 21]

$$L_C = \frac{G_f \cdot E}{\sigma_f^2} \ [m] \tag{12.18}$$

The characteristic length L_C should be as large as possible.

Laboratory experiments with a large number of refractory materials have shown that the fracture energy passes through a maximum at approx. 1000 °C and that

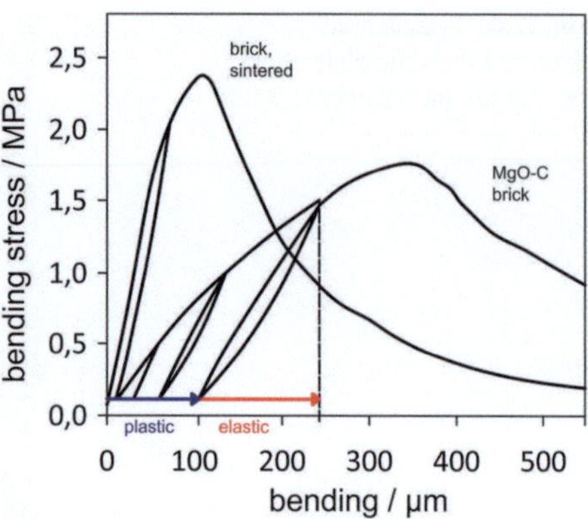

Fig. 12.8 Comparison of the part-elastic bending behavior of two materials under increasing stress [24]

there is no close correlation between the thermal shock resistance and the characteristic length [15–19]. According to these results, R_{St} seems to be more suitable as a guide for the thermal shock resistance. Brochen [20] comes to a comparable result. Conversely, Harmuth (priv.Mitt. [4]) refers to his experience, according to which L_C, when comparing strung together product spectra (e.g. differently fired brick types but also including carbon-bonded brick types) according to their characteristic length, agrees very well with the thermal shock resistance known from practical observations. This is especially true if the crack initiation, i.e. L_C is decisive.

Another point is the determination that in Eq. (12.16) no direct influence of temperature can be seen. In the case of stationary crack progression (Eq. 2.17), on the other hand, the denominator contains the square of the linear thermal expansion coefficient α_T [K^{-1}]. α_T should be as small as possible, which practice confirms. If $\alpha_T = 0$, there is no thermal stress in the component (heat-resistant glass).

12.2.1   Example of Calculation of Damage and Damage Tolerance Parameters

A standard brick ($0.240 \cdot 0.175 \cdot 0.065$ m^3) out of corundum cools rapidly from 1500 °C to room temperature. Let the characteristic length be $R_x = 0.0325$ m. The heat transfer coefficient is $h = 300$ W/m^2 K, density $\rho = 3300$ kg/m^3, elasticity modulus E $= 35$ GPa, thermal expansion coefficient $\alpha_T = 9 \times 10^{-6}$ K^{-1}, specific heat $C_P = 1050$ J/kg K, thermal conductivity $\lambda = 5.5$ W/m K, thermal conductivity $a = 1.6 \times 10^{-6}$ m^2/s, transverse contraction coefficient $\nu = 0.22$, critical fracture stress $\sigma_f = 0.15$ GPa, work of fracture $G_f = 300$ J/m^2. Thus, from Eq. 12.11, the permissible constant cooling rate is approximately calculated to be

$$\left(\frac{\partial T}{\partial t}\right)_C = \frac{a \cdot \sigma_f \cdot (1 - \nu)}{\alpha_T \cdot E} \cdot \left[\frac{-3}{R_x^2}\right] \leq -2K/s \tag{12.19}$$

The calculated damage and damage tolerance parameters are

$$
\begin{aligned}
R &= 371 \ [\text{K}] \ (12.12)\\
R' &= 2043 \ [\text{W/m}] \ (12.13)\\
R'' &= 6 \times 10^{-4} \ \left[\text{K m}^2/\text{s}\right] \ (12.14)\\
R'' &= 2 \times 10^{3} \ [1/\text{Pa}] \ (12.15)\\
R''' &= 3 \times 10^{-4} \ [\text{m}] \ (12.16)\\
R_{\text{St}} &= 8.2 \ \left[\text{K m}^{1/2}\right] \ (12.17)\\
L_C &= 5 \times 10^{-4} \ [\text{m}] \ (12.18)
\end{aligned}
\tag{12.20}
$$

If these values are exceeded, there is a risk of cracks forming due to thermal stresses.

This simple interpretation is doubtful, because it is not clear which parameter provides the most reliable information in which case. Nevertheless, in the same group of materials there exist useful comparisons using the same parameter that are sufficiently possible. In steel mill practice, one prefers to use the "characteristic length" L_C to compare materials. This takes into account, at least to some extent, the nonlinear behavior of the refractory material used, in particular MgO-C.

12.3 Improved Thermal Shock Quality Values

All thermal shock quality values have in common that the temporal change of the stress field in the entire component is not considered. For this purpose numerical models are created, which, however, do not result in a parameter that can be calculated quickly [16, 19]. In his dissertation, Brochen [20] measures and calculates this stress distribution in Al_2O_3 and MgO products of simple design (geometry) in order to develop enhanced parameters.

For evaluation, he measured the temperature field during both rapid heating-up and quenching and described it in agreement with an analytical model. This was then used to calculate the thermal stresses as a function of location and time. In summary, improved damage and damage tolerance parameters are obtained [20, 21].

12.3.1 Critical Temperature Difference

A critical temperature difference can occur between the environment and the component, at which the strength limit σ_f of the body is reached (damage parameter R) [20]:

$$\Delta T_C = (T_u - T_0)_c = \frac{\sigma_f \cdot (1 - v)}{\alpha_T \cdot E \cdot \sigma^*_{\max}(0, t_C)} \; [^\circ C] \qquad (12.21)$$

σ_f [MPa] is the fracture strength of the material, preferably measured at the temperature being considered. $\sigma^*_{\max}(0, t_c)$ is the dimensionless stress calculated from the time-dependent local temperature distribution, which reaches its maximum at the surface of the body ($x = 0$) at time t_C and leads to fracture:

$$\sigma^*_{\max}(0, t_c) = 0.155 \cdot \ln(\text{Bi}) + 0.1 [-] \; (1 < \text{Bi} < 100) \qquad (12.22)$$

ΔT_C is, consequently, the improved critical temperature jump by taking the stress field into account, which Kingery (Eq. 12.12) calculated in a simplified manner by comparing the surface temperature with the mean integral average of the internal temperature at a given time [4].

12.3.2   Temporal Changes of the Surface Temperature

The critical, temporal change of the surface temperature $\dot{T}_C$, which is the maximum tolerable heating-up rate, is calculated as (E. Brochen [20]):

$$T_C = \frac{a \cdot (T_{Ziel} - T_u)}{L^2} \cdot \left(\frac{\alpha_T \cdot E \cdot (T_{Ziel} - T_u)}{(1 - v) \cdot \sigma_f} \right)^{-2} \; [K/s]. \qquad (12.23)$$

With good approximation, the following applies for this damage parameter:

$$T_C = 19 \cdot V \cdot S^{-2} \; [K/s] \qquad (12.24)$$

L [m] is the characteristic length of the body. (In many cases it equals R_X.) V [K/s] and S [−] are two parameters which, in addition to material properties, take into account the difference between the ambient temperature T_U and the target temperature $T_{Ziel \, (\text{target})}$ specified for the calculation:

$$V = \frac{a \cdot (T_{Ziel} - T_u)}{L^2} \; [K/s], \quad S = \frac{\alpha_T \cdot E \cdot (T_{Ziel} - T_u)}{\sigma_c \cdot (1 - v)} \; [-] \qquad (12.25)$$

12.3.3   Damage Tolerance Parameters R''''

The damage tolerance parameters derived by Hasselman [8, 9] (Eqs. 12.16 and 12.17) take into account the amount of stored energy in a thermally homogeneous body, i.e.

Hasselman, like Kingery [3], neglects the actual temperature distribution present. The resulting distortion is considered to be suppressed.

Brochen [20] calculates the temperature and stress distribution in the entire body at the time t_C, at which the critical fracture stress σ_f [MPa] is just being reached and forms the integral mean value from it:

$$\sigma_m = \frac{\alpha_T \cdot E \cdot (T_U - T_0)}{1 - \nu} \cdot \sigma_m^* \ [\text{MPa}] \tag{12.26}$$

σ_m represents the stress distribution resulting from the inhomogeneous temperature distribution in one value. $\sigma_m^*[-]$ is the calculated mean value of the dimensionless stress. Its approximation is

$$\sigma_m^* = 2.7 \cdot \text{Bi}^{-n} S^{-m} \ [-] \tag{12.27}$$

with

$$n = 1.1 - 0.007 \cdot \text{Bi}; \ (10 < \text{Bi} < 50)$$

and

$$m = 2.2 - 0.08 \cdot V; \ (1.7 < S < 15)$$

The kinetic crack progression is thus calculated to be

$$R''''_{\text{Model}} = \frac{G_f \cdot E}{2 \cdot \sigma_m^2 \cdot (1 - \nu)} \ [\text{m}] \tag{12.28}$$

Accordingly, the following applies to the enhanced "characteristic length":

$$L_{c/\text{Model}} = \frac{G_f \cdot E}{\sigma_m^2} \ [\text{m}] \tag{12.29}$$

An improved calculation of the stationary crack progression index R_{St} (Eq. (6.2.1.7)) requires knowledge of the number, location and length of the incipient cracks and their interaction with each other, which is not given. Therefore, it had to be omitted.

(**Remark**: *The detailed calculations of Brochen [20] are reproduced here as approximate equations sufficiently accurate for practical use*).

12.3.4   Examples for the Calculation of Enhanced Thermal Shock Quality Values

The application of the above-mentioned parameters to describe the thermomechanical behavior of refractory materials does not lead to absolutely accurate results. However, it does allow similar products to be compared with each other. The reason for this is the heterogeneous structure of the refractory materials. It is optimized for a variety of service-related properties, of which thermal shock resistance is only one. The discussion of the properties of different materials using the example of very rapid heating-up, e.g. of a slide gate plate, is carried out with the aid of the thermal shock quality values listed above. The cracks run from the inside to the outside and are deliberately accepted. The material values used are compiled in Table 12.1 [18–21]. σ_F [GPa] denotes the compressive (crushing) strength. In addition, the heat transfer coefficients required for the calculation are given in Table 12.2.

The calculated key figures are given in Table 12.3.

The damage parameters given by Kingery (R) and Büssem (R', R'') do not take into account the design (geometry) of the body, which is why their statement is limited. The influence of the material values is not very clear. The damage tolerance parameters R'''', L_C and R_{St} (Hasselman) are calculated taking into account the fracture energy. However, they do not show clear dependencies consistent with practical experience. The corresponding correction calculations of Brochen [20] prove the significant influence of design (geometry) and heat transfer. Thus, they are an improvement. For water quenching (DIN 51068), i.e. tensile stress at the surface,

Table 12.1 Material values of different refractory materials [18–20]

Data	Al$_2$O$_3$	MgO	MgCr$_2$O$_4$	MgO–C
L [m]	0.25	0.25	0.25	0.25
ρ [kg/m^3]	3300	3100	3100	2950
C_P [J/kg K]	1050	1300	1000	1360
λ [W/m-K]	6	15	6	17
$a \times 10^{+6}$ [m^2/s]	1.7	3.7	1.9	2.4
$\alpha_T \times 10^{+5}$ [K^{-1}]	0.8	1.35	0.9	1.2
σ_F [GPa]	0.15	0.08	0.05	0.03
E [GPa]	35	40	25	20
G_F [J/m^2]	200	200	250	180

Table 12.2 Thermal shock conditions [20]

Heat-transfer-coefficient	Calm air ~ 1000 °C	Natural gas-burner ~ 1300 °C	Contact with SiC—plate			Liquid pig iron ~ 1500 °C
			500 °C	1000 °C	1500 °C	
h [W m^{-2} K]$^{-1}$	~ 60	~ 150	~ 250	~ 300	~ 350	> 1000

Table 12.3 Calculated thermal shock quality values

Result	Dimension	Equation	Al_2O_3	Al_2O_3	MgO	$MgCr_2O_4$	MgO-C
Lsign.	[m]	–	00325	0.25	0.25	0.25	0.25
Bi	[–]	2.2.66	1.63	12.5	5	12.5	4.4
R	[K]	12.12	418	418	116	173	97
R'	[W/m]	12.13	2510	2510	1733	1040	1658
R''-10^4	[K m^2/s]	12.14	7	7	4	3	4
R'''-10^{-3}	[1/Pa]	12.15	2	2	8	13	28
R''''	[m]	12.16	0.2	0.2	0.8	1.6	2.6
R_{St}	[K m$^{1/2}$]	12.17	6.7	6.7	3.7	7.9	5.6
L_C	[m]	12.18	0.3	0.3	1.3	2.5	4
$-(dT/dt)_C$	[K/s]	12.19	2	0.034	0.02	0.016	0.02
ΔT_C	[K]	12.21	2,384	850	331	353	295
$-(dT/dt)_C$	[K/s]	12.24	3.6	0.06	0.01	0.01	0.01
R_m''''-10^3	[m]	12.28	1	97	1535	6394	5680
L_{Cm}	[m]	12.29	1.6	151	2394	9975	8860

calculated for $\Delta T = 1500\ °C$, $\nu = 0.22$, $h = 300$ W/m K^2

Brochen [20] calculated the ratios R_{St} and R'''' for various refractory materials and presented them graphically according to Harmuth [22]. A clear tendency can be seen in Fig. 12.7, which reflects what is seen in practice well. (The tensile strength is only approx. 10% of the compressive (crushing) strength [23]).

Graphite-bonded materials are not included. Table 12.3 shows that their thermomechanical behavior should be similar to that of $MgO\cdot Cr_2O_3$. However, this is not the case. Figure 12.8 shows the comparison between MgO and MgO-C using an example from Hampel and Aneziris [24]. The course for magnesia indicates a comparatively low microstructural flexibility [2], whereas MgO-C has a high one, i.e. the risk of brittle fracture is lower. This is due to the graphitized binder phase, which leads to crack branching (spreading) along the grain/matrix boundary, so that trans-granular cracks are prevented. This pronounced thermoplastic behavior is not taken into account in any key (indicator) figure. Plasticity follows the increase in cyclical stress, whereas elastic behavior hardly changes. The plastic behavior is a "mechanical reserve" of the material and increases the structural flexibility which is so important for the thermal shock resistance.

It becomes clear that the material values compiled in Table 12.3 for MgO-C are questionable and the calculation of the key (indicator) figures does not allow for any statement. Much more, examinations taking plasticity and its dependence on temperature and time into account are required to characterize this group of materials.

References

1. Routschka, G., Krause, O. (ed.): Refractory materials and refractory engineering, DIN-Standards, 2. Vulkan-Verlag (2010): DIN 51068 Testing of ceramic raw materials and materials: determination of resistance to sudden thermal shock and water quenching process for refractory bricks (2010)
2. Harmuth, H., Tschegg, E.K.: Fracture mechanics characterization of coarse grain refractory materials. Veitsch-Radex Rundschau **1–2**, 4 (1994)
3. Tschegg, E.K.: Test equipment for the determination of fracture mechanics characteristic data. Austrian Patent Specification AT 390328 (1986)
4. H. Harmuth: Modelling and Simulation in Building Materials Technology, Lecture Montanuniversität Leoben, Austria Professorship Ceramic Engineering and Materials/Personal Communication (2013)
5. Kingery, W.D.: Factors affecting thermal stress resistance of ceramic materials. J. Am. Ceramic Soc. **38**(1), 3–15 (1955)
6. Buessem, W.R.: Thermal shock resistance of ceramic mixes. Sprechsaal für Keramik-Glas-Email **6**, 137–141 (1960)
7. Griffith, A.A.: Theory of rupture. 1. In: International Congress for Applied Mechanics, Delft, Netherlands, pp. 55 – 63 (1924)
8. Irwin, G.R.: Analysis of stress and strains near the end of a crack traversing a plate. J. Appl. Mech. **24**(3), 361–364 (1957)
9. Hasselman, D.P.H.: Unified theory of thermal shock fracture initiation and crack propagation in brittle ceramics. J. Am. Ceramic Soc. **52**, 600–604 (1969)
10. Woijsa, J., Wrona, A.: Prediction of thermal shock resistance of refractory castables. In: 47th International Colloquium on Refractories, Aachen, Germany 35–37 (2004)
11. Jansen, H.: Wear due to thermomechanical treatment and abrasion in steelmaking processes. Stahl EisenEisen **125**(11), 43–48 (2005)
12. de Anchieta Rodrigues, J., dos Santos, S.F.: The energy of fracture of refractories at high temperatures. In: 48th International Colloquium on Refractories, Aachen, Germany, pp. 190–193 (2005)
13. Schickle, B., Telle, R., Tonnesen, T., Miyaji, D.Y., de Anchieta Rodrigues, J.: Change of mechanical and elastic properties of castables as a function of thermal shock cycles. In: 53rd International Colloquium on Refractories, Aachen, Germany, pp. 86–89 (2010)
14. Buchebener, G., Pirker, S.: New High-Performance Products for Application in the Converter. Veitsch–Radex–Rundschau **2**, 3–14 (1996)
15. Alapin, B., Ollig, M., Pötschke, J.: Thermomechanical Properties of selected refractory materials up to 1500°C. In: 46th International Colloquium on Refractories, Aachen, Germany, pp. 107–112 (2003)
16. Mittler, G., Klima, R., Alapin, B., Pötschke, J.: New method for the thermo-mechanical durability analysis of refractory constructions. In: 46th International colloquium on refractories, Aachen, Germany, pp. 112–114 (2003)
17. Alapin, B., Pötschke, J.: Investigation of the high-temperature properties of some industrial refractories with the wedge-splitting method. In: 48th International colloquium on refractories, Aachen, Germany, pp. 183–189 (2005)
18. Brochen, E., Brückmann, C., Pötschke, J., Mittler, G., Vollenberg, I., Mudersbach, D.: Thermomechanical characterization of magnesia-based refractory products by means of wedge-splitting test above 1,000°C. In: 53rd International Colloquium on Refractories, Aachen, Germany, pp. 82–85 (2010)
19. Vollenberg, I., Mittler, G., Brochen, E., Brückmann, C., Pötschke, J., Mudersbach, D.: Optimization of refractories by combining laboratory investigations with numerical simulation using the example of a MgO-brick in steel ladle and converter. In: 53rd International Colloquium on Refractories, Aachen, Germany , pp. 152–154
20. Brochen, E.: Measurement and modelling of the thermal shock resistance of refractory materials. Dr.-Ing. Dissertation. Bergakademie TH Freiberg, Germany (2011)

21. Brochen, E., Pötschke, J.: A contribution to the measurement and modelling of the thermal shock resistance of refractories—investigation of Al_2O_3- and MgO-based materials. In: 54th International Colloquium on Refractories, Aachen, Germany, pp. 75–79 (2011)
22. Harmuth, H.: A fracture mechanics approach for the development of refractory materials with reduced brittleness. Fatigue Fracture Eng. Mater. Structure **20**, 1585–1603 (1997)
23. Eschner, A.: Refractory seminar GHI–RWTH, Aachen, Germany, 28 Mar to 2 Apr 2011
24. Hampel, M., Anezieris, C.G.: Thermal shock performance of carbon bonded materials. In: 48th International Colloquium on Refractories, Aachen, Germany, pp. 28–32 (2005)

Chapter 13
Thermomechanical Behavior
of Carbon-Containing Refractories

13.1 MgO–C in Steel Metallurgy

13.1.1 Experiments

Mag-carbon bricks consist of approx. 90% by weight MgO and 10% graphitized carbon. They are preferably used in steel metallurgy. The examinations described here serve to systematically determine their properties.

Pure iron (2 kg) is inductively melted (10 kHz.) in a crucible made of sintered corundum and let dwell at 1600 °C. The vacuum vessel contains pure argon (800 mbar). A test bar (20×20 mm^2) is immersed 40 mm deep in the melt for up to 5 h. Two different materials are examined: MgO–C (14% flake graphite, resin-bonded) and MgO–C with an addition of 3% metallic aluminum as an antioxidant, both carbonized at 450 °C for 6 h. To control the oxygen content, it is monitored electrochemically without interruption during the entire test period by means of an EMF probe permanently installed in the crucible.

Samples are taken from the melt during the test and analyzed for their dissolved carbon or aluminum content [1].

13.1.2 Results

After immersion of the MgO–C sample material, gas bubbles are observed rising to the sample surface. This is not the case when using the MgO–C–Al material. Figure 13.1 shows the quantitative results graphically [1, 2]. The continuous red curve describes the increase in dissolved carbon over time when using the MgO–C sample material. The dissolution of the graphite comes to a halt after about 30 min. When the sample is pulled out, a white coating is noticed, which turns out to be MgO (refer to Fig. 13.2). Iron has infiltrated only in very few places.

J. Pötschke, *Refractory Fundamentals in Metallurgical Practice*,
https://doi.org/10.1007/978-3-031-63709-4_13

Fig. 13.1 Time progression of the dissolved carbon and aluminum contents in the molten iron [1]

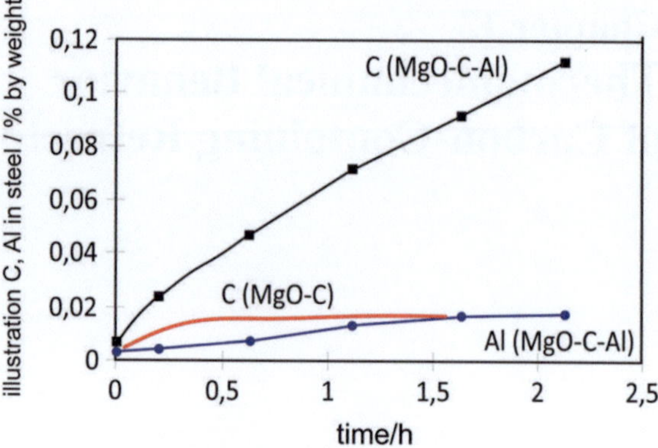

Fig. 13.2 Microstructures: MgO–C (top) and MgO–C–Al (bottom), both after the reaction with iron at 1600 °C

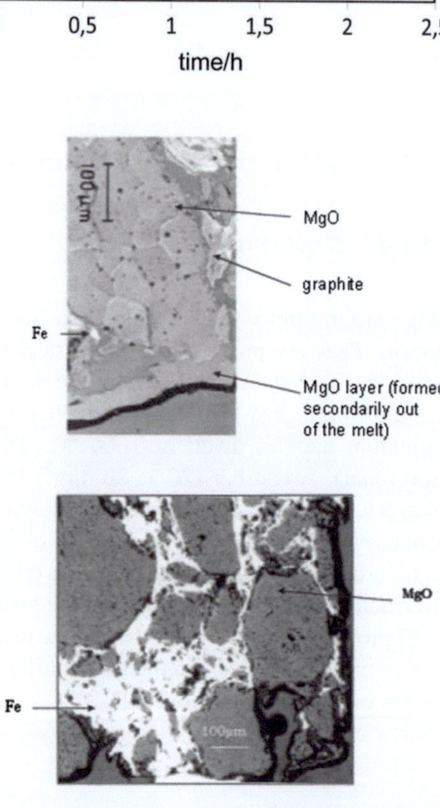

The MgO–C–Al material behaves completely differently. The carbon and aluminum contents increase continuously. The surface remains black, i.e. it is not covered by any MgO layer. The microstructure is strongly infiltrated by iron to a depth of approx. 4 mm (Fig. 13.2).

Figure 13.3 shows the distribution of dissolved carbon and aluminum in the infiltrated iron. Both contents increase with increasing distance from the surface.

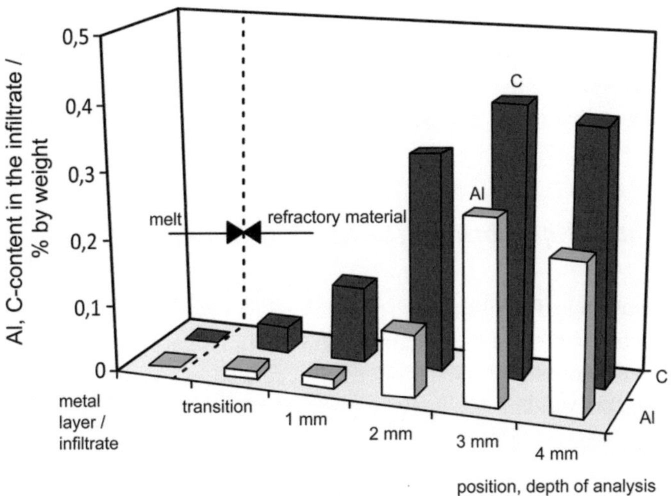

Fig. 13.3 Carbon (black) and aluminum (white) contents dissolved in the infiltrated iron [2]

For a variety of such experiments, the measured dissolved oxygen content is plotted against the carbon content analyzed in the infiltrate and shown double logarithmically in Fig. 13.4. A relatively narrow scatter plot is obtained in which the MgO–C values tend to be at lower contents of carbon and higher oxygen values than the MgO–C–Al value pairs. The envelope can be assigned to CO partial pressures between 0.01 and 0.1 atm.

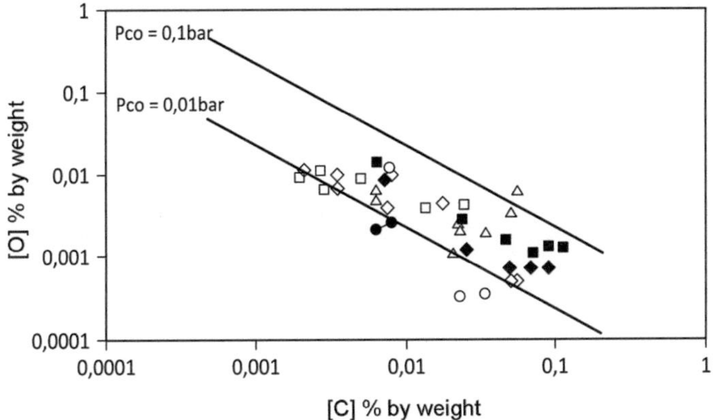

Fig. 13.4 Vacher—Hamilton equilibrium in liquid iron (1600 °C) measured in dissolution experiments taking different MgO–C and MgO–C–Al materials [2]

Based on these results, the behavior of MgO–C and MgO–C–Al materials in steel metallurgy will be systematically investigated and explained in the following chapters. It is a sum of physicochemical influences which makes the use of this group of materials possible in the first place.

13.1.3   Phase Equilibria in the MgO–C Material (Gibbs)

The Gibbs phase rule describes phase equilibria. It is a thermodynamically based law, which is only often referred to as a rule, because its practical application, especially in the determination of phases and components, can lead to uncertainties which falsify the result [3]. It is valid.

$$F = K + 2 - P - R - B \tag{13.1}$$

F Number of degrees of freedom, i.e. number of state variables (pressure, temperature, concentration) necessary for unambiguous determination of equilibrium.
K Number of all chemical components.
P Number of coexisting phases at equilibrium.
R Number of linearly independent reactions.
B Number of further constraining conditions, e.g. stoichiometric linkages upon the transition from an initial material state to the equilibrium state.

Mag-carbon bricks consist of approx. 90% by weight MgO and 10% graphitized carbon. At high temperature, both components (MgO and C) react with each other, producing equal amounts of gaseous Mg and CO:

$$\langle MgO \rangle + \langle C \rangle = \{Mg\} + \{CO\}. \tag{13.2}$$

Therefore, the following applies

$$K = 4 \ (MgO, C, \{Mg\}, \{O\})$$
$$P = 3 \ (MgO, C, gas)$$
$$R = 1 \ D(13.2)$$
$$B = 1 \ \left(P_{Mg} = P_{CO}\right),$$

i.e. $F = 4 + 2 - 3 - 1 - 1 = 1$ (temperature or gas pressure).

From Fig. 13.5 [4], it can be taken that at a total pressure of 1 atm ($P_{CO} = P_{Mg} = 0.5$ atm) the reaction takes place at 1770 °C. If the total pressure drops to 0.1 atm ($P_{CO} = P_{Mg} = 0.05$ atm), the periclase and graphite already react at 1560 °C.

$P_{CO} = P_{Mg} = 1$ atm, i.e. $P_{total=}$ 2 atm, are not reached until 1835 °C. Due to the stoichiometric excess of MgO, C is completely converted to CO.

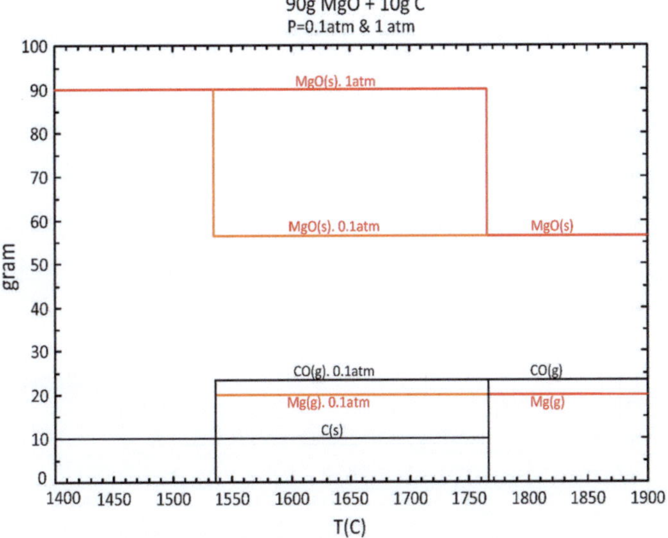

Fig. 13.5 Reaction between periclase and graphite at total pressures 1 atm and 0.1 atm (FactSage)

Figure 13.6 illustrates the dependence of the reaction Eq. (13.2) on temperature and pressure.

As an approximation, the equation can be used:

$$T_{\mathrm{Mg,CO}} = 1757 \cdot P_{[\mathrm{atm}]}^{0.056} \; [^\circ\mathrm{C}]. \tag{13.3}$$

Fig. 13.6 Dependence of the reaction Eq. (13.2) on temperature and pressure

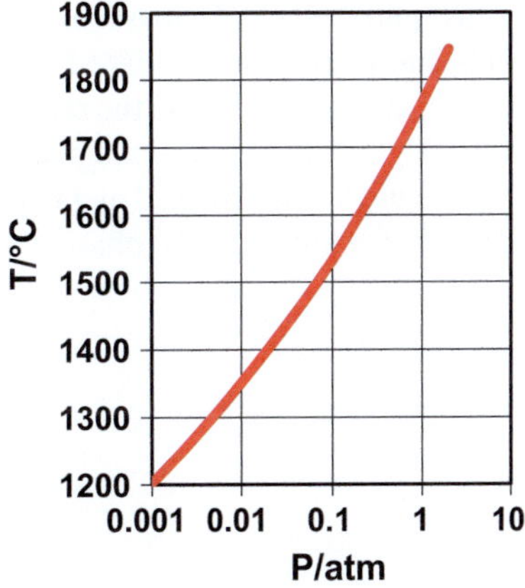

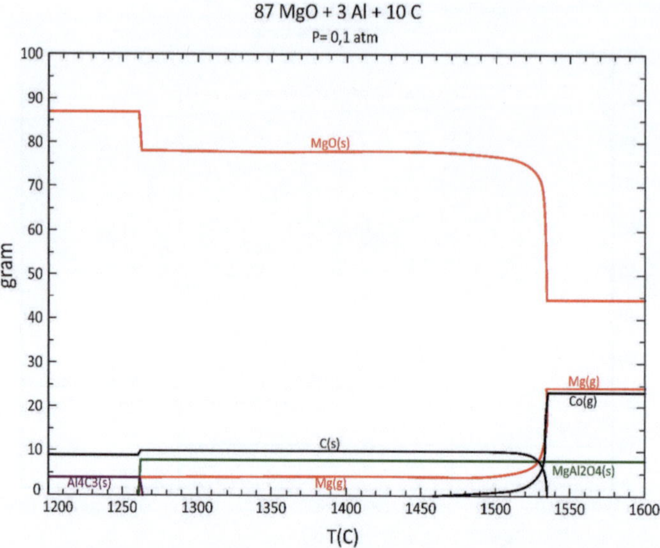

Fig. 13.7 Reactions between periclase, graphite and metallic aluminum at total pressure $P = 0.1$ atm

The addition of the antioxidant Al_{met} increases both the number of components (Al) as well as the number of phases (Al_4C_3 respectively MgO-Al_2O_3) by 1. However, the ratio {CO}/{Mg} is no longer at constant one, i.e. $B = 0$ and therefore $F = 2$. In Fig. 13.7, at a total pressure of $P_{total} = 0.1$ atm, both pressures reach the same value (0.05 atm) only once 1535 °C is reached. Up to 1260 °C, Al_4C_3 is the stable phase, along with graphite and MgO. Above 1260 °C, aluminum carbide disappears completely and spinel is stable along with MgO.

In liquid steel, Al_4C_3 dissolves completely. The solution equilibrium of all components and phases is established with $F = 3$. When the temperature of 1535 °C is reached at $P_{total} = 0.1$ atm, the remaining graphite gasifies completely with reduction of the MgO. Spinel is a stable phase as long as MgO is present (Fig. 13.8).

It should be emphasized that in practice these reactions may be delayed if kinetic impairments influence the phase transformations.

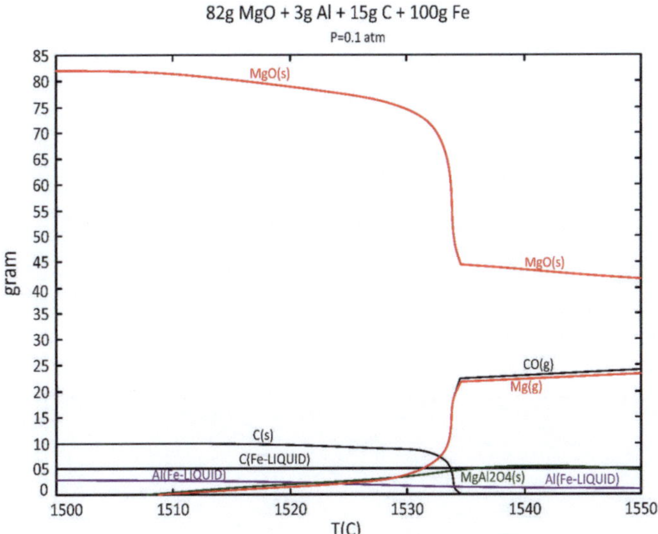

Fig. 13.8 Reaction between periclase, graphite, aluminum and molten iron at total pressure $P =$ 0.1 atm

13.1.4 *Stability of Various Oxides in a CO Atmosphere (Richardson)*

13.1.4.1 Solid Oxides (MgO)

Figure 13.9 shows the thermochemical stability of various oxides of interest for the durability (service life) of refractory materials at different partial pressures of oxygen [3]. The stabilities of MgO (violet line) and MnO (brown line) are graphically highlighted.

Since a compound is more stable the more negative its free enthalpy of formation is ΔG°, the metal of a compound further down reduces the oxide arranged above it to metal, itself being oxidized. For example, metallic aluminum at one temperature reduces the silica (SiO_2) above it and is itself oxidized to Al_2O_3.

According to Gibbs, there is only **one** degree of freedom for all reactions of solid carbon with condensed phases under formation of gaseous carbon monoxide and of further condensed phases. Thus, the gas pressure at a given temperature is fixed. For example, for the reaction

$$\langle MnO \rangle + \langle C \rangle = \langle Mn \rangle + \{CO\} \tag{13.4}$$

the equilibrium constant $K = P_{CO}$, since the activities of the solid components are one. At the intersection of the straight lines of both partial reactions

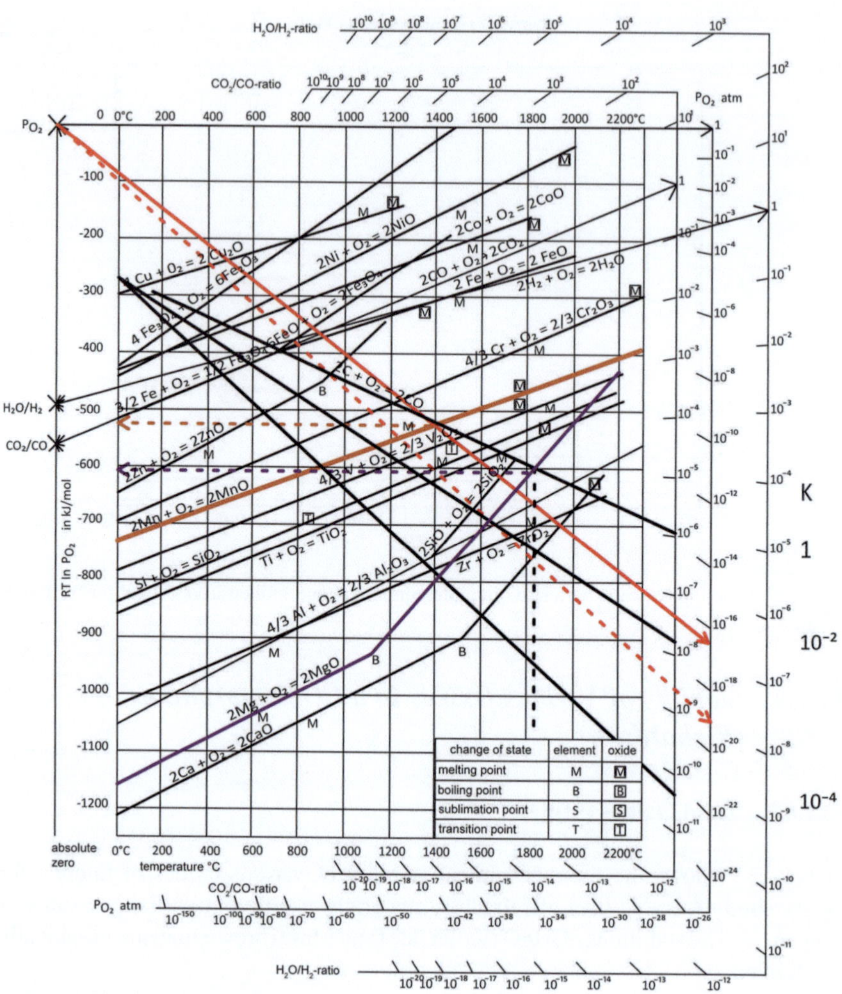

Fig. 13.9 Richardson—Ellingham plot of the thermochemical stability of various oxides at different O₂ partial pressures [3]

$$2\langle Mn\rangle + \{O_2\} = 2\langle MnO\rangle \tag{13.5}$$

and

$$2\langle C\rangle + \{O_2\} = 2\{CO\} \tag{13.6}$$

their free enthalpies of formation

$$\Delta G^{\circ} = RT \cdot \ln P_{O2} \cdot [kJ/mol] \tag{13.7}$$

are equal, i.e. equilibrium exists at this temperature and the corresponding O_2 partial pressure. Figure 13.9 shows the value $\Delta G^\circ = -530$ kJ/FU for $P_{CO} = 1$ atm and 1400 °C (brown, dashed horizontal line). The oxygen partial pressure is read by starting from the point "$\mathbf{P_{O2}}$" (oxygen) and drawing a straight line through the intersection point of the reaction in question at the given temperature (red line) and extending it to the edge scale ($\mathbf{P_{O2}}$). One reads at 1400 °C $P_{O2} = 5 \times 10^{-17}$ atm (red line). The O_2 partial pressure is directly connected to the CO partial pressure via Eq. (13.6). The lines drawn for the CO partial pressures $K = P_{CO,}$ therefore, directly indicate the stability of the initial oxide over against carbon.

If the pressure, especially the CO partial pressure, drops, the free enthalpy ΔG of the reaction changes according to Eq. (13.6)

$$\Delta G = \Delta G^\circ + 2RT \cdot \ln K, \tag{13.8}$$

where the equilibrium constant in the case under consideration is defined as $K \equiv P_{CO}$. Consequently, this equation answers the question that is important in practice for the durability (service life) of refractory material, e.g. in the vacuum process: below what CO pressure does the refractory material react intensely with the graphite once a certain temperature is reached? For example, graphite at $P_{CO} = K = 0.1$ atm reduces MnO already above 1150 °C.

In Fig. 13.9, it can be seen that at $P_{CO} = 1$ atm above 1835 °C MgO is reduced by C. The gases (CO) and (Mg) are formed in equal parts:

$$\langle MgO \rangle + \langle C \rangle = \{Mg\} + \{CO\}; \quad K_{MC} = \frac{P_{Mg} \cdot P_{CO}}{a_{MgO} \cdot a_C}. \tag{13.2}$$

The partial pressure of both gases is equal ($P_{CO} = P_{Mg}$), which is why the equilibrium constant $K_{MC} = P_{CO} \cdot P_{Mg}$ of this reaction can be defined as $K_{MC} \equiv P_{CO}^2$. The sum of the gases gives the total pressure $P_{total} = P_{CO} + P_{Mg}$. As explained above, there is only one degree of freedom. The intersection of the straight lines of both partial reactions

$$2\{Mg\} + \{O_2\} = 2\langle MgO \rangle \tag{13.9}$$

and

$$2\langle C \rangle + \{O_2\} = 2\{CO\} \tag{13.6}$$

is 1835 °C and its free enthalpies of formation

$$\Delta G^\circ = RT \ln P_{O2} = -605 \, \text{kJ/mol} \quad (1.4.18) \tag{13.10}$$

are then of the same size. This value can be taken from Fig. 13.9 [3, 5] or relevant tables [6, 7]. The gas constant is $R = 8.31$ J/mol ·K. For a given partial pressure, e.g. $P_{CO} = 0.1$ atm, corresponding to $K_{MC} = 0.01$, one obtains from Eq. (13.8) $\Delta G = -$

761 kJ/mol. Drawing this value at 1835 °C in Fig. 13.9 and draws a line connecting it to the intersection of the reaction line Eq. (13.6) with the ordinate at $T = 0\ K$ ($-$ 222 kJ/mol), we obtain the intersection $\Delta G = -690$ kJ/mol with the reaction Eq. (13.9) at 1620 °C. The total pressure of the system should, therefore, not fall below $P_{\text{total}} = 0.2$ bar ($P_{CO} = P_{MG} = 0.1$ bar) at 1600 °C. Otherwise the refractory material MgO–C, consisting of magnesia and carbon, will decompose. MgO–C is unsuitable for vacuum metallurgy.

If the partial pressure of the magnesium vapor were $P_{Mg} = 1$ atm, the CO partial pressure in Fig. 13.9 would correspond exactly to the equilibrium constant K (refer to Fig. 13.11), since $K = P_{CO}$ would then apply.

With each power of ten of pressure drop, the "decomposition temperature" of many oxides at steel mill temperatures drops by about 150–200 °C in the presence of graphite (rough rule of thumb).

It should also be remembered that the reactions

$$2\{CO\} + \{O_2\} = 2\{CO_2\} \tag{13.11}$$

and

$$2\{H_2\} + \{O_2\} = 2\{H_2O\} \tag{13.12}$$

describe the redox reactions shown in the Richardson-Ellingham diagram as auxiliary gas equilibria. The CO_2/CO and H_2O/H_2 ratios are read from the scales shown on the right, starting from the respective origins at absolute zero (refer to Chap. 2.1).

13.1.4.2   Molten Slags (SiO_2)

The Richardson-Ellingham diagram also answers the question as to the extent to which an oxide component, whose activity is less than 1, is reduced by carbon. At 1600 °C, in a slag of composition (% by weight) 18% Al_2O_3, 30% CaO, 10% MgO, 27% SiO_2 the activity of silica $a_{SiO_2} = 0.01$ (Fig. 13.10) [8], i.e. only 1% of the value used in the Richardson-Ellingham diagram $a_{SiO_2} = 1$.

How does the pure carbon ($a_c = 1$) present in a MgO–C brick react with the silica content of this slag at 1600 °C?

The two partial reactions are

$$2\langle C\rangle + \{O_2\} = 2\{CO\} \tag{13.13}$$

and

$$(Si) + \{O_2\} = \langle SiO_2\rangle. \tag{13.14}$$

Fig. 13.10 SiO_2 activity in the slag CaO–Al_2O_3–SiO_2–10% by weight MgO (1600 °C) [8]

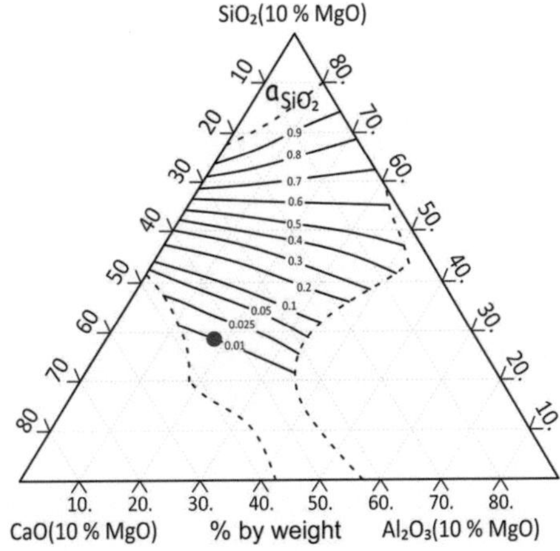

Equation 13.13 is shown for activities $a_C = 1$ and $P_{CO} = 1$ atm, black in Fig. 13.11, and Eq. 13.14 for $a_{SiO_2} = 1$ is blue. Both degrees intersect at about 1650 °C. At 1650 °C and 1 atm CO, pure SiO_2 is still barely stable.

However, if the free enthalpy $\Delta G° = RT \ln p_{O2}$ of reaction Eq. (13.13) is less than or at least equal to that of reaction Eq. (13.14), the carbon reduces the silica to metallic (liquid) silicon, forming CO:

$$\langle SiO_2 \rangle + 2\langle C \rangle = (Si) + 2\{CO\} \tag{13.15}$$

This reaction takes place above 1650 °C (Fig. 13.11).

However, the CO partial pressure in the MgO–C material at 1600 °C is not 1atm, but approx. 0.05 atm (refer to Sect. 13.1.2). The black, roughly dotted straight line intersects the blue one at 1450 °C, i.e. at higher temperature the silica is reduced. The oxygen partial pressure is 10^{-18} atm (red arrow).

Analogous to Eq. (13.8), in the slag the free enthalpy of the "silica reduction" Eq. (13.14) changes as a consequence of the decreased activity a_{SiO_2} to

$$\Delta G = \Delta G° = RT \ln a_{SiO_2}. \tag{13.16}$$

From Fig. 13.11, at about 1650 °C, we take on the ordinate $\Delta G° = -560$ kJ/mol. With $R = 8.31$ J/mol · K, $T = 1923$ K and $a_{SiO_2} = 0.01$, it follows from Eq. (13.16) $\Delta G = -634$ kJ/mol, i.e. the "SiO_2 line" turns downward around its zero point (0 K, -890 kJ/mol) (dashed blue line). The thermochemical stability of the silica dissolved in the slag has become greater than that of the pure silica. Its reduction by carbon at a CO partial pressure of 1 atm is first possible only above 1850 °C, i.e. only 200 °C higher. Now, in the MgO–C material at 1650 °C, the CO partial pressure P_{CO} ≈ 0.05 atm, so that the reaction line Eq. (13.13) also turns downward (coarse dotted black straight line). One notices that at this temperature the dissolved silica (dotted blue straight line) is barely still stable.

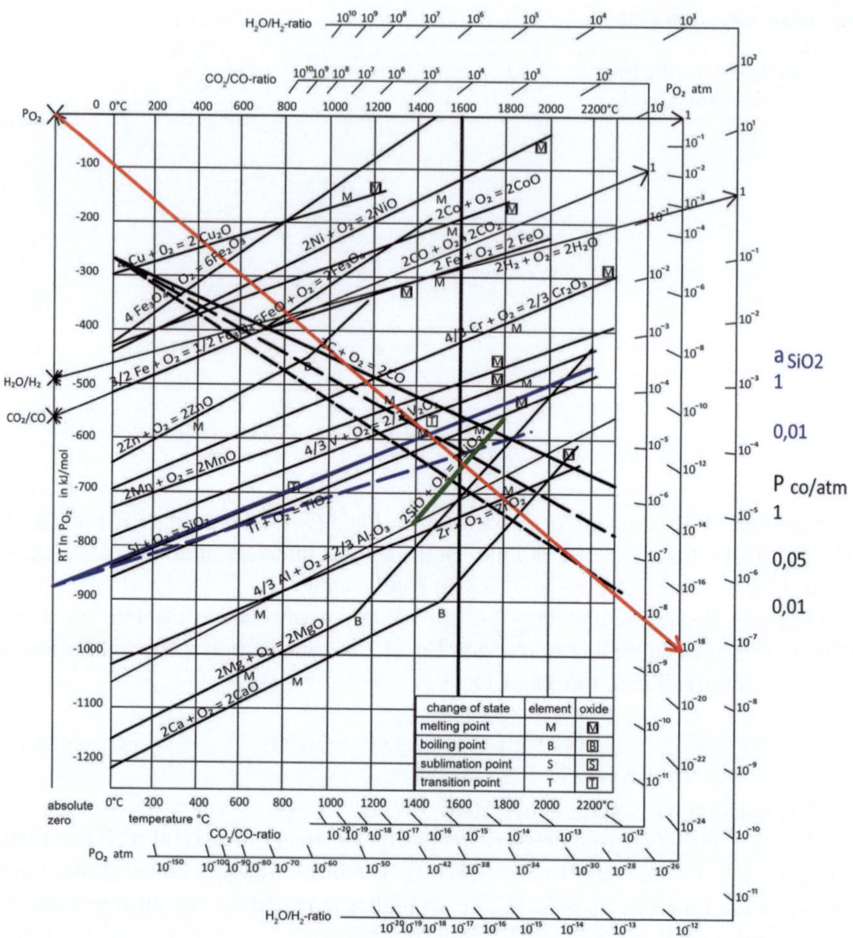

Fig. 13.11 Graphite-silica reaction in the Richardson-Ellingham illustration [3]

However, strong "smoke generation" is to be expected because $< SiO_2 >$ can be reduced to its gaseous suboxide $\{SiO\}$ if, at a given temperature, the partial pressure of CO is low enough. The reaction $2\{SiO\} + \{O_2\} = 2 < SiO_2 >$ is highlighted in green in Fig. 13.11. Given $P_{CO} = 1$ atm, pure silica above 1750 °C is reduced by carbon to $\{SiO\}$ and $\{CO\}$. Formally, the ratios are similar to those of the reduction of MgO by graphite in Eq. (13.2), here in the form $< SiO_2 > + < C > = \{SiO\} + \{CO\}$. Consequently, above 1535 °C, $< SiO_2 >$ is reduced to $\{SiO\}$ by graphite given $P_{CO} = \sqrt{0.01}$ atm $= 0.1$ atm.

Less stable oxides, such as FeO, MnO and Cr_2O_3, are reduced by graphite at all circumstances at steel mill temperatures. Since the stability of oxide compounds is between those of the oxides forming them, it can be concluded that, for example, mullite ($2SiO_2$-$3Al_2O_3$) is more stable than silica but more easily reduced by graphite than corundum (Al_2O_3). Chrome-spinel (MgO-$3Al_2O_3$, picrochromite) is formed from the less stable chromia (Cr_2O_3) and the very stable magnesia (MgO) and is less stable than silica (refer to Fig. 2.8).

13.1.5 Solution Equilibria

13.1.5.1 Carbon in Molten Iron

The solubility of carbon in liquid iron follows the relationship

$$[C]_{sat} = 1.3 + 2.57 \times 10^{-3} \cdot \vartheta \, [\% \text{ in wt.}]. \; (1152 - 2000\,^{\circ}\text{C}) \tag{13.17}$$

ϑ [°C] is the temperature along the C'-D' line in the iron-carbon diagram [5]. Consequently, at 1600 °C $[C]_{sat} = 5.41\%$ in weight are soluble. From this, it can be concluded first of all that the MgO–C material is unsuitable for the steel mill sector from a thermochemical point of view.

For control by kinetics, the dissolution rate of graphite is estimated in molten iron at 1600 °C. In the simplest case, the equation of Nernst applies

$$\dot{m} = \frac{D}{\delta} \cdot \frac{F \cdot \rho_{St}}{100} (C_s - C_0). \; \left[g/s\right] \tag{13.18}$$

$D = 10^{-4}$ cm^2/s is the mean diffusion coefficient of carbon in iron at steel mill temperature [9]. The mean density of the molten steel is $\rho_{St} = 7$ g/cm^2. The concentration difference is $(C_s - C_0) = (5.4 - 0)\%$ by weight. The thickness of the diffusion boundary (interfacial) layer depends on the mean flow velocity of the melt u[cm/s] [9]:

$$\delta_{Steel} = 10^{-2} - 2 \times 10^{-4} \cdot u \; [\text{cm}] \tag{13.19}$$

Within the diffusion boundary (interfacial) layer, the concentration curve of the dissolved carbon is linear.

In the laboratory crucible of an induction furnace, the mean velocity of the molten steel is about 30–50 cm/s [9]. Thus, $\delta_{Steel} \simeq 3 \times 10^{-3}$ [cm]. The dissolution rate is

$$\frac{\Delta x}{\Delta t} = \frac{\dot{m}}{F \cdot \rho_{refr.}} \; [\text{cm/s}] \tag{13.20}$$

i.e. the reaction area F [cm^2] truncates. The density of the graphite is $\rho_{refr} = 2.9$cm^3/g.

$$\beta \equiv \frac{D}{\delta} = \frac{10^{-4}}{3 \times 10^{-3}} = 0.03 \,\text{cm/s} \tag{13.21}$$

is the defined mass transfer coefficient of the dissolved carbon [9]. It follows with

$$\frac{\Delta x}{\Delta t} = \beta \cdot \frac{\rho_{St}}{\rho_{refr}} \cdot \frac{C_s - C_0}{100} = 0.03 \cdot \frac{7}{2.9} \cdot \frac{5.4}{100} = 4 \times 10^{-3} \, [\mathrm{cm/s}] \tag{13.22}$$

the dissolution of 14 cm/h. Consequently, other effects must prevent the catastrophic dissolution and disintegration of the MgO–C refractory material in the steel. It will be shown later what these are precisely.

13.1.5.2 Carbon in Molten Slag

Lime-aluminate slags (CA slags) are very frequently used in steel production. The wear of the MgO–C refractory lining is greatest in the slag area, which is initially surprising due to the supposed low solubility of graphite in slag. Consequently, this must be examined.

Not much is known about the solubility of carbon in liquid slags. The few experiments suggest that carbon is dissolved in acid slags as C^{2-}[10] and in basic slags as C_2^{2-} [11]. In general, therefore, the following applies:

$$2C\langle\rangle + \left(O^{2-}\right) = \left(C^{2-}\right) + \{CO\} \text{ (acidic slags)} \tag{13.23.1}$$

$$3C\langle\rangle + \left(O^{2-}\right) = \left(C_2^{2-}\right) + \{CO\} \text{ (basic slags)} \tag{13.23.2}$$

Since the activity $a_{<C>}$ of pure and solid graphite <C> is one, the following applies for the equilibrium constant in basic CA slag according to Eq. (13.23.2)

$$K = \frac{a\left(C_2^{2-}\right)}{a\left(O^{2-}\right)} \cdot P_{CO} \tag{13.24}$$

Since $a_{(O^{2-})}$ is almost constant in the slag with small contents of dissolved carbon, the oxygen ion activity is taken into the constant. For the activity of the carbon dissolved as anion, its chemically analyzed concentration (C) is used, taking an activity coefficient $\gamma_{(C)}$ into account:

$$a_{\left(C_2^{2-}\right)} = \gamma_{(C)} \cdot (C). \tag{13.25}$$

In a narrow concentration range, likewise $\gamma_{(C)} = $ const and can be taken into the constant. Then the equilibrium constant is

$$K' = (C) \cdot P_{Co}. \tag{13.26}$$

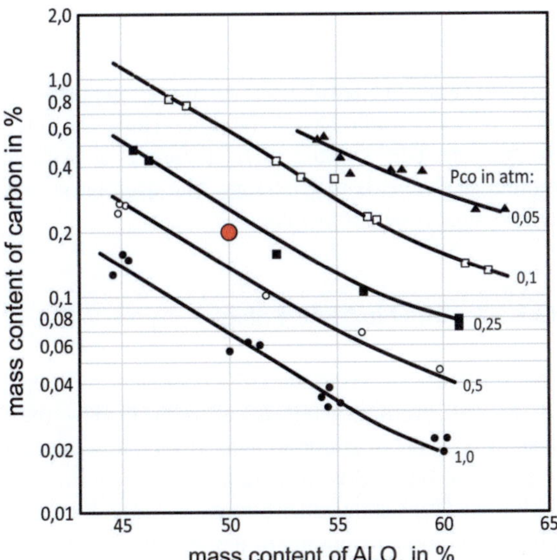

Fig. 13.12 Solubility of carbon in molten CA slag [11, 12]

Consequently, as the CO partial pressure increases, less and less carbon is dissolved into the slag. Figure 13.12 shows this exemplarily for CaO-Al$_2$O$_3$ slags at 1600 °C [11]. Up to 1% by weight carbon is soluble.

The carbon concentration of an iron melt in equilibrium with the slag can be used to determine the activity coefficient $\gamma_{(C)}$.

A practical example is the three-phase area steel/slag/MgO–C of a ladle lining [12]. Assuming the local thermodynamic equilibrium slag/steel, the following applies in this case $a_{(C_2^{2-})} = a_{[C]}$.

meaning

$$\gamma_{(C)} = (C) = \gamma_{(C)} \cdot [C] \tag{13.27}$$

(C) is carbon dissolved in the slag and [C] is carbon dissolved in the steel.

For the standard applicable condition of the 1% solution, which is very common in steel, $\gamma_{[C]} = 1$ applies and when 1% C is reached, $a_{[C]} = 1$. Then

$$\gamma_{(C)} = [C]/(C). \tag{13.28}$$

Example [12]: In acid slag (40% SiO$_2$, 45% CaO, 5% Al$_2$O$_3$, 5% MgO, balance FeO, MnO), when MgO–C was dissolved in contact with steel, the values [C] = 0.4% and (C) = 0.08 % were determined analytically (1650 °C, P_{CO} = 0.5 bar), i.e. γ (C) (acidic slag) = 5.

In basic slag (40% CaO, 40% Al$_2$O$_3$, 5% MgO, 5% SiO$_2$, balance FeO + MnO), [C] = 0.4% and (C) = 0.2% were determined analytically (1650 °C, P_{CO} = 0.5 bar), i.e. $\gamma_{(C)}$ (basic slag) $= 2$. This agrees quite well with the results of [11] (refer to Fig. 13.12, red dot).

To estimate the dissolution rate of graphite in molten slag, a calculation according to Eq. (13.22) would have to be done. However, since under comparable conditions the mass transfer coefficient β in slags is at least one power of ten lower than in molten steel, this is unnecessary. The usually observed premature wear in the slag region must have another cause.

13.1.5.3   MgO in Molten Slag

The solubility of magnesia (MgO) in steel mill slags has been frequently studied [8, 13–16]. From a metallurgical point of view, the Al_2O_3, CaO, MgO, SiO_2 slag system is of greatest importance. It includes calc-silicate (CS) and calc-aluminate (CA) slags. Basic CA slags are preferred specifically in secondary steel metallurgy.

Figure 13.13 shows measurement results for the solubility of MgO in CA slags. There is a close correlation to the basicity $B \equiv CaO/(Al_2O_3 + SiO_2)$ [15].

Recently, the saturation of MgO in CS and CA slags was calculated with consistent data using FactSage [2] and presented graphically [17]. Figure 13.14 shows the result for CA slags at 1600 °C.

The x-axis shows the CaO content of the slag normalized to 100% by weight of the Al_2O_3-CaO-SiO_2 ternary system, as CaO. The y-axis shows the corresponding standardized Al_2O_3-content as Al_2O_3. The normalized SiO_2 content SiO_2, which is not explicitly shown, results from the difference

$$SiO_2^\circ = 100 - \left(Al_2O_3^\circ + CaO^\circ\right) \ [\text{wt}\%] \tag{13.29.0}$$

Each point (CaO, Al_2O_3) is assigned an interpolated MgO saturation value in % by weight. Sites of different composition but the same MgO saturation are connected by lines with an increment of 0.25 % by weight. Double saturation of MgO and CaO is highlighted from the rest of the melt area by grayish graduated coloring. Locations

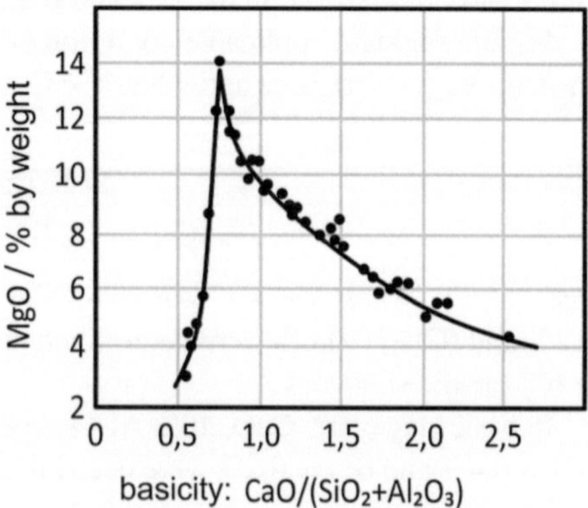

Fig. 13.13 Saturation concentration of MgO in calc-aluminate slags [15]

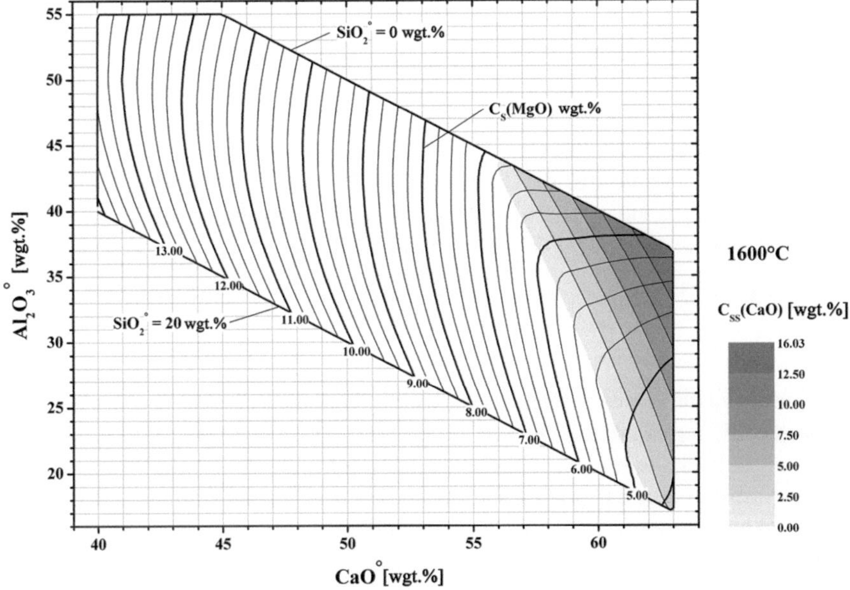

Fig. 13.14 MgO saturation in calc-aluminate slags at 1600 °C [17]

of different compositions but equal CaO saturation are shown with an increment of 2.5 % by weight.

The approximate calculation of the MgO saturation concentration C_S (MgO) of calc-aluminate slags in the limits $0 < SiO_2^{\circ} < 10$ % by weight and $0.7 < B < 1.4$ enables the equation

$$C_S(\text{MgO}) = 20.7 - 11.5 \cdot B + 0.02 \cdot \left(T\left[°C\right] - 1600\right)\left[\%\,\text{by weight}\right] \quad (13.29.1)$$

and for $10 < SiO_2^{\circ} < 20$ % by weight the following applies approximately

$$C_S(\text{MgO}) = \frac{9.5}{B} + 0.02 \cdot \left(T\left[°C\right] - 1600\right)\left[\%\,\text{by weight}\right] \quad (13.29.2)$$

$B = CaO^{\circ}/\left(Al_2O_3^{\circ} + SiO_2^{\circ}\right)$ is the defined basicity of the slag.

For $B < 0.67$, MA spinel is formed. The calculation was also performed for 1650 and 1700 °C [18].

Calculation example [18] at 1600 °C is a slag of the following composition (m_S = 1000 kg):

10 kg FeO or MnO, 30 kg MgO, 540 kg CaO, 300 kg Al_2O_3, 120 kg SiO_2.

Standardization ($m_{S\text{-}N}$ = 960 kg) results in:

$$CaO^{\circ} = (540/960) \cdot 100 = 56.25\%\,\text{by weight}$$

$$Al_2O_3^\circ = (300/960) \cdot 100 = 31.25\% \text{ by weight and}$$
$$SiO_2^\circ = (120/960) \cdot 100 = 12.50\% \text{ weight.}$$

The basicity is $B = 1.29$.

Figure 13.14 reads C_S (MgO) ≈ 6.80 % by weight. The approximate calculation with Eq. (13.29.2) gives for C_S (MgO) ≈ 7.40 % by weight, i.e. the relative error made is < 10%. Practically, the slag is then weakly supersaturated in MgO.

The soluble amount of MgO, $m_{\text{MgO-S}}$ turns out to be

$$m_{\text{MgO-S}} = \frac{m_{S-N} \cdot C_S(\text{MgO})}{100 - C_S(\text{MgO})} \; [\text{kg}] \tag{13.30}$$

By filling in the values one gets for $m_{\text{MgO-S}} \approx 70.0$ and 76.7 (3.2b) kg, respectively. Consequently, for example, 40 or 46.7 Eq. (13.29.2) kg MgO per 1000 kg slag can be dissolved from the refractory lining containing MgO.

The corresponding calculation and illustration for CS slags (Figure 13.15) can be expressed by the following approximate equation:

$$C_S(\text{MgO}) = 3.5 \cdot X^2 - 17 \cdot X + 31 + 0.02 \cdot \left(T\left[^\circ C\right] - 1600\right) \left[\% \text{ by w\%}\right] \tag{13.31}$$

With $X = \left(60 - SiO_2^\circ\right)/(65 - CaO^\circ)$.

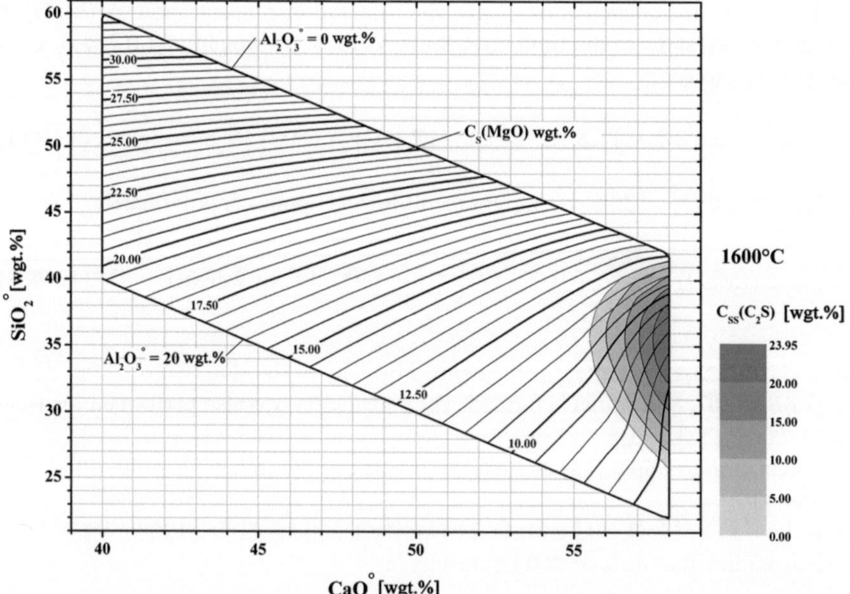

Fig. 13.15 MgO saturation in silicate-calc slags at 1600 °C [17]

A temperature increase of 50 °C boosts the MgO solubility by about 1% by weight.

An experimentally [19] confirmed relationship (by the author) for calculating the maximum solubility of MgO in slags containing FeO and CaF_2 is

$$(MgO)_{sat} = \frac{8.2}{B} + 0.06 \cdot (FeO) + 0.2 \cdot (CaF_2) + 0.02 \cdot \left(T/^\circ C - 1550\right) \left[\% \text{ by weight}\right]$$

with basicity $B \equiv (CaO)/(Al_2O_3 \cdot SiO_2)$ (13.32)

Thereby, (FeO) and (MnO) are largely interchangeable, since their thermochemical behavior is similar. Both are seamlessly mixable into each other [5].

The dissolution rate of magnesia in a CA slag should be estimated by using Eq. (13.33) (W.Nernst):

$$\frac{\Delta x}{\Delta t} = \beta \cdot \frac{\rho_{Slag}}{\rho_{refr}} \cdot \frac{C_s - C_0}{100} \ [cm/s]$$ (13.33)

The mass transition coefficient $\beta \equiv D/\delta$ in slags of this type at steel mill temperatures is $\beta \approx 10^{-3}$ cm/s [20]. The densities of the refractory material and slag are approximately equal. The concentration difference is chosen to be $(C_S - C_0) = 10\%$ by weight. Then, $\Delta x/\Delta t \approx 10^{-4}$ cm/s, i.e. 4 mm/h.

This wear, while being remarkable, is comparatively small to the calculated dissolution of the graphite in the molten steel.

13.1.5.4 Deoxidation Equilibria

The consideration of the solubility of MgO in steel melts is identical with their deoxidation by Mg. Figure 13.16 shows the effect of some common elements used

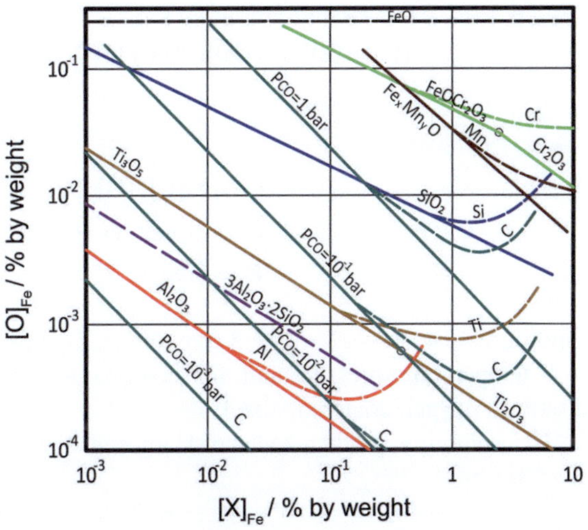

Fig. 13.16 Deoxidation diagram for steel melts at 1600 °C [20]

for deoxidation of steel melts at 1600°C [20, 21]. Highlighted is the deoxidation by manganese (dark brown), silicon (blue), aluminum (red) and carbon (dark green). As already shown in the Richardson-Ellingham diagram, the partial pressure of carbon monoxide has a significant influence on the result: the lower P_{CO} is, the greater the reduction in dissolved oxygen in the steel bath having the same initial carbon content. Vacuum metallurgy, in particular, makes use of this. If the partial pressure of carbon monoxide drops to $P_{CO} \leq 10^{-1}$ atm at steel mill temperatures, for example, silica (blue) and, at $P_{CO} \leq 10^{-3}$ atm, additionally aluminum oxide (red) are noticeably reduced. In the for metallurgically interesting concentration range of the deoxidation elements, their effect is almost linear if plotted double logarithmically. Only at high contents does the oxygen concentration deviate from the activity toward higher concentration values, a consequence of the influence of the "interaction parameters" [3] (refer to Sect. 2.1.3.3).

Some equilibria important for understanding the function of MgO refractory material in steel metallurgy are discussed here. The basis for the theoretical consideration of steel deoxidation is Gibbs' phase law (refer to Sect. 2.1.1 **Basics**).

$$F = K + 2 - P - R - B \tag{13.34}$$

With

K Components
P Phases
R Number of independent linear reactions
B Stoichiometric limitations.

Iron-Carbon (Vacher-Hamilton Equilibrium)

For the deoxidation with carbon, the reaction equation applies

$$[C] + [O] = \{CO\}, \ \Delta G°(1600\ C°) = -93\ kJ/mol \tag{13.35}$$

With

$$K = 4([C], [O], Fe, CO)$$
$$P = 2(melt, gas\ phase)$$
$$R = 1(eq.13.35),$$

follows: $F = 4 + 2 - 2 - 1 = 3$ (pressure, temperature, one concentration).

If temperature, pressure and, for example, the carbon content are specified, the dissolved oxygen content is fixed.

The solubility product (Vacher-Hamilton equilibrium) is calculated from Eq. (13.35) [20]:

$$[C] \cdot [O] = 2.4 \times 10^{-3} \cdot P_{CO}, \; (1600°C) \tag{13.36}$$

Consequently, if a steel melt contains more than 25 ppm oxygen dissolved, a CO partial pressure of more than 1 atm develops when it comes into contact with the graphite of the MgO–C material, the activity of which is $a_C = 1$. Such oxygen contents are quite common, so that a gas cushion is formed on the refractory material, which repels the melt and hinders its penetration into the structure.

This explains the gas bubbles observed in the experiment (Chap. 1.2) and could be an initial **protective mechanism** for the MgO–C lining.

From Fig. 13.4, the CO partial pressure can be calculated via the Vacher-Hamilton equilibrium Eq. (13.36). As shown, it is always between $0.01 \leq P_{CO} \leq 0.1$ atm in the microstructure. The mean value $P_{CO} \approx 0.05$ atm just barely corresponds to the temperature range of 1550–1600 °C. If the molten steel were to enter the microstructure with a dissolved oxygen content of 25 ppm, only $[C] = 2.5 \times 10^{-3} \cdot 0.05/[2.5 \times 10^{-3}] = 0.05\%$ by weight carbon would be soluble there at $P_{CO} = 0.05$ atm. Consequently, the graphitic **binder phase** would be quite well **protected** from strong dissolution.

Iron-Aluminum/(Si)

For deoxidation with aluminum, the reaction equation is as follows

$$2[Al] + 3[O] = \langle Al_2O_3 \rangle, \; \Delta G^0(1600\,°C) = -470\,kJ/mol \tag{13.37}$$

With
 $K = 4$ ([Al], [O], Fe, Al$_2$O$_3$)
 $P = 2$ (melt and deoxidation product)
 $R = 1$ (Eq. 13.37)
 follows: $F = 4 + 2 - 2 - 1 = 3$.

Since the pressure dependence is negligible, the oxygen content results when the temperature and the aluminum content are specified. The same applies to any deoxidation without the participation of a gas phase.

The solubility product of aluminum deoxidation is [20]:

$$K_{Al_2O_3} = [Al]^2 \cdot [O]^3 = 8 \times 10^{-14}(1600\,°C) \tag{13.38}$$

(Note: Oeters [20] gives, among others, $K_{Al_2O_3} = 2.5 \times 10^{-14}(1600\,°C)$).

In practice, this value is not reached during deoxidation due to supersaturation effects, but $3 \times 10^{-13} < K_{Al_2O_3} < 5 \times 10^{-13}$ has been experimentally confirmed [22] (refer also to [23]).

In the experiment, it was observed that metallic aluminum, as an antioxidant, provides for a strong infiltration of the molten iron. Calculating with the mean aluminum content observed in the infiltrate of $[Al] = 0.1\,\%$ by weight and $K_{Al_2O_3} =$

3×10^{-13}, the dissolved oxygen content is $[O] = 3$ ppm (3×10^{-4} %). Placing this value at 1600 °C into the Vacher-Hamilton product of Eq. (13.36) and noting that the CO partial pressure in the microstructure is approximately $P_{CO} = 0.05$ atm, the dissolved carbon concentration is $[C] = 0.4$ % by weight. This corresponds exactly to the observation and means that in the infiltrate not 5.4% carbon, but only the tenth part can go into solution. This can also be seen as a **protective mechanism of** the MgO–C material.

For deoxidation with silicon, [20] applies:

$$[Si] + [O]^2 = SiO_2, \Delta G^0(1600°C) = -168.4\,\text{kJ/mol} \tag{13.39}$$

$$K_{SiO_2=}[Si] \cdot [O]^2 = 2 \times 10^{-5}(1600\,°C) \tag{13.40}$$

Iron–Magnesium/(Ca)

The sum of two reaction equations must be considered:

$$\{Mg\} = \left[Mg\right] \quad \Delta G^0 = -58\,\text{kJ/mol} \tag{13.41.1}$$

$$\underline{\left[Mg\right] + [O] = \langle MgO\rangle} \quad \Delta G^0 = -204\,\text{kJ/mol} \tag{13.41.2}$$

$$\{Mg\} + [O] = \langle MgO\rangle \quad \Delta G^0 = -262\,\text{kJ/mol} \tag{13.41.3}$$

With
$K = 4(\{Mg\}, Fe, [O], \langle MgO\rangle$
$P = 3$ (gas phase, melt, oxide)
$R = 1$ (Eq. 13.41.3),
$F = 4 + 2 - 3 - 1 = 2$ (pressure, temperature). If the temperature and the partial pressure of the magnesium vapor are specified, the oxygen content is fixed.

However, the solubility of magnesium in the melt is very low. At 1600 °C and 1 atm magnesium vapor, it is only 0.023 % by weight [20]. For this reason, magnesium is added to the melt as an alloy or with an immersion bell.

The solubility product at 1600 °C is [20]:

$$\left[Mg\right] \cdot [O] = 2 \times 10^{-6}(1600\,°C). \tag{13.42}$$

The low solubility of magnesium in steel and its very small deoxidation product indicate good durability (service life) of magnesia refractory material in steel metallurgy. This is one further **protective mechanism**.

The same applies to deoxidation with calcium. Its solubility in molten steel is also very low and amounts to $P_{Ca} = 1$ atm only 0.03 % by weight [20]. It is, therefore, used for deoxidation exclusively as an alloy or compound, e.g. CaSi and CaC_2. From the Richardson-Ellingham diagram, Figure 13.9, it can be seen that CaO is thermochemically even more stable than MgO. Its effect as a deoxidizing agent is sufficiently described as

$$[Ca] + [O] = (CaO), \ \Delta G^0(1600\,^\circ C) = -217.7\,kJ/mol \tag{13.43}$$

$$[Ca] \cdot [O] = 9 \times 10^{-7}(1600\,^\circ C) \tag{13.44}$$

13.1.6 Wetting, Infiltration and Nucleation

13.1.6.1 Wetting

The wetting of the refractory material by the molten steel (Fe) or the slag (Sl) also determines the infiltration into the structure.

The wetting between the precipitation products, such as MgO, Al_2O_3 and $MgAl_2O_4$ decisively determines the nucleation of these oxides in the melt and, thus, the reaction of the refractory material along with it.

The mechanical equilibrium between a solid phase and a liquid phase is described by the equation of Th. Young (1804):

$$\sigma_{SG} - \sigma_{LS} = \sigma_{LG} \cdot \cos\theta \ \left[J/m^2\right] \tag{13.45}$$

The assignment of the specific interfacial energies σ_{xy} to the phases (solid, liquid and gas) is explained in Fig. 13.17. The wetting angle θ is decisive for the interaction of the phases with each other.

The mentioned oxides, just like graphite, are not wetted by liquid steel, $\theta > 90°$. Molten slag, on the other hand, wets the oxides, $\theta < 90°$ but not the graphite $\theta > 90°$.

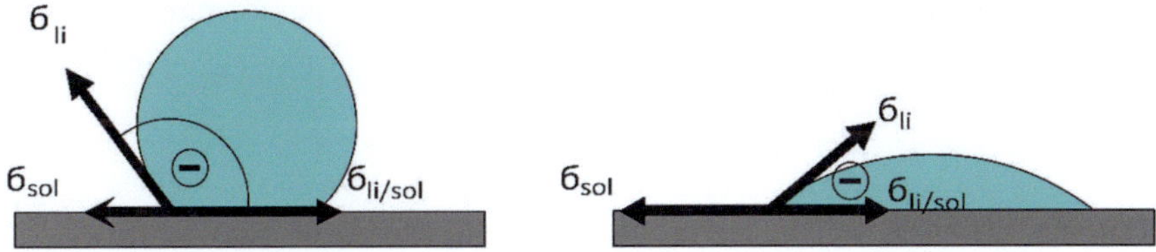

Fig. 13.17 Non-wetting (left) and wetting (right)

13.1.6.2 Infiltration

Slag

The equilibrium between the gravitational pressure and the capillary pressure is important for infiltration (refer to Chap. 5):

$$\rho \cdot g \cdot h = \frac{4\sigma \cdot \cos\theta}{d} \left[\text{dyn/cm}^2\right] \qquad (13.46)$$

If the melt wets the refractory material ($\theta < 90°$), it infiltrates such pores whose diameter d [cm] is sufficiently large:

$$d > \frac{4\sigma \cdot \cos\theta}{\rho \cdot g \cdot h} \text{ [cm]} \qquad (13.47)$$

Infiltration by slag into graphite-containing refractory material is possible as soon as the wetting angle becomes less than 90°. In fact, it varies depending on whether the slag is in contact with MgO ($\theta < 90°$) or graphite ($\theta > 90°$). In practice, one observes on the average $\theta \approx 90°$ and the infiltration of the slag. This promotes wear as the binder phase is weakened by oxidation of the graphite.

The infiltration depth is calculated from the flow velocity of the slag in the pores:

$$\frac{dl}{dt} = \frac{d \cdot \sigma \cdot \cos\theta}{8 \cdot \eta \cdot l} \text{ [cm/s]} \qquad (13.48)$$

and the labyrinth factor [24, 25]

$$\lambda = \frac{x}{l} = 1.84 \cdot \sqrt{\varepsilon} - 0.3 \text{ [−]} \qquad (13.49)$$

It describes the relationship between the externally measured penetration depth x and the entangled (interlaced) actual path l of the slag inside the structure of porosity ε. Equation (13.49) is put into Eq. (13.48) and integrated after separation of the variables [17]:

$$x = \lambda \cdot \left(\frac{d \cdot \sigma \cdot \cos\theta}{4 \cdot \eta} \cdot t\right)^{0.5} \text{ [cm]} \qquad (13.50)$$

As an example, the following values are valid [17]: Open porosity $\varepsilon = 0.2$ (20% by volume), labyrinth factor $\lambda = 0.52$, mean pore diameter $d = 8\ \mu\text{m} = 8 \times 10^{-4}$ cm, dynamic viscosity of the slag: $\eta = 1.7$ g/cm $\cdot$ s, surface tension of the slag: $\sigma_S = 520$ mN/m and wetting angle: $\theta = 60°$, infiltration time $t = 60$ s. Inserting these values into Eq. (13.50) gives $x = 0.7$ cm for the infiltration depth.

Steel

Since MgO–C is not wetted by steel ($\theta > 90°$), no infiltration is to be expected at low bath heights, which is usually confirmed in laboratory tests. Although the metallic aluminum (MgO–C–Al), added as an antioxidant, does not change the wetting conditions at first glance, the melt does infiltrate. Two explanations can be offered:

1. The missing oxide layer (MgO) facilitates infiltration but this should not play a major role at $\theta > 90°$, or
2. The aluminum melted above 660 °C, which pushes out of the refractory material into the steel bath, wets the molten steel in contact and, thus, acts as a "wick," i.e. overcomes the repulsive effect of the large wetting angle.

Due to the lack of oxygen on the one hand and kinetic retardation on the other, presumably only a small fraction of the aluminum reacts to form spinel ($MgAl_2O_4$) and the major part is, therefore, present in the microstructure in liquid form and dissolves in the penetrating liquid iron. This is indicated by the above-mentioned test results (refer to Sect. 13.1.1).

13.1.6.3 Nucleation

Nucleation is important for the observed excretion of MgO on the surface of the MgO–C material and is, therefore, looked at in more detail. The maximum solubility of magnesium at a partial pressure of 1 atm is 0.023% by weight. In the microstructure, the partial pressure of magnesium is $P_{Mg} \approx 0.05$ atm (refer to Fig. 1.2.4). The vapor diffuses to the surface and dissolves in the steel. That is $[Mg]_a = 0.05 \cdot 0.023 = 1.2 \times 10^{-3}\%$ by weight. About 25 ppm oxygen is dissolved in the steel at the same time ($[O]_a = 2.5 \times 10^{-3}\%$ by weight) and both components react to form $< MgO >$.

This is not self-evident, because the formation of a new phase requires supersaturation α_{KB} of its components to apply the necessary nucleation work ΔG_N [26]:

$$\Delta G_N = \frac{4}{3}\pi r_K^2 \cdot \sigma \cdot f(\theta) \ [\text{J}] \tag{13.51}$$

If the wetting angle between the melt and the new phase is $\theta < 180°$, nucleation is easier. The measure is the factor [26]

$$f(\theta) = \frac{1}{4}(2 + \cos\theta)(1 - \cos\theta)^2 \tag{13.52}$$

For example, for $\theta = 120°, f(\theta) =$ is 0.8. Nucleation is easier (facilitated) by 20%.

The critical nucleation radius r_K is calculated from the relation [26]

$$r_K = \frac{2 \cdot \sigma_{MgO/Fe} \cdot V_{MgO}}{R \cdot T \cdot \ln \alpha_{KB}} \; [\text{cm}] \tag{13.53}$$

The frequency of excretion (nucleation rate) is

$$J = J_0 \cdot \exp\left(-\frac{\Delta G_N}{kT}\right) \; [\text{cm}^{-3} \cdot \text{s}^{-1}] \tag{13.54}$$

In liquid metals, the abundance factor is $J_0 \approx 10^{37}$ [nuclei/cm$^3 \cdot$ s] [27–29].

For the reaction [Mg] + [O] = < MgO > , the supersaturation α_{KB} is calculated from the ratio of the current concentration product $K_a = [Mg]_a \cdot [O]_a$ to $K_{gl} = [Mg]_{gl} - [O]_{gl}$ given in thermodynamic equilibrium. According to Fig. 13.4, approx. 0.02% by weight oxygen is dissolved in the steel at the beginning. This gives

$$\alpha_{KB} = \frac{K_a}{K_{gl}} = \frac{0.0012 \cdot 0.02}{2 \times 10^{-6}} = 12. \tag{13.55}$$

The critical nucleation radius is obtained from Eq. (13.53) with $\sigma_{MgO/Fe} \approx 2 \text{N/m}$ [30] and $V_{MgO} = 12 \times 10^{-6} \text{m}^3/\text{mol}$ to give

$$r_K = \frac{2 \cdot 2.0 \,\text{N/m} \cdot 12 \times 10^{-6} \text{m}^3/\text{mol}}{8.31 \,\text{J/mol} \cdot \text{K} \cdot 1.873 \,\text{K} \cdot \ln 12} = 12 \times 10^{-10} \; [\text{m}] \mathrel{\hat{=}} 12 \tag{13.56}$$

It follows from Eq. (13.51) for the nucleation work of a particle. $\Delta G_N = 9 - 10^{-18}$ [J] . This value is so large that Eq. (13.54) calculates the nucleation frequency to be $J \to 0 \left[\text{nuclei/cm}^3 \times \text{s}\right]$.

This means that MgO deposits only in heterogeneous nucleation on the rough surface of the refractory material. Especially narrow gaps facilitate nucleation significantly, since only a fraction of the spherically imagined nucleation surface has to be formed to achieve stability [31].

The question remains to be clarified how the metallic aluminum present in the structure of the MgO–C–Al material influences this process. It is assumed that the aluminum emerging from the pores on the surface of the refractory material, forcibly wetted by the liquid steel, deoxidizes the bath to such an extent that the formation of MgO is not possible. A prerequisite is that Al_2O_3 or $MgO \cdot Al_2O_3$ is formed. This requires the nucleation of the oxide, and this could be deposited protectively on the surface of the refractory material, just like the MgO.

However, the excess of dissolved aluminum over against magnesium is comparatively large, so that the formation of spinel should be negligible in comparison with Al_2O_3.

The solubility product of Al_2O_3 in liquid iron at 1600 °C is $K_{gl} = 8 \times 10^{-14}$ [20]. Let the initial oxygen content again be [O] = 0.02% by weight (Fig. 13.4) and that of aluminum [Al] = 0.08% by weight (Fig. 13.3). The actual concentration product in the steel is then

$$K_a = \left[\text{Al}\right]_a^2 \cdot \left[\text{O}\right]_a^3 \approx \left[0.08\right]^2 \cdot \left[0.02\right]^3 = 5 \times 10^{-8}$$

Consequently, Eq. (13.55) calculates the supersaturation to be $\alpha_{KB} = 6 \times 10^5$. The critical nucleation radius Eq. (13.56) follows with $\sigma_{Al_2O_3/Fe} = 1.9\,\text{N/m}$ [30] and $V_{Al_2O_3} = 25,5 \times 10^{-6}\,\text{m}^3/\text{mol}$ to $r_K = 5 \times 10^{-10}$ [m] $\overset{\wedge}{=} 5\,\text{Å}$. The nucleation work is given by Eq. (13.51) $\Delta G_N = 1.6 \times 10^{-18}$ [J]. From Eq. (13.54), the nucleation frequency $J = 2 \times 10^{+10}$ $\left[\text{nuclei/cm}^3 \cdot \text{s}\right]$ is obtained. Thus, nucleation occurs homogeneously in the melt and does not require a helping support. This explains why no oxide was deposited on the sample and the observed low oxygen content in the melt of about 10 ppm.

In summary, the laboratory tests show that aluminum as an antioxidant increases the wear on the refractory material:

1. Aluminum lowers the dissolved oxygen content in the bath and prevents the formation of a protective CO–cushion.
2. Aluminum prevents the formation of a protective MgO layer.
3. Aluminum increases the solubility of carbon in the infiltrate due to its strong deoxidizing effect.

13.1.7 Premature Wear of MgO–C

13.1.7.1 Experiments

Over the course of 10 years, the corrosion properties of different refractory linings looking at alternating steel-slag combinations at steel mill temperatures were tested in a large number of corrosion tests in the induction furnace at the German Institute for Refractories and Ceramics (DIFK) in Bonn, Germany. Figure 13.18 schematically illustrates the test setup and the observed zones of premature wear. They are always located along the three-phase lines refractory/slag/air or rather to a significantly greater extent along refractory/slag/steel.

The flow course assumed in the model in the bath and the slag is drawn. The relative velocity of both is zero at their contact surface by definition. The same applies for contact with the wall. Nevertheless, as the flow velocity increases, the mass transfer increases because the diffusion boundary layer δ becomes thinner [32].

Figure 13.19 shows a summary of all results of carbon-bonded AC, MC and AMC materials [33]. The mean corrosion rate is plotted against the equilibrium distance of the corrosion-determining component(s) in the slag, i.e. the driving concentration difference $(c_{sat} - c_0)$ [% by weight]. The scatter results from the different materials and test conditions [33]. The individual values can be calculated approximately taking into account the conditions present in the experiment [33–35] (refer to Sect. 2.2.5.2 and 7.6.3).

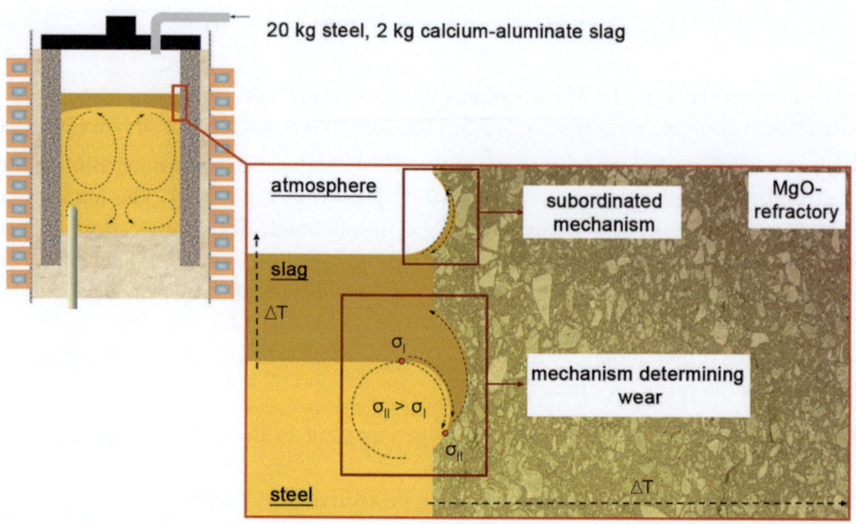

Fig. 13.18 Schematic illustration of the zones of premature wear in the induction crucible test (10 kHz, 1550–1650 °C, Ar + 5% H_2)

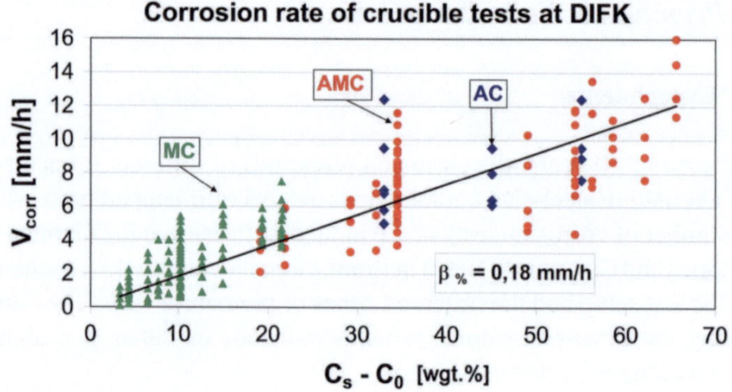

Fig. 13.19 Corrosion of the AMC linings of a laboratory induction furnace (20 kg steel, 2 kg slag) at approx. 1650 °C [33–35]

13.1.7.2   Evaluation and Discussion

An explanation of the premature wear is offered by the so-called Marangoni convection. It is an interfacial convection of two adjacent liquid phases driven by gradients of concentration and/or temperature [36] (refer to Sect. 2.2.5.2). The interfacial convection occurs at high velocity, which significantly increases the mass transfer between the two phases or an adjacent solid phase [32].

The driving force is mainly the concentration gradient of the capillary-active oxygen dissolved in the steel along the steel/slag (St/Sl) phase boundary. In direct contact with the graphite of the refractory material, the concentration of oxygen drops, i.e. $\sigma^{II} > \sigma^{I}$, so that both melts flow toward the circulating line II (Fig. 13.18). Conversely, the temperature distribution ($T^{II} > T^{I}$) causes the flow to weaken. The following applies [37]:

$$\frac{d\sigma_{St/Sl}}{dC_{[O]}} \simeq -\frac{220}{C_{[O]_I}}\left[\frac{mN}{m \cdot \%}\right] \quad \text{and} \quad \frac{d\sigma_{St/Sl}}{dT} = -0.2\left[\frac{mN}{m \cdot K}\right] \tag{13.57}$$

The corrosion rate is calculated under the assumption of a linear boundary (interfacial) layer diffusion according to Nernst:

$$V_{CORR} = \beta_M \cdot (C_S - C_0) \text{ [mm/h]} \tag{13.58}$$

where the mass transfer coefficient refractory/slag is given by

$$\beta_{FF/SL} = 360 \cdot \frac{\rho_{Sl}}{\rho_R} \cdot \left(\frac{4 \cdot D_{Sl} \cdot \tan(180 - \theta_{R/St,Sl})}{\pi^3 \cdot h \cdot \eta_{St}} \cdot \left(\frac{D \cdot \rho}{\eta}\right)^{1/3}_{Sl}\right)^{1/2} \cdot$$

$$\left(\frac{d\sigma_{St/Sl}}{dC_{[O]}} \cdot (C^I_{[O]} - C^{II}_{[O]}) + \frac{d\sigma_{St/Sl}}{dT} \cdot (T^I - T^{II})\right)^{1/2} \text{ [mm/h} \cdot \%] \tag{13.59}$$

Pötschke and Brüggmann [38]. The height h [cm] of the corrosion zone in the laboratory test can be calculated using Laplace's equation [39] for the slope of the phase boundary $\psi = 0$:

$$h = 2 \cdot \sqrt{\frac{\sigma_{St/Sl}}{\rho_{St} \cdot g}} \cdot \sin\big(0.5 \cdot \big(\psi - \theta_{R/St/Sl}\big)\big) \text{ [cm]} \tag{13.60}$$

Table 13.1 lists all data required for the calculation. The attached micrograph [38] shows the typical conditions along the three-phase line refractory/steel/slag.

The result agrees quite well with the laboratory experiment. However, it must be pointed out that the model is only applicable to the refractory—steel/slag system to a limited extent, since the boundary conditions are less precisely defined than for refractory—glass or slag melts [38]. In addition, there is the uncertainty of the broad spectrum of material data required for the calculation [7, 8, 20, 40, 41].

Finally, it should be noted that although Marangoni convection explains the advanced wear of submerged nozzles [42] and steel ladles in the steel/slag/refractory contact area, theoretical prediction based on laboratory tests is not possible because the operational boundary (interfacial) conditions are subject to too intensive fluctuations.

In practice, the crucial question is how to counter this effect without operational restrictions. There is currently only one answer: the refractory material must be as chemically inert as possible. For this reason, zirconium dioxide, which reacts only slightly with both liquid phases, steel and casting powder slag, has been used for years in the critical area of the submerged nozzle.

Table 13.1 Corrosion of the AMC lining of a laboratory induction furnace by steel and CA slag at 1650 °C (refer to Fig. 13.19) [35]

T	[°C]	1650
$C_{[0]}^{\mathrm{I}}$	% by weight	0.005
$C_{[0]}^{\mathrm{II}}$	% by weight	0.0025
$\sigma_{\mathrm{st/sl}}$	[mN/m]	1135
$\sigma_{\mathrm{FF/st}}$	[mN/m]	1900
σ_{FF}	[mN/m]	900
$\theta_{\mathrm{FF/st/sl}}$	[°]	150
ρ_{st}	[g/cm^3]	7
ρ_{sl}	[g/cm^3]	2.7
ρ_{FF}	[g/cm^3]	3.3
η_{st}	[g/(cm s)]	0.01
η_{sl}	[g/(cm s)]	1
D	[cm^2/s]	2.4×10^{-7}
ΔC	% by weight	10
initial values/7,12,20,40,41/		

Equation	Result	Value	Dimension
1	ψ	0	[°]
2	h	-0.79	Cm
5	θ	150	°
9	$\mathrm{d}\sigma/\mathrm{d}C$	-4.4×10^4	mN/(m % by weight)
–	$\mathrm{d}\sigma/\mathrm{d}T$	-0.2	mN(m K)
10	gradC	-2.0×10^3	% by weight/cm
10	gradT	$+16.2$	K/cm
8	u_{M}	244	cm/s
11	β	7.2×10^4	cm/s
–	$\beta_{\%}$	0.21	mm/(h %)
12	V_{corr}	2.1	mm/h

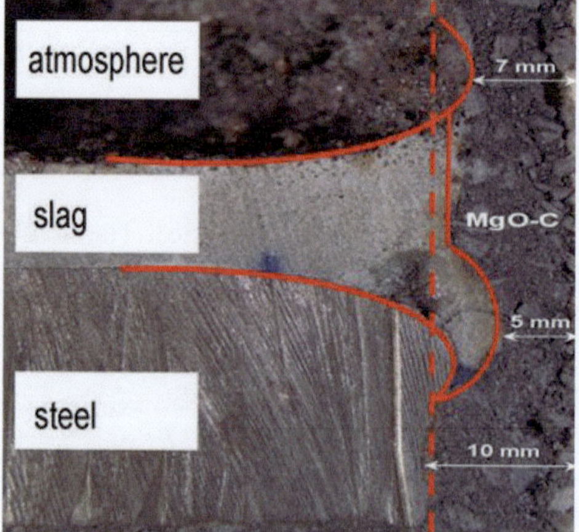

13.1.8   Oxidation of MgO–C

13.1.8.1   Experiments

The decarburization of carbon-bonded refractory material during the heating of steel mill ladles by means of the ladle fire in air to approx. 1100°C is a serious operational problem because the edge (boundary) layer decarburizes and, consequently, corrodes. Laboratory tests were conducted at DIFK/FGF in Bonn, Germany, to describe the process quantitatively [43–45]. For this purpose, among other things, cylinders out of industrially produced MgO–C materials (36 mm Ø × 100 mm) were oxidized at elevated temperature in air and their weight loss continuously recorded. In addition, 3% by weight metallic aluminum and/or boron carbide (B$_4$C) were used as antioxidants. Figure 13.20 shows the results for three examples.

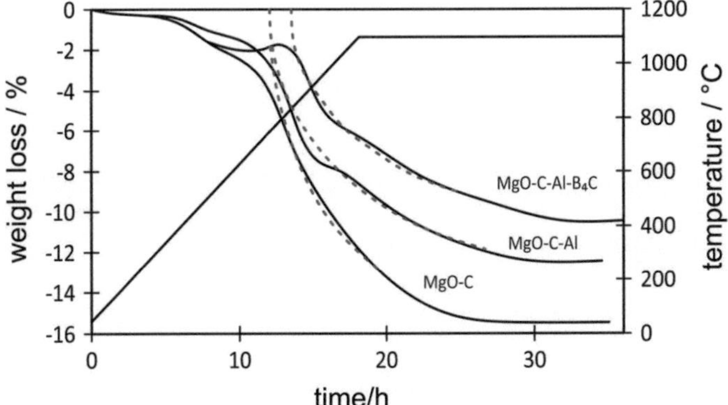

Fig. 13.20   Decarburization of three MgO–C materials in air at 1100 °C

The samples were heated at 5 K/min. A significant reaction starts above 600 °C. Discontinuities in the course of time signal the influence of the antioxidants. They oxidize and cause an increase in weight, which seems to delay the combustion reaction.

If the experiment is stopped before the specimen is completely decarburized, a sharp color transition of the light area decarburized from the outside to the black core is already visible to the naked eye in the saw cut. This points to a transport-controlled topochemical reaction.

13.1.8.2   Evaluation and Discussion

As has been shown earlier, slag and steel infiltrate the open pores of the refractory material. Its dissolution thus takes place virtually "from the inside out." The same applies for furnace gases and air. The oxidation reaction of the graphite is

$$2\langle C\rangle + \{O_2\} = 2\{CO\} \tag{13.61}$$

If the temperature and the CO partial pressure are specified, the partial pressure of the oxygen is fixed. From the Richardson-Ellingham diagram, Fig. 13.11, the value $P_{O2} = 10^{-18}$ atm (red line) is taken for 1100 °C and $P_{CO} = 1$ atm.

The pore diffusion coefficient is one of the factors that determines the timing of corrosion.

$$D_p \equiv D \cdot \lambda \cdot \varepsilon \ \left[\text{cm}^2/\text{h}\right] \tag{13.62}$$

In this, λ is the labyrinth factor. It describes the ratio between the externally measured penetration depth x and the entangled actual path l of the gas inside the structure of porosity ε.

In many cases, [24, 46] is approximately valid.

$$\lambda = \frac{x}{l} = 1.84 \cdot \sqrt{\varepsilon} - 0.3 \; [-] \tag{13.63}$$

As an example, the selective oxidation of carbon during the heating of a ladle lined with MgO–C in a steel mill is considered. The diffusion coefficient of oxygen in the O_2-N_2-CO gas is [43]:

$$D = 2.85 \times 10^{-2} \cdot T_K^{1,79} \; \left[cm^2/h\right] \tag{13.64}$$

Summarizing Eqs. (13.62–13.64), the pore diffusion coefficient is

$$D_P = 2.85 \times 10^{-2} \cdot T_K^{1,79} \cdot \varepsilon \cdot \left(1.84 \cdot \sqrt{\varepsilon} - 0.3\right) \left[cm^2/h\right] \tag{13.65}$$

With an apparent porosity of $\varepsilon = 0.37$ after complete burnout and temperature $T_K = 1.373 \, \mathrm{K}(1100 \,^\circ C)$, the result is $D_P = 3572 \left[cm^2/h\right]$.

The diffusion of oxygen from the outside to the reaction front [46, 47] determines the rate. The linear oxidation rate follows the law of time

$$\frac{dx}{dt} = -V \cdot D_P \cdot \frac{n_0 - n_i}{x_0 - x_i} \; [cm/h] \tag{13.66}$$

$V = 5 \; cm^2/mol$ is the molar volume of graphite.

$$n_0 = 1.23 \times 10^{-4} \cdot C_{Ox}/T_K \; \left[mol/cm^3\right] \tag{13.67}$$

is the oxygen concentration in the gas. C_{Ox} in the air is 21% by volume, i.e. $n_0 = 1.88 \times 10^{-6} \; [mol/cm^3]$. $n_i = 10^{-19}$ atm, on the other hand, is negligibly small (refer to Fig. 2.5).

The solution of the **differential** Eq. (13.66) is [44, 48]:

$$(1 - \phi)^2 = \tau . \tag{13.68}$$

This is the so-called $\sqrt{t}$ − Gesetz (law), where the following applies:

$$\phi = \frac{x_i}{x_0} \quad \text{(Location of reaction front/wall thickness)}, \tag{13.69}$$

$$\tau = \frac{t}{t_0} \quad \text{(time/total duration),} \quad \text{with} \tag{13.70}$$

$$t_0 = \frac{x_0^2}{z \cdot 2 \cdot V_O \cdot D_p(n_a - n_i)} \tag{13.71}$$

t_0 is the time which would be required for one-sided oxidation of the entire wall thickness x_o according to reaction (13.61). z stands for the geometry (design) of the specimen: 2 (plate), 4 (cylinder), 6 (sphere). With $T_K = 1373$ K and the brick thickness $x_0 = 12$ cm, the calculation from (13.71) gives us the one-sided through-oxidation $t_0 = 1072$ h.

In operational practice, approx. 1 cm burns out during heating-up. Thus, Eq. (13.70) for the reduced time gives

$$\tau = \left(1 - \frac{11}{12}\right)^2 = 0.008 \tag{13.72}$$

The time for decarburization of one centimeter of MgO–C brick is according to Eq. (13.70)

$$t = \tau \cdot t_0 = 0.008 \times 1072 \approx 9\,\text{Std,} \tag{13.73}$$

which agrees well with operating results considering that heating-up was not taken into account in the calculation. It was found experimentally that the type of binder phase, resin or Carboplast, has no significant influence on the decarburization rate [49].

The model offers the possibility of a simple and very accurate evaluation of experiments involving decarburization. For this purpose, the measured time progression of the weight change is simulated mathematically, whereby one obtains t_0 as a result [46]. The red-dotted lines in Fig. 13.20 are the result of such calculations. It can be seen that the adjustment is most successful if no antioxidants delay the course of decarburization.

The result is given in Table 13.2.

The following facts become obvious from the quantitative evaluation:

1. Aluminum delays decarburization by approx. 25% which is not much compared to the disadvantages:
1.1 Prevention of a CO buffer protecting against infiltration at the material surface in contact with steel
1.2 Prevention of a MgO layer protecting against infiltration on the material surface in contact with steel
1.3 Increasing the carbon solubility in the infiltrated steel by deoxidizing it
2. Aluminum makes the recycling of used material more difficult (Refer to Sect. 13.1.5)
3. The addition of approx. 3% boron carbide has a negligible effect.

Table 13.2 Influence of antioxidants on decarburization (refer to Fig. 13.20)

	MgO–C	MgO–C–Al	MgO–C–Al–B$_4$C
t_0 [h]	14	19	19.5
m_0 [%]	16.5	12.7	10.5

4. Because of the weight gain, due to the oxidation of antioxidants, their (positive) effect shown in Fig. 13.20 is greatly exaggerated. This leads to the assumption that boron carbide, for example, considerably enhances the effect of aluminum.

To facilitate the calculation in practice, Eq. (13.71) can be rewritten to the so-called oxidation coefficient [45, 46, 48]:

$$\alpha_{Ox} = 2 \cdot V \cdot D_P \cdot n_0 = \frac{x_0^2}{z \cdot t_0} \left[cm^2/h \right] \tag{13.74}$$

From the results of numerous experiments at different temperatures, α_{Ox} was calculated [46]. The function is

$$\alpha_{Ox} = 4.5 \times 10^{-5} \cdot T_C + 0.015 \left[cm^2/h \right] \tag{13.75}$$

For $T_C = 1100\ °C$ and $\varepsilon = 0.37$, we get $\alpha_{Ox} = 0.064$. In summary, for $1173\ K < T_K < 1643\ K$ and $0.16 < \varepsilon < 0.37$, the relation holds

$$\alpha_{Ox} = 3.5 \times 10^{-5} \cdot \varepsilon \cdot \left(1.84 \cdot \sqrt{\varepsilon} - 0.3 \right) \cdot C_{Ox} \cdot T_K^{0.79} \left[cm^2/h \right] \tag{13.76}$$

C_{Ox} is the oxygen content in the gas, e.g. 21% by volume in the air. If α_{Ox} has been measured or calculated from Eq. (13.76), t_0 can be calculated immediately from it if the geometry and dimension x_0 are known. With Eqs. (13.72) and (13.73), the time t [h] required to decarburize the length x[cm] is then obtained.

13.1.9   Recycling of MgO–C–Al

13.1.9.1   Development Pressure of Methane (CH$_4$)

Metallic aluminum is one of the antioxidants used in MgO–C material. When the refractory material is heated, aluminum carbide is formed in the brick:

$$4Al + 3C = Al_4C_3, \left(\Delta G^0 (1000°C) = -1.4 \times 10^5\ J/FU \right). \tag{13.77}$$

Depending on the partial pressure of the magnesium vapor, usually above about 1400 °C, Al_4C_3 reacts further:

$$Al_4C_3 + 8MgO = 2MgO \cdot Al_2O_3 + 3C + 6\{Mg\} \tag{13.78}$$

At $P_{Mg} = 0.05\ atm$, spinel is formed above 1250 °C (refer to Fig. 13.7).

Both reactions have only one degree of freedom, i.e. their equilibrium is solely determined by temperature.

If such material is stored and reprocessed after use, methane is formed in moist air:

$$Al_4C_3 + 6H_2O = 3CH_4 + 2Al_2O_3; \quad \Delta G^0(25\,°C) = -1.74 \times 10^6 \text{ J/FU.} \quad (13.79)$$

This reaction has two degrees of freedom (temperature and a partial pressure) and is strongly exothermic ($\Delta H = -1.9 \times 10^6$ J/FU), so that the material can heat up by about 100 °C (refer to Sect. 13.1.5.2), which kinetically accelerates the reaction. The equilibrium constant is ($R = 8.31$ J/Mol $\cdot$ K; $T = 300$ K):

$$K = \frac{P_{CH_4}^3}{P_{H_2O}^6} = \exp\left(-\frac{\Delta G^\circ}{RT}\right) = 1 \times 10^{+300} \quad (13.80)$$

Assuming an atmosphere saturated with steam at 25 °C. $\left(P_{H_2O} = 0.035 \text{ atm}\right)$, then follows:

$$P_{CH_4} = \left(10^{300} - 0.035^6\right)^{1/3} = 10^{97} \text{ atm.} \quad (13.81)$$

The refractory material disintegrates as a result of this high development pressure of methane (Table 13.3).

Table 13.3 Data for the calculation of heat generation during the reaction of aluminum carbide with water vapor [6]

Reactants (298 K)	n_i	H_i (Joule/mol)	
MgO	360	$-601,701$	
N_2	114	0	
O_2	30	0	
H_2O	6	$-241,856$	
Al_4C_3	1	$-209,200$	
$\Sigma n_i H_i$ (298 K)		$-218,272,696$ [Joule]	
Products (T)	n_i	H_i (Joule/Mol)	
		300 K	400 K
Al_2O_3	2	$-1,675,545$	$-1,666,526$
CH_4	3	$-74,807$	$-71,005$
O_2	30	$+53$	$+3029$
N_2	114	$+52$	$+2984$
MgO	360	$-601,632$	$-597,557$
$\Sigma n_i H_i$ (T)		$-220,155,513$	$-218,235,541$ (Joule)

13.1.9.2   Heating of the MgO–C–Al$_4$C$_3$ Material

After its use, the refractory material contains about 1% by weight Al$_4$C$_3$. Its molecular weight is $M_{Al_4C_3} = 144$ g/mol. If the molecular weight of the total refractory material is set equal to that of MgO ($M_{MgO} = 40$ g/mol), 360 mol of refractory material (MgO) is involved in the conversion of 1 mol of Al$_4$C$_3$.

Air saturated in steam (H$_2$O) at room temperature (RT $= 25$ °C) contains 0.035% by volume H$_2$O. The gas composition is 0.76% N$_2$, 0.20% O$_2$, 0.04% H$_2$O. Consequently, 150 mol of air consist of 114 mol of N$_2$, 30 mol of O$_2$ and 6 mol of H$_2$O.

Both the entire refractory material and the air are heated from RT to T[K] as a result of the exothermic transformation of the aluminum carbide with steam (The reaction equations Eq. (13.82) and Eq. (13.79) are identical):

$$(360\, MgO + 114\, N_2 + 30\, O_2 + 6H_2O + Al_4C_3)_{R,T}$$
$$= (2Al_2O_3 + 3CH_4 + 30\, O_2 + 114\, N_2 + 360\, MgO)_{P,T} \tag{13.82}$$

Neglecting possible heat losses, the reaction is considered adiabatic, i.e. the heat content of the reactants (R) and the products (P) is equal: $\Delta H = 0$. Consequently, the temperature T of the products is sought at which their enthalpy is equal to that of the reactants:

$$\Delta n_i H_i(R) - \Delta n_i H_i(P) = 0 \tag{13.83}$$

The enthalpy values are taken from the tables of Knacke et al. [6]. Fig. 13.20

In a narrow temperature range, the enthalpy changes linearly with temperature. Therefore, the following applies for the products

$$\Sigma n_i H_i(P) = 19200\, T - 225915429\, \big[\text{joules}\big]. \tag{13.84}$$

For the reactants, at room temperature [6] $\sum n_i H_i(R) = -218272696$ joules.

Putting $\sum n_i H_i(P) = \sum n_i H_i(R) = -218272696\,$j into Eq. (13.84) gives $T = 398\, K = 125$ °C.

Consequently, the temperature increase of the reacting refractory material can be more than 100 °C.

13.1.10 Elasto-plastic Behavior of MgO–C

As already discussed in Sect. 12.3.4, MgO–C material exhibits excellent elasto-plastic behavior (refer to Fig. 12.8). The consequence is its outstanding thermal shock resistance. However, the current state of knowledge does not suggest the quantitative application of mechanical models to the thermomechanical behavior of this material. Further experiments are required to make the theory more precise.

13.1.11 Summary

Carbon-bonded refractory materials used most of all in the steel industry. This is astonishing at a superficial glance, if one thinks of the excellent solubility of carbon in liquid iron and of the two oxides alumina and magnesia in slags. The main arguments for their use are the low wetting of this refractory material and its outstanding thermal shock resistance.

In this work, the corrosion and thermal shock resistance properties of AMC materials (Chap. 12) are subjected to a thorough analysis. The aim is to understand why their properties are of such great importance, especially for the steel industry.

The following connections emerge.

Chemical corrosion is kept within tolerable limits mainly by thermodynamic effects:

- From a thermodynamic point of view, MgO is one of the most stable oxides.
- The non-wetting of the refractory materials by both melts, steel and slag, hinders their infiltration and protects the binder phase from its dissolution.
- According to Vacher-Hamilton, if the graphite comes into contact with the molten steel, a pressure of the carbon monoxide of more than 1 bar can occur which prevents the melt from penetrating the pore structure.
- At approx. 1600 °C, a total pressure of approx. 0.1 atm exists in the structure, consisting of 0.05 atm CO gas and 0.05 atm Mg steam. According to Vacher-Hamilton, penetrating iron can, therefore, only dissolve about 0.04% by weight carbon.
- The magnesium steam diffuses at the refractory material/steel interface, dissolves in the steel and reacts by heterogeneous nucleation to form periclase, which can form a protective layer on the surface of the refractory material that impedes infiltration.
- Antioxidants, which, on the one hand, impair the combustion of the graphite (within limits), especially metallic aluminum, but, on the other hand, promote infiltration by deoxidizing the steel. No CO cushion and no MgO protective layer can develop, i.e. infiltration is favored.

In the infiltrate, the oxygen content is lowered so that the solubility of the carbon increases by a factor of about 10–0.4% by weight.

- Another disadvantage of using metallic aluminum is its reaction to Al_4C_3. Reprocessing is limited by its reaction with the steam in the air to form methane, whose enormously high development pressure destroys the microstructure and can lead to a heating of more than 100 °C.

- The premature wear along the three-phase line refractory material/steel/slag, observed in the laboratory and in practice, can be explained by Marangoni convection. This is caused by gradients of the interfacial energy as a result of chemical reactions and temperature gradients.
- The thermal shock resistance of AMC material is excellent due to the thermoplastic properties of the graphitic binder phase. But even without this, a good service is already foreseeable that is superior to that of chromium magnesite.

13.2   SiC in Copper Metallurgy

13.2.1   Introduction

Silicon carbide (SiC) is used for the pyrometallurgical production of conductive (line) copper as a lining for cathode shaft furnaces [50]. There are essentially two completely different types of bonding: SiO_2 bonding (10% oxide) and Si_3N_4 bonding (20% nitride). The service temperature ranges up to approx. 1650 °C. The density of the materials is 2.6 g/cm^3, and the open (apparent) porosity is 18%.

13.2.2   Experiments

Laboratory tests of the materials as crucible test [50] and finger test [50, 51] lead to the same result and correspond well with practical experience.

Quantitative results are obtained in particular with the finger test, in which a test rod (20×20 mm^2) is immersed approx. 90 mm deep in the molten copper. The change in the weight of the specimen over time is determined via its buoyancy and the change in the oxygen content of the melt is determined electrochemically via a continuous EMF measurement. The operating principle is described in detail in Sect. 9.4.1, Figs. 9.5 and 9.6. The EMF measurement is evaluated using Eq. 9.30. Finally, the microstructure of the specimens is examined microscopically in order to precisely describe the reactions with the melt.

13.2.3   Results

Figure 13.21 shows a summary of the results of several tests. The graphs illustrate the relative change in buoyancy $B = [(A-A_0)/A_0]$ on the one hand and the oxygen content [O] % by weight dissolved in the copper at 1150 °C on the other. Once the specimen material is immersed, the formation of CO bubbles begins immediately which can be seen by the unsteady buoyancy measurement. After a few minutes, however, the measurement calms down as silicic acid (SiO_2) is deposited on the specimen surface. Microscopic analysis of the specimens shows that this thin layer repeatedly spalls (chips) off the surface of the nitride-bonded material and soon only forms and remains, to a small extent, preferably in indentations. In contrast to this, the layer adheres to the oxide-bonded material and effectively hinders the further dissolution of the refractory material (red line). Accordingly, the buoyancy of the oxide-bonded material is constant over time while the nitride-bonded material continues to dissolve very slowly with the slight formation of gas bubbles (green line). Material, which is infiltrated by the melt, shows a reduction in buoyancy as its density increases on average.

A comparable observation can be made if monitoring the oxygen content of the melt using EMF measurement: starting with $10^{-2}\%$ by weight, the oxygen content initially drops to less than $10^{-6}\%$ by weight when using nitride-bonded material, only to stabilize at $2 \times 10^{-6}\%$ by weight after a few minutes (blue line). If using silica-bonded material, the oxygen content remains almost constant over time at $10^{-2}\%$ by weight after an initial fluctuation (yellow line).

The initial oxygen content plays a key role here: if it is less than $10^{-3}\%$ by weight, no silicate excretion occurs on either material, i.e. the SiC refractory material is fully exposed to the melt and, therefore, dissolves slowly. Conversely, if the oxygen content is initially high and amounts to 0.1% by weight, for example, a cover layer is formed on both materials which is so dense that dissolution does not occur.

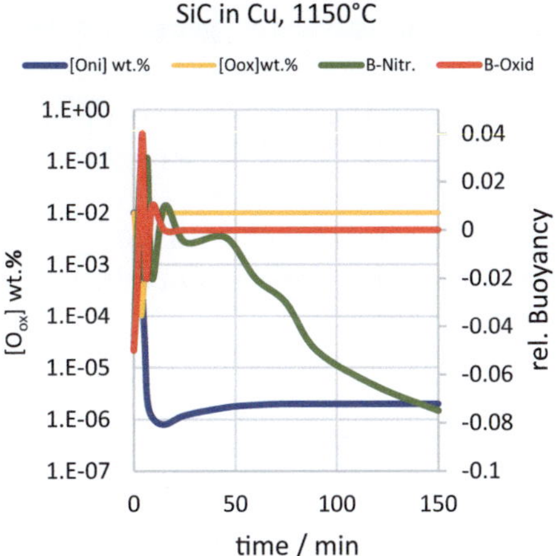

Fig. 13.21 Dissolution behavior of silicon carbide (SiC) with oxide bond (SiO_2) and nitride bond (Si_3N_4) in liquid copper at 1150 °C. ([**O**] denotes the oxygen content dissolved in the melt and **B** the relative buoyancy of the immersed specimen)

13.2.4　Discussion

The experimentally determined results can be confirmed theoretically. The cover layer is formed according to the reaction equation [51]

$$\langle SiC \rangle + 3[O] = \langle SiO \rangle_2 > +\{CO\}. \tag{13.85}$$

At 1400 K, the free enthalpy has the value. $\Delta G_1^0 = -112283\,\text{cal/FU}$ At a CO pressure of 1 atm, the following can be calculated from the law of mass action

$$[O] = \left[P_{CO} \cdot \exp\left(\frac{+\Delta G_1^0}{RT} \right) \right]^{1/3} \tag{13.86}$$

the oxygen content of the melt is $2 \times 10^{-6}\%$ by weight. The calculated content agrees with the measured content. This means that the development pressure of the carbon monoxide at the surface of the SiC is 1 atm. According to this, a silicate layer formed there is easily blown off (Fig. 13.17).

The reaction equation [51]

$$\{CO\} = [C] + [O]; \ \Delta G_2^0 = 73850\,\text{cal/FU} \tag{13.87}$$

according to $5 \times 10^{-5}\%$ by weight dissolved oxygen is the carbon content (Fig. 13.18).

$$[C] = P_{CO} \left[[O] \cdot \exp\left(\frac{+\Delta G_2^0}{RT} \right) \right]^{-1} \tag{13.88}$$

at equilibrium. At $P_{co} = 1$ atm, this is only $2 \times 10^{-6}\%$ by weight carbon.

The dissolved silicon content is calculated from the oxygen content using the reaction equation [51] (Fig. 13.19).

$$\langle SiO_2 \rangle = [Si] + 2[O]; \ \Delta G_3^0 = 105530\,\text{cal/FU} \tag{13.89}$$

and the law of mass action to

$$[Si] = \left[[O]^2 \cdot \exp\left(\frac{+\Delta G_3^0}{RT} \right) \right]^{-1} \tag{13.90}$$

This is also only $[Si] = 10^{-5}\%$ by weight.

The dissolved carbon and silicon contents in thermodynamic equilibrium are, therefore, below their detectability limit of approx. $10^{-3}\%$ by weight which corresponds to the experimental findings.

The different behavior of the two materials is due to their bonding phase. The silicate binder phase is thermodynamically very stable, whereas the nitride phase is comparatively not. It consists of Si_3N_4, and this applies at 1400 K:(Figs. 13.20 and 13.21).

$$\langle Si_3N_4 \rangle = 3[Si] + 4[N]; \ \Delta G_4^0 = 73294 \, cal/FU/51 \tag{13.91}$$

Consequently, at 1400 K

$$[Si]^3 \cdot [N^4] = \exp\left(\frac{-\Delta G_4^0}{RT}\right) = 4 \times 10^{-12}. \tag{13.92}$$

In addition to the maximum soluble nitrogen content of approx. $10^{-3}\%$ by weight in the molten copper, approx. 1.4% by weight of silicon is soluble at equilibrium. That is a lot.

Further experiments illustrate that the pure nitride phase alone reacts strongly with the melt with the initial oxygen content dropping from 0.1% by weight to $2 \times 10^{-2}\%$ by weight in half an hour. Consequently, the nitride significantly determines the behavior of the material. At the same time, it was seen that Si_3N_4 is susceptible to cracking and breaks easily. This is another reason why an oxidic protective layer repeatedly spalls and the dissolution of the nitride-bonded material cannot be stopped.

References

1. Ollig, M.: Reactions of Carbon-Containing MgO—Materials with Liquid Iron, Dr.-Ing. Dissertation. RWTH—Aachen, Germany (2000)
2. Pötschke, J., Beimdiek, K., Ollig, M.: Reaction between MgO–C bricks and steel melts. In: Proceedings. 6th Biennial Worldwide Congress of Refractories, pp. 166–169. Berlin, Germany (1999)
3. Frohberg, M.G.: Thermodynamics for Materials Engineers and Metallurgists, 2nd edn. Deutscher Verlag für Grundstoffindustrie, Leipzig, Germany (1994)
4. Hack, K.M.: FACTSage, Thermfact & GTT Technologies (2007). www.factsage.com
5. Hütte.: Pocket book for ironworks specialists. 5th Edition. Verlag W. Ernst & Sohn, Berlin, Germany (1961)
6. Knacke, O., Kubaschewski, O., Hesselmann, K.: Thermochemical Properties of Inorganic Substances I/II, 2nd edn. Springer-Verlag, Verlag Stahleisen (1991)
7. Elliott, J.F., Gleiser, M., Ramakrishna, V.: Thermochemistry for Steelmaking I/II, p. 63. Addison-Wesley Publishing Company Inc. (1960)
8. VDEh: Slag Atlas, 2nd Edition, Verlag Stahleisen GmbH Düsseldorf, Germany (1995)
9. Kalvelage, L., Markert, J., Pötschke, J.: Measurement of the dissolution of graphite in liquid iron by tracking Buoyancy. Archiv. Eisenhüttenwesen 50, 107–110 (1979)
10. Swisher, J.H.: Thermodynamics of carbide formation and graphite solubility in the $CaOSiO_2Al_2O_3$ system. Trans. Metal Society. AIME 242, 2033–2037 (1968)
11. Schubert, H.G., Schwerdtfeger, W.: Solubility of carbon in ESU—Slags. Archiv. Eisenhüttenwesen. 45, 437–439 (1974)

12. Pötschke, J., Deinet, T., Routschka, G., Simmat, R.: Properties and corrosion of AMC-refractories II. In: Proceedings UNITECR'03 Osaka, pp. 584–586, Japan (2003)
13. Koch, K., Trömel, G., Heinz, G.: The blast furnace slag system Al_2O_3-CaO-MgO-SiO_2 at 1600, 1500 und 1400 °C. Archiv. für das Eisenhüttenwesen. **46**(3), 165–171 (1975)
14. Janke, D.: Physical Properties of Slag Melts in Slags in Metallurgy, K. Koch und D. Janke (Publisher), Verlag Stahleisen mbH
15. Park, J.M., Lee, K.K.: Reaction equilibria between liquid iron and CaO-Al_2O_3-$MgO_{sat.}$-SiO_2-Fe_tO-MnO-P_2O_5 slag. In: Proceedings of the 79th Steelmaking Conference, 24 Mar 1996, Pittsburgh, Pennsylvania, USA (1996)
16. Reisinger, P., Presslinger, H., Hiebler, H., Zednicek, W.: MgO-solubility in steel mill slags. BHM Berg Hüttenmänn. Monatsh. **144**(5), 196–203 (1999)
17. Brüggmann, C.: A Contribution to Slagging of MgO in Secondary Metallurgical Slags, Dr.-Ing. Dissertation. TU Bergakademie Freiberg, Saxony, Germany (2011)
18. Brüggmann, C., Pötschke, J.: MgO Saturation in secondary lime-aluminate and lime-silicate slags. Steel Res. **4, 82**(1011), 422–427
19. Wöhrmeyer, C., Elorza-Ricart Jolly, E., Guichard, S., Brüggmann, C., Sax, A.: The impact of synthetic slags on steel ladle refractory life time. In: Proceedings of the 51st International Colloquium on Refractories, pp. 80–83, Aachen, Germany, 15–16 Oct (2008)
20. Oeters, F.: Metallurgy of Steelmaking. Verlag Stahleisen, Springer (1989)
21. Gatellier, C., Olette, M.: Aspects Fondametaux des reactions des elements Metallurgiques dans les acieres liquids. Rev. Met. **76**, 377–386 (1979)
22. Smeets, L.: Examination of Formation Conditions of Oxidic and Sulfidic Inclusions Upon the Solidification of Steel, Dr.-Ing. Dissertation RWTH—Aachen, Germany (1972)
23. Bannenberg, N.: Interactions between the refractory material and steel and their influence on the degree of purity. Stahl u. Eisen **115**(9), 78–86 (1995)
24. Jeschke, P.: Texture analysis of basic refractory brick. J. Am. Ceram. Soc. **49**(7), 360–363 (1966)
25. Pötschke, J., Routschka, G., Simmat, R.: The evaluation of oxidation experiments on carbon-containing refractories. In: Proceedings 45th International Colloquium on Refractories, pp. 36–39. Aachen, Germany (2002), 16/17 Oct 2002
26. Volmer, M.: Kinetics of Phase Formation. Th. Steinkopf, Dresden, Germany (1939)
27. Cahn, R.W.: Physical Metallurgy. North Holland Publishing Company Amsterdam, Holland (1965)
28. Pötschke, J., Frohberg, M.G.: Oxide nucleation in liquid steel—Homogenous or heterogenous? Archiv. Eisenhüttenwesen **41**(8), 723–729 (1970)
29. Pötschke, J.: The periodic precipitation of oxide in liquid copper. Metall 24, Heft 2 + 3 (1970)
30. Eustahopoulos, N., Drevet, B.: Wettability at High Temperatures. Pergamon (1999)
31. Pötschke, J.: Emergence and Behavior of Excretions During the Solidification in "Solidification of Metallic Melts". DGM Seminar (81), 221–231
32. Levich, V.G.: Physicochemical Hydrodynamics, pp. 373–390. Prentice Hall Inc. (1962)
33. Pötschke, J., Deinet, T.: Premature Corrosion of Refractories by Steel and Slag. Millennium Steel 109–113 (2005)
34. Pötschke, J., Deinet, T., Routschka, G., Simmat, R.: Properties and corrosion of AMC—Refractories Part II: Corrosion by Steel/Slag, pp. 584–587. In: Proceedings UNITECR—2003—Congress Osaka, Japan (2003)
35. Pötschke, J., Deinet, T.: The corrosion of AMC—Refractories by Steel and Slag, pp. 75–79. UNITECR Congress, Orlando, Florida, USA (2005)
36. Marangoni, C.: Spread of droplets of one liquid on the surface of another one. Ann. d. Phys. Chem. **143**(1871), 337
37. Borgmann, F.O.: Influence of Surface Tension of Carbon-Saturated Iron Melts by Additional Elements. Dr.-Ing. Dissertation, TU – Berlin, Germany (1971)
38. Pötschke, J., Brüggmann, C.: Premature Wear of Refractory Material as Result of Marangoni Convection. Steel Res. **82** (2011) in print

39. Langbein, D.: Capillary Surfaces—Shape—Stability—Dynamics, in Particular under Weight-lessness, Springer Tracts in Modern Physics 178 (2002)
40. Turkdogan, E.T.: Physical Chemistry of High Temperature Technology. Academic Press (1980). ISBN 0-12-704650-x
41. Ogino, K., Hara, S., Miwa, T., Timoto, S.: The Effect of Oxygen Content in Molten Steel on the Interfacial Tension Between Molten Steel and Slag. Tetsue-to-Hagane **65**, 2012–2021 (1979)
42. Hauck, F.G., Pötschke, J.: The wear of submerged nozzles during continuous casting of steel. Archiv. für das Eisenhüttenwesen. **53**(4), 133–138 (1982)
43. Bogdandy, L.V., Engell, H.J.: The Reduction of Iron Ore, pp. 49–60, Verlag Stahleisen (1967)
44. Friedrichs, H.-A., Knacke, O.: Process Engineering Basic Types of Isothermal Reactions in Blown Through and Overflown Fills. Research Reports of the State of North Rhein-Westphalia No. 2240, (1973)
45. Pötschke, J., Deinet, T., Routschka, G., Simmat, R.: Properties and Corrosion of AMC-Refractories. In: Proceedings UNITECR'03, pp. 580–583
46. Pötschke, J., Routschka, G., Simmat, R.: The Evaluation of Oxidation Experiments on Carbon-Containing Refractories; Proceedings 45th International Colloquium on Refractories, pp. 36–39. Aachen, Germany (2002)
47. Ghosh, N.K., Gosh, D.N., Jangannathan, K.P.: Oxidation mechanism of MgO–C in air at various temperatures. Brit. Ceram. Trans. **99**, 124–128 (2003)
48. Routschka, G., Pötschke, J.: The Oxidation of Carbon-Containing Refractories. In: Proceedings UNITECR 01 Cancun, Vol. II, pp. 1092–1099, Mexico (2001)
49. Simmat, R., Pötschke, J.: Oxidation experiments on different types of carbon bond of refractories. In: Proceedings 46th International Colloquium on Refractories, pp. 11–12. Aachen, Germany (2003)
50. Brückner, H.W., Völker, W., Beimdick, K., Pötschke, J.: Proceeding 41. International Feuerfest-Kolloquium, pp. 100–102, Aachen (1998)
51. Maier, R., Pötschke, J.: Reaction of technical silicon carbide with oxygen-containing copper melts. Metall **32**(11), 1109–1111 (1978)

Chapter 14
Thermal Wear

Exceeding the service (application) temperature, together with insufficient thermal shock resistance (cf. Chap. 12), is one of the most frequent causes for failure. One example is the operation of a furnace for reheating slabs in the rolling mill of a steel mill [1].

Knowledge of heat transfer is, therefore, of great importance.

14.1 Heat Transfer

14.1.1 Radiation

Thermal radiation follows the

$$P = A \cdot \varepsilon \cdot \sigma \cdot T_K^4 \ [\text{W}]. \tag{14.1}$$

In the future, the consideration will be focused on the area $A = 1\ \text{m}^2$. The emission constant of refractory oxide materials can be assumed to be $\varepsilon = 0.8$. The radiation constant is according to Stefan–Boltzmann $\sigma = 5.7 \times 10^{-8}\ \text{W/m}^2\ \text{K}^4$ [2]. A temperature difference is calculated from the approximation

$$\dot{Q} \simeq 4 \cdot \varepsilon \cdot \sigma \cdot T_K^3 \cdot (T_1 - T_2)\ [\text{W/m}^2] \tag{14.2}$$

Note: Integration over T in the limits T/T$_{12}$ gives Eq. (14.1).

If the values in front of the parenthesis are combined to the so-called heat transfer coefficient α, the value $\alpha \approx 6\ \text{W/m}^2\ \text{K}$ is obtained for 300 K, and for approx. 1500 K $\alpha_1 \approx 600\ \text{W/m}^2\ K$. In practice, this value depends, among other things, on the design and contents of the furnace and can vary considerably.

© The Author(s), under exclusive license to Springer Nature Switzerland AG 2024

J. Pötschke, *Refractory Fundamentals in Metallurgical Practice*,

https://doi.org/10.1007/978-3-031-63709-4_14

A typical value for the heat flux density in industrial furnaces is $\dot{Q} = 5000$ W/m^2 (A. Eschner, personal communication). From this, the temperature drop ΔT toward the inner furnace wall is calculated to $\Delta T = \dot{Q}/\alpha_1 = 5000/600 = 8$ K. The gas temperature of 1300 °C drops to 1292 °C toward the wall.

14.1.2   Convection

The calculation of the heat transfer as a result of gas convection is usually only possible in an approximate way, since the operating conditions are subject to fluctuations which are only recorded inaccurately. The heat flux·density follows the known relation [3]:

$$\dot{Q} = \alpha_K \cdot (T_1 - T_2) \ [\text{W/m}^2] \tag{14.3}$$

The temperature difference between gas and wall is ΔT [K], and the heat transfer coefficient is α [W/m^2 K]. In furnaces, the flow is forced because of the use of burners. The gas velocity exceeds very possibly the value $w = 10$ m/s. In the simplest case then applies [3]:

$$\alpha_K = 7.5 \cdot w^{0.78} = 45 \ \text{W/m}^2 \ \text{s} \tag{14.4}$$

In the transition region from laminar to turbulent flow, the relationship [4] can be applied.

$$\alpha_K = \frac{\lambda}{L} \cdot \left(0.037 \cdot \text{Re}^{0.8} - 871\right) \cdot \text{Pr}^{0.3} \tag{14.5}$$

The transition takes place when the Reynolds number of the gas is $\text{Re} \equiv \frac{w \cdot L}{v} \geq 5 \times 10^5$ [4]. The Prandtl number $\text{Pr} \equiv \frac{v}{a} 0.7$ of gases is negligible dependent on temperature, because both, the kinematic viscosity v and the thermal diffusivity a are determined by the momentum (impulse) exchange [4]. The thermal conductivity of the gas at high temperature is approx. $\lambda = 0.1$ W/m K [3]. With these figures, taking Eq. 14.5 you get $\alpha_K = 88$ W/m^2 s.

Calculating the Reynolds number taking into account the operational estimate $w = 10$ m/s for the length $L = 1$ m subjected to flow and the kinematic viscosity of the gases at high temperatures $v \simeq 2 \cdot 10^{-4}$ m^2/s [4], we obtain $\text{Re} = 5 \times 10^4$. The flow would be laminar and [4] would apply.

$$\alpha_K = \frac{\lambda}{L} \cdot \left(0.664 \cdot \text{Re}^{0.5} \cdot \text{Pr}^{0.3}\right) = 30 \ \text{W/m}^2 \cdot \text{s} \tag{14.6}$$

However, if the flow length L and/or the flow velocity w increase by more than a factor of 10, which may well be the case in large furnaces, turbulence will occur.

For Re $= 10^6$, this results in [4]

$$\alpha_K = \frac{\lambda}{L} \cdot \left(\frac{0.037 \cdot \text{Re}^{0.8} \cdot \text{Pr}}{1 + 2.443 \cdot \text{Re}^{-0.1} \cdot (\text{Pr}^{0.67} - 1)} \right) = 185\,\text{W/m}^2\,\text{s} \qquad (14.7)$$

14.1.3 Radiation and Convection

Both processes add up to a total heat transfer. It is questionable at which temperature the radiation predominates. If both processes are equally strong, the driving temperature difference is equally big and $\alpha_K = \alpha_{Str}$ is valid. From Eqs. 14.2 and 14.5, we obtain ($\varepsilon = 0.8$; $\sigma = 5.7 \times 10^{-8}$ W/m^2 K^4 [2]):

$$T_{K/S} = \left(\frac{\lambda \cdot (0.037 \cdot \text{Re}^{0.8} - 871) \cdot \text{Pr}^{0.3}}{4 \cdot L \cdot \varepsilon \cdot \sigma} \right)^{1/3} = 783\,\text{K} \stackrel{\triangle}{=} 510\,^\circ\text{C} \qquad (14.8)$$

In the event that the gas flow is in the "laminar/turbulent" transition range radiation above 510 °C begins to dominate the heat transfer. Practically, this is the temperature range at which radiation becomes visually visible. With laminar flow, radiation already dominates at lower temperatures; with turbulence, only at higher temperatures.

14.2 Linear Heat Conduction

In many cases, it can be assumed that the temperature drop in the lining is linear. Then, the following applies for the heat flux density [3]

$$\dot{Q} \simeq \lambda \cdot \frac{(T_1 - T_2)}{L} \; [\text{W/m}^2]. \qquad (14.9)$$

λ [W/m K] is the temperature-dependent thermal conductivity of the material. If the inner wall of the furnace is made of mullite, then at high temperature. $\lambda_1 \approx 1.65$ W/m K. For a thickness of $L = 0.2$ m the temperature drop in the wall is

$$\Delta T = \frac{\dot{Q} \cdot L}{\lambda_1} = \frac{5000 \cdot 0.2}{1.65} = 606\,\text{K} \qquad (14.10)$$

i.e. the temperature drops from 1292 to 686 °C. The mullite layer is followed by the insulation layer. Its thermal conductivity is only $\lambda_2 \approx 0.2$ W/m K. With a thickness of 0.03 m, the temperature drop in this layer is

$$\Delta T = \frac{\dot{Q} \cdot L}{\lambda_2} = \frac{5000 \cdot 0.025}{0.2} = 625 \text{ K}. \tag{14.11}$$

Consequently, the outside temperature of the lining is 61 °C.

The outer steel jacket has the thermal conductivity $\lambda_3 \approx 20 \, \text{W/m K}$ and a thickness of 0.005 m. Inside it, the temperature drops by 1 K to 60 °C.

This is followed by external cooling with, $\alpha_2 \approx 30 \, \text{W/m}^2 \cdot K$ a value that depends strongly on the external conditions, so that room temperature is already reached at a short distance from the sheet metal jacket.

The individual coefficients are usually combined into the heat transfer coefficient k [W/m^2 K]. This is a sequence of individual resistances, so that [3] applies:

$$k = \frac{1}{\frac{1}{\alpha_1} + \frac{L}{\lambda_1} + \frac{L}{\lambda_2} + \frac{L}{\lambda_3} + \frac{1}{\alpha_2} \dots} \, [\text{W/m}^2 \text{ K}] \tag{14.12}$$

with $\dot{Q} \simeq k \cdot (T_A - T_E) \, [\text{W/m}^2]$. In the present case, k $\approx$ 4 W/m^2 K and $\Delta T \approx 1250$ K.

The calculation suggests that the insulation should be done on the inside. If the insulation were to be carried out on the outside, the temperature in the lining would rise and it could be damaged. For example, care must be taken that the classification temperature of the insulation layer in contact with the wear layer is not exceeded. Not taken into account are the temperature jumps between two material transition points, which cannot be excluded.

14.3 Measurement of Thermal Conductivity with the Calorimeter Method

The measurement of the heat conduction coefficient λ [W/m K] according to ASTM is carried out with the aid of a calorimeter. The refractory material, in particular ceramic mats, is heated to a maximum of 1300 °C on one side. On the opposite side, the heat flux density

$$\dot{Q} = \lambda \times \frac{T_j - T_k}{x_j - x_k} = C_{P,\text{H}_2\text{O}} \times \dot{m}_{\text{H}_2\text{O}} \times \Delta T / A \, [\text{W/m}^2] \tag{14.13}$$

is determined by means of a flow calorimeter [5]. Inside the mat, five thermocouples are located at equal intervals, which measure the temperature when the stationary state is reached (refer to Fig. 14.1). From the sequence of temperatures, differences ΔT are formed. The result is ten linear temperature gradients $\Delta T / \Delta$.

As agreed, the thermal conductivity is related to the mean value of all temperatures: $\overline{T} = (T_j + T_k)/2$. However, this is only correct if in the specimen the temperature curve between the measuring points x_j and x_k is linear. Often this is not the case.

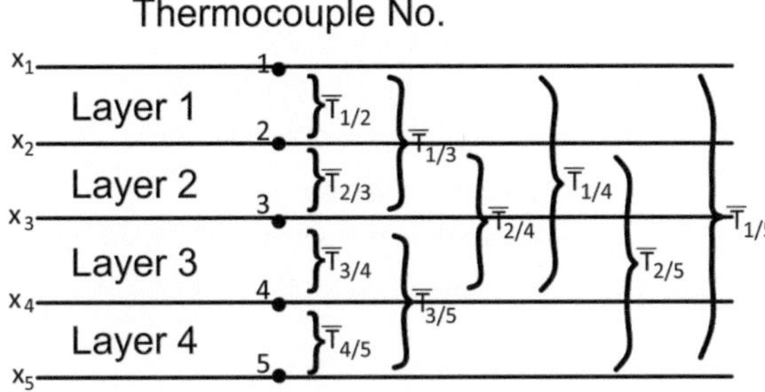

Fig. 14.1 Ten arithmetic mean values of the measured temperatures [5]

Figure 14.2 shows an example of the measurement result on a fiber mat made of aluminosilicate. The temperatures $T(x)$ lie on a curve. The reference to the mean value $\overline{T}_{1/5}$ is far away from the measured curve.

The measured temperatures as a plot over their measurement location are, therefore, represented as a polynomial to improve accuracy [5]:

$$T(x) = Ax^3 + Bx^2 + Cx + D \tag{14.14}$$

From this the integral mean value is calculated

$$\overline{T}_{j,k} = \frac{1}{x_j - x_k} \int_{x_j}^{x_k} T(x)\, \mathrm{d}x \tag{14.15}$$

Term-wise integration results in

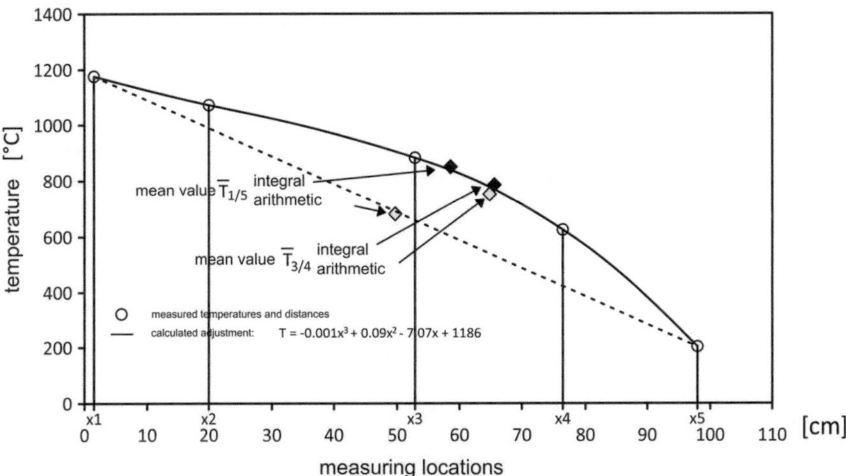

Fig. 14.2 Temperature distribution in the aluminosilicate specimen [5]

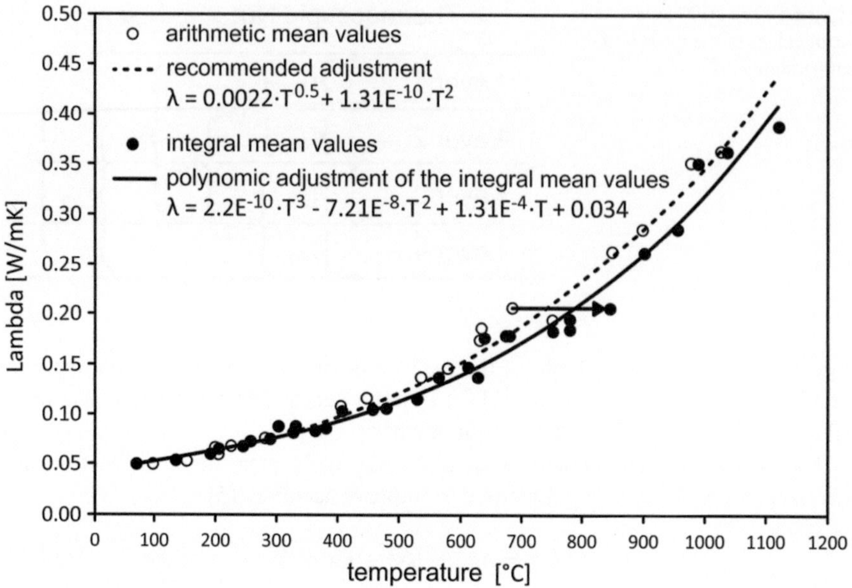

Fig. 14.3 Comparison of the results with classical evaluation by formation of arithmetic mean values oooo with the formation of integral mean values •••• [5]

$$\overline{T}(x_{j,k}) = \frac{1}{x_j - x_k}\left[\frac{A}{4}\left(x_j^4 - x_k^4\right) + \frac{B}{3}\left(x_j^3 - x_k^3\right) + \frac{C}{2}\left(x_j^2 - x_k^2\right) + D\left(x_j - x_k\right)\right]$$

$$(14.16)$$

By means of Eq. 14.13, $\lambda_{j,k}$ is calculated for each measuring point and assigned to the mean temperature $\overline{T}_{j,k}$ calculated according to Eq. 14.16.

The result is shown in Fig. 14.3 [5]. The temperature dependence of the thermal conductivity constant $\lambda\ (T)$ [W/m K] of the measured fiber mat is shown. Above 400 °C, the values assigned to the arithmetic mean (••••) differ increasingly clearly with increasing temperature from those assigned to the integral mean (• • ••). The latter are shifted toward higher temperatures. In practice, this means a possibility to save insulation material if referring to the same temperature, since the thermal conductivity related to the integral mean value is comparatively lower.

14.4   Measurement of Thermal Diffusivity by Monotonic Heating-Up

If a specimen heats up linearly in time, i.e. $\dot{T} = $ const., the solution of the unsteady temperature conduction equation is [6]

$$\frac{\partial T}{\partial t} = -a\frac{\partial^2 T}{\partial x^2} \; [\text{K/s}] \tag{14.17}$$

$$\dot{T} = -z \cdot a \cdot \frac{\Delta T}{x^2} \; [\text{K/s}]. \tag{14.18}$$

a [m^2/s] is the sought temperature-dependent thermal diffusivity. The temperature difference between the surface and the core of the specimen is measured continuously and the mean value is calculated: $\Delta T = ((T_{+x} + T_{-x})/2) - T_0$ (refer to Fig. 14.4). The same is true for cooling, but it cannot be controlled so well experimentally. z is a geometry factor: $z = 2$ (plate), $z = 4$ (cylinder) [7].

Via the relation $a = \frac{\lambda}{C_P \cdot \rho}$, a [m^2/s] is connected to the thermal conductivity [W/ m K], the specific heat C_P [J/kg K] and the density ρ [kg/m^3] [8, 9].

An optimal result is obtained if in the experiment the condition $\frac{\dot{T}}{\Delta T} \leq z \cdot \frac{a}{x^2} [\text{s}^{-1}]$ is fulfilled. If the temperature change is too rapid, the core $\dot{T}$ can no longer follow the external temperature change $\Delta T(t)$ and a is too low. For example, with the estimated thermal diffusivity a $\approx 0.4\,\text{m}^2/\text{s}$ and the radius of the cylindrical specimen $x = 0.02$ m, the cooling rate of the core $\dot{T} \approx 0, 25\,\text{K/s}$ if $\Delta T \approx 60$ K is set.

Figure 14.5 shows an example. It is noticeable that the measurement shows enormous deflections after 4500 s. In this range, a microstructural change occurs as a result of a chemical reaction. The reaction heat decouples the two measured quantities $\dot{T}$

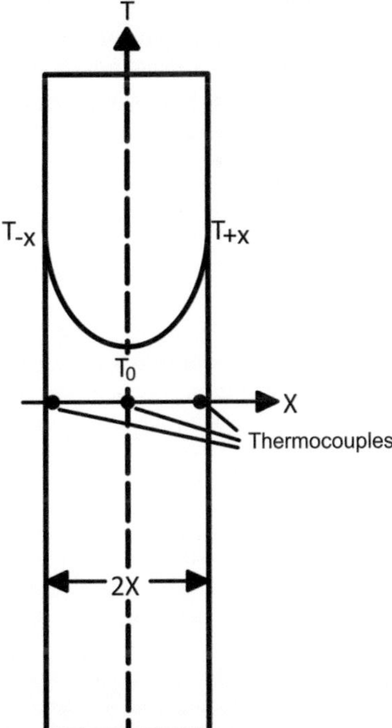

Fig. 14.4 Arrangement of the thermocouples and temperature profile in the cylindrical specimen during dynamic measurement of thermal conductivity by monotonic heating [6]

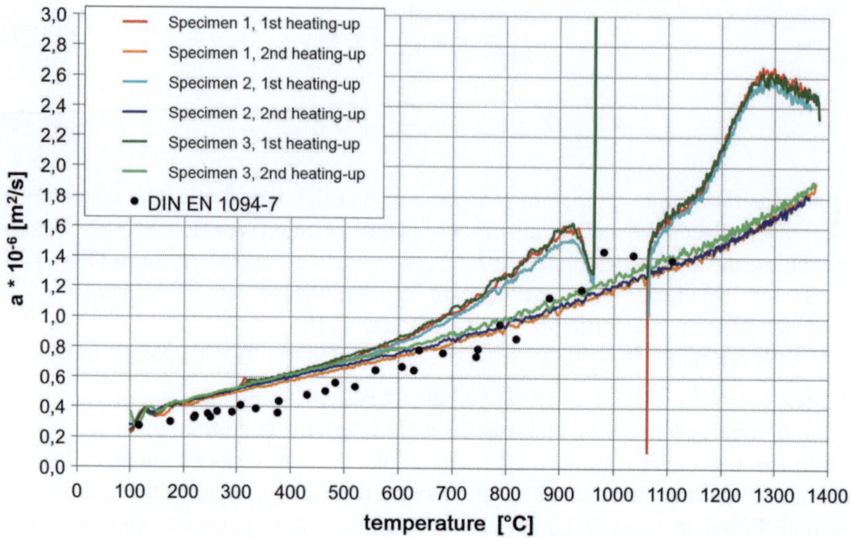

Fig. 14.5 Measurement of the thermal conductivity of three different aluminosilicate fiber mats with short-term microstructural transformation by monotonic heating [6]

and ΔT from each other, i.e. Equation 14.18 is no longer applicable. It is noticeable that the reaction is indicated very distinctly. The method can, therefore, be used advantageously to make such reactions visible and to assign them to a temperature.

An undisturbed determination of the thermal diffusivity is then advantageously carried out in a second round, since the microstructural change is reversible only in rare cases and in practice occurs completely during the first heating up.

References

1. Wuthnow, H., Pötschke, J., Buhr, A., Boßelmann, D., Pozun, F., Gerhaz, N., Golder, P., Grass, J.: Experiences with Microporous calcium hexaluminate insulating materials in steel reheating furnaces at Hoesch Hohenlimburg and Thyssen Krupp Stahl AG Bochum, Germany. In: 47th International Colloquium on Refractories, Aachen, Germany, 13 & 14 Oct 2004, pp 198–204
2. Gerthsen, C., Meschede, D.: Physics. Springer, Berlin (2005)
3. Sass, F., Bouché, Ch., Leitner, A.: Dubbels Pocket Book for Mechanical Engineering, p. 443ff. Springer, Berlin (1966)
4. Groß, U.: Heat and Material Transfer I. Institute Wärmetechnik. u. Thermodynamik Script, TU Bergakademie Freiberg, Germany (2010) (Chapter 30)
5. Pötschke, J., Simmat, R.: Improved evaluation of calorimetric thermal conductivity measurements according to PREN1094-7, pp. 60 – 61. In: 48. International Colloquium on Refractories, Aachen, Germany (2005)
6. Pötschke, J., Simmat, R., Litovski, E., Kleimann, J.: A Method to determine the thermal conductivity in the range of 200–1,600°C. In: Proceedings 49th International Colloquium on Refractories, Aachen, Germany, pp. 14–18 (2006)

7. Simmat, R., Pötschke, J.: Determining the high temperature thermal conductivities of Al_2O_3-rich insulating materials. In: Presented in International Science Conference. "Refractories, Furnaces and Thermal Insulation" Strebske, Slovakia, 20–22 April 2008
8. Krause, O., Simmat, R., Pötschke, J., Quirmbach, P.: Newest development in the field of testing hot properties of refractory products. In: Proceedings 10th ECerS Conference Göller Verlag, pp. 1997–2000 (2007)
9. Simmat, R., Sokoll, T., Pötschke, J., Quirmbach, P.: Thermal conductivity in the range of 200 to 1,600°C due to the monotonic heating method. CFI/Reports DKG 9,84, E115–E119 (2007)

Chapter 15
Mechanical Wear of Refractory Material

Basically, a distinction is made between four wear mechanisms:

1. Adhesion, as a result of local welding (galling) with subsequent fracture.
2. Abrasion, as a result of microcutting.
3. Surface spalling, as a result of fatigue (refer to Sect. 3.1.1.5.3.2.2).
4. Ablation, as a result of chemical decomposition, evaporation, etc.

For mechanical wear of refractory materials, abrasive wear plays the main role. Two examples, friction wear and abrasive blast/impact wear, are considered here:

15.1 Friction Wear

To carry out annealing treatments in an inert or reducing atmosphere, ceramic annealing boxes filled with the material to be treated are placed on the push plates (skillets) arranged in a row and slowly pressed through a push plate (skillets) furnace. The pusher plates (skillets) stand on a ceramic slide rail. Both materials are subjected to wear. The task is to find the material combination with the least wear. The number of revolutions of the pusher plates (skillets), i.e. the number of load and temperature changes they can withstand, should be as large as possible. In addition to the materials used, temperature, atmosphere, annealing material, weight, continuous or jerky movement, speed and duration are the most important influences. Any disturbance can lead to a long standstill, i.e. loss of production.

© The Author(s), under exclusive license to Springer Nature Switzerland AG 2024   355
J. Pötschke, *Refractory Fundamentals in Metallurgical Practice*,
https://doi.org/10.1007/978-3-031-63709-4_15

Of great importance for the evaluation of the most favorable material combination is the so-called relative friction coefficient μ. It combines the frictional force K_R with the normal force K_N:

$$K_R = \mu \cdot K_N \, [N]. \tag{15.1}$$

The friction force is independent of the dimension of the contact surface of both bodies between which the normal force acts. If the contact surface is inclined against the horizontal by the angle α, the normal force is less than the force of gravity acting on the body, i.e. its weight $P = m \cdot g$:

$$K_N = \cos \alpha \cdot P. \tag{15.2}$$

The friction force is calculated from Eqs. (15.1 and 15.2) (refer to Fig. (15.1)). The force required to move the body on the inclined plane is

$$K_R = \sin \alpha \cdot P. \tag{15.3}$$

A distinction is made between static friction K_{RH} and dynamic (sliding) friction K_{RG}, whereas $K_{RH} > K_{RG}$. The static friction can be determined, for example, with a particularly illustrative experiment. The opposite body (push plate/skillet) to be tested is loaded and placed on the base body (ceramic rail). The angle of inclination α is slowly increased. The last setting **before** the start of sliding gives the coefficient of friction for the static friction μ_H:

$$\mu_H = \frac{P \cdot \sin \alpha}{P \cdot \cos \alpha} = \tan \alpha. \tag{15.4}$$

The coefficient of sliding friction μ_G is determined by determining the tensile force K_R as a function of the weight P for $\alpha = 0°$ after adhesion has been overcome:

Fig. 15.1 Forces at the inclined plane

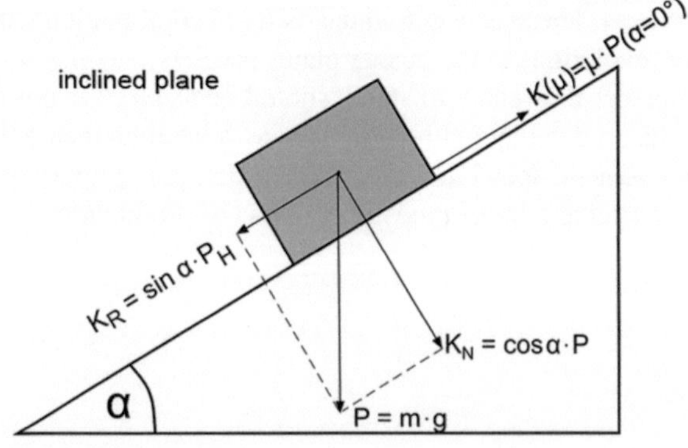

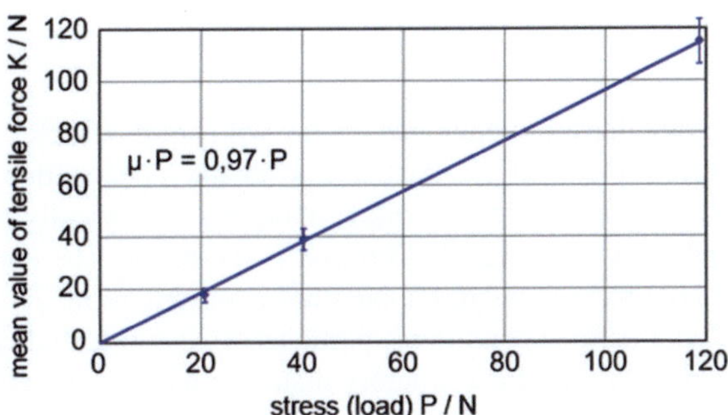

Fig. 15.2 Measurement of the coefficient of friction μ

$$K_R = \mu_R \cdot P. \tag{15.5}$$

Figure 15.2 shows an example of the determination of the coefficient of friction of a push plate/skillet (mullite) on a ceramic/SiC-based slide rail. The result is $\mu_R = 0.97$. The denser the structure of the push plate/skillet and the smoother its surface, the lower the coefficient of friction will be. If, for example, polished sintered corundum is used, the coefficient of friction can drop to $\mu_R = 0.5$.

If the abrasion is determined in the laboratory by continuously axially rotating a flat cylindrical wear body (round plate, radius r_a) by a motor and an opposite body, with a constant force P, acts face to face against it, it is possible to determine the motor power $L[W]$ and at the same time the torque $D[Ws]$ (R. Simmat, DIFK, Bonn, Germany, 2004). By definition, the area-weighted radius $r = r_a/\sqrt{2}$ is the radius at which the outer circular area equals the inner circular area. The rotational speed of the wear body is v [1/s]. It is valid

$$L = 2\pi r \cdot v \cdot K_R \cdot t/t \,[W] \text{ and } D = r \cdot K_R. \tag{15.6}$$

Consequently, the friction force is obtained from two simultaneously measured results which can be compared. According to Eq. (15.5), the coefficient of friction μ_R is calculated.

Example A flat ceramic cylinder ($r_a = 0.071$ m, $r = 0.05$ m) rotates with a speed of $v = 2/s$. The motor power is $L = 50$ W. One gets $K_R = 80$ N. The measured torque is $D = 4$ Nm. This also corresponds to $K_R = 80$ N. The contact force is $P = 70$ N, from which Eq. (15.5) calculates the coefficient of friction to be $\mu_R = 1.14$. The coefficient of friction can, therefore, be greater than one.

15.2   Beam Wear/Impact Wear

In pipelines, in which gas loaded with solid particles is transported, e.g. in cement calcination or in hot blast stoves, abrasion occurs due to jet (blast) wear. The particles carried in the gas collide obliquely or vertically with the refractory wall lining and chip the material. Abrasion is the result. The quantitative measurement of jet (blast) wear at room temperature is described in EN ISO 16822:2008 (Fig. 15.3). Under defined conditions, particles of silicon carbide (SiC–opposite body) are shot vertically onto the refractory material (base body) with a strong air stream and the abrasion/time is determined [1].

This method was transferred to temperatures up to 1200 °C at DIFK, Bonn, Germany, starting in 2003 by using a furnace setup as a measuring chamber and coupling the Venturi-nozzle with an additional burner [2]. In this way, the standard-compliant application was approximately maintained. The sample reaches the specified temperature, measured in the immediate vicinity of the corrosion area. The duration of the experiment is 7.5 min.

For a better understanding of the experiment, it is calculated step by step. If not especially mentioned, all data apply for room temperature. As a first step, it is clarified which state the flow has [3].

For the Reynolds number $Re = u \cdot d \cdot \rho/\eta > 2300$, turbulence is present. The diameter characterizing the measurement system is $d \sim 5 \cdot 10^{-3}$ m. Under normal conditions, the density of air is $\rho_0 = 1.3\,\text{kg/m}^3$ and its viscosity is $\eta = 1.38 \times$

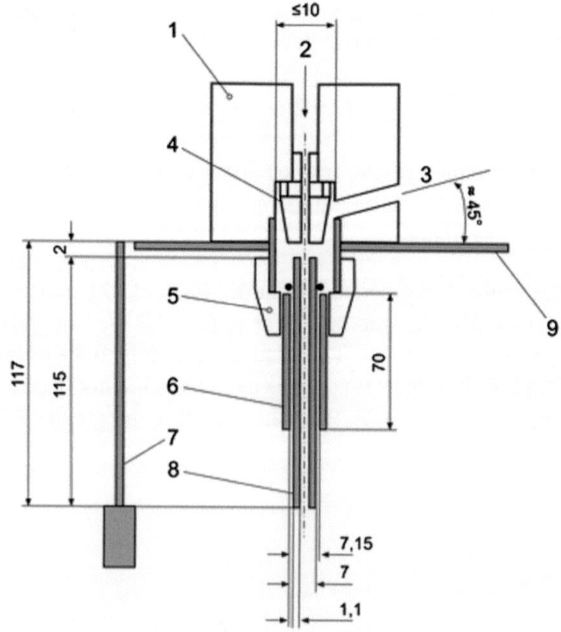

Fig. 15.3 Measurement of jet abrasion according to EN ISO 16282:2008(D), Explanation: 1—Venturi nozzle, 2—Air supply, 3—Abrasive agent supply, 4—Air supply nozzle, 5—Union (swivel) nut, 6—Steel sleeve, 7—Brass rod for positioning glass tube, 8—Glass tube with sealing ring, 9—Top side of test chamber

10^{-5} kg/m-s. Consequently, if the flow velocity exceeds $u = 5$ m/s, turbulence is present. This is always the case here.

Secondly, the type of flow, formed or unformed, is clarified. To do this, calculate the length L_S of the "flow inlet" until the "formed flow" is reached: $L_S = 50 \cdot d$ for Re > 2320. In the glass tube, Re $= u \cdot d \cdot \rho / \eta \sim 50 \cdot 5 \times 10^{-3} - 1.3 / 1.38 \times 10^{-5} = 24{,}000$ at RT. It follows that $L_S = 0.5 \cdot 50 = 25$ cm. This just corresponds to the length of the guided flow, which is why there is largely an as yet unformed piston flow.

Another orientation value for the calculation is the speed of sound $u_S = (402 \cdot T)^{0.5} = 346$ m/s. This is the maximum air velocity achievable here at room temperature. At 1200 °C, it is 770 m/s.

The air is at pressure $P_1 = 4.5$ atm in front of the Venturi-nozzle. The density of the air at RT and 4.5 atm is $\rho_1 = 5.85$ kg/m^3. Its viscosity is $\eta = 1.38 \times 10^{-5}$ kg/m s. The maximum pressure allowed in the measurement chamber is 310 Pa. This is a counter-pressure that is negligible and helps to avoid the decay of the free jet (blast) (refer to Remark 1). In the Venturi-nozzle, the free diameter decreases from $d_1 = 2.9 \times 10^{-3}$ to 2.4×10^{-3} m. The inlet velocity of the gas into the nozzle is calculated with the given mass flow $\dot{m}_L = 1.25 \times 10^{-3}$ kg/s and the equation

$$u_1 = \frac{4 \cdot \dot{m}_L}{\pi \cdot \rho_L \cdot d^2} = \frac{4 \cdot 1.25 \times 10^{-3}}{\pi \cdot 5.85 \cdot \left(2.9 \times 10^{-3}\right)^2} = 32 \, \text{m/s}. \tag{15.7}$$

The pressure drop ΔP_{12} in the Venturi-nozzle is calculated from Bernoulli's equation (refer to Sect. 3.2.1.4, Eq. 3.91) with the side condition that the static height is equal on both sides [3]. The gas is therefore approximately considered to be incompressible. The pressure drop in the Venturi-nozzle is

$$\Delta P_{12} = \frac{\rho_L}{2 \cdot 101330} \cdot u_1^2 \left(\left(\frac{d_1}{d_2}\right)^4 - 1\right)$$

$$= \frac{5.85}{202660} \cdot 32^2 \cdot \left(\left(\frac{2.9}{2.4}\right)^4 - 1\right) = 0.03 \, \text{atm}. \tag{15.8}$$

This low drop in pressure is the technically exploited advantage of measuring the flow velocity of gases with the aid of the Venturi-nozzle. Consequently, the calculation continues with the inlet pressure $\Delta P = 4.5$ atm (4.56×10^5 N/m^2). In the Venturi-nozzle, the gas relaxes and its gas velocity increases: $u_2 = u_1 \cdot (d_2/d_1)^2 = 47$ m/s. After leaving the nozzle, the air flows through a glass tube with length $L_3 = 117$ mm and inner diameter $d_3 = 4.8$ mm. Following this, the gas flows 203 mm freely into the measuring chamber. The bundling of the jet (blast) is largely maintained due to the low chamber pressure (310 Pa) (refer to Remark 1).

If the burner increases the temperature in the furnace vessel, there is a chamber pressure of approx. 1 atm, that is not 310 Pa. The gas is fed in a tube made of corundum $L_3 = 200$ mm far to 3 mm in front of the specimen. The inner diameter remains unchanged $d_3 = 4.8$ mm. The mass flow is constant because of the continuity of the flow and the gas hits the sample surface with velocity

$$u_3 = \frac{\dot{m} \cdot T_3}{\rho_0 \cdot F_3 \cdot T_0} [\text{m/s}]. \tag{15.9}$$

The temperature increase from T_0 to T_3 and the impact area

$$F_3 = 0.25 \cdot \pi \cdot \left(4.8 \times 10^{-3}\right)^2 = 1.81 \times 10^{-5}\,\text{m}^2$$

of the grains are taken into account. Starting from the room temperature $T_0 = 300$ K, $T_3 = 1473$ K results in

$$u_3 = \frac{1.25 \times 10^{-3} \cdot 1473}{1.3 \cdot 1.81 \times 10^{-5} \cdot 300} = 261\,[\text{m/s}].$$

At room temperature, i.e. without temperature increase of the air, $u_3 = 53$ m/s. (refer to Remark 2).

The grain of SiC (300–600 μm°) is fed into the chamber (approx. 1 cm^3) surrounding the Venturi-nozzle and accelerated from there by the gas flow. The average velocity $\bar{u}$ [m/s] of the gas/solid mixture is calculated as follows:

$$\dot{m}_G \cdot u_{23} + \dot{m}_P \cdot u_0 = (\dot{m}_G + \dot{m}_P) \cdot \bar{u}[\text{kg\,m/s}]. \tag{15.10}$$

In 7.5 min, 1 kg of SiC (2.2×10^{-3} kg/s) and 0.6 kg of air (1.25×10^{-3} kg/s, approx. 1 l/s) are consumed. The initial velocity of the SiC grains is $\dot{m}_P = 0\,$m/s, that of the air $u_{23} = 53\,$m/s. The average velocity of the batch is obtained at RT $\bar{u} = 19\,$m/s and at 1200 °C $\bar{u} = 95\,$m/s.

However, it must be checked whether the air/solid mixture is heated that much at all. For this purpose, the heat flux density required for heating is calculated and compared with the heat content/time required for heating [5]:

$$\dot{Q} = k \cdot F \cdot \Delta T = c_P \cdot \dot{m} \cdot \Delta T\,[\text{W}]. \tag{15.11}$$

$$k \cdot F = \frac{\pi \cdot L}{\frac{1}{\alpha_i \cdot d_i} + \frac{\ln(d_a/d_i)}{2\lambda_1} + \frac{1}{\alpha_a \cdot d_a}}[\text{W /K}]. \tag{15.12}$$

k[W/m^2 K] is the heat transfer coefficient (refer to Sect. 14.2). The heat transfer coefficients on the outside and inside of the guide tube α_i and α_a are combined for simplicity since turbulence is present on both sides. The diameters $d_a = 7.2$ mm and $d_i = 4.8$ mm are approximately averaged to $d_m = 6$ mm. The following applies to the averaged heat transfer coefficient [5]:

$$\bar{\alpha} = 0.029 \cdot \left(\frac{\dot{m} \cdot c_P}{d_m \cdot \lambda}\right)^{0.78} \cdot \frac{\lambda}{d_m}[\text{W/m}^2 \cdot \text{K}]. \tag{15.13}$$

The mass current is calculated as the sum of the partial currents to $\dot{m}_0 = \dot{m}_L + \dot{m}_P = 1.25 \times 10^{-3} + 2.2 \times 10^{-3} = 3.5 \times 10^{-3}$ kg/s.

The specific heat is averaged. Since it is about the same for both components, it follows that $c_P = 1.1 \times 10^3$ J/kg K. The thermal conductivity of the corundum tube is $\lambda = 3.5$ W/m K. Putting this into Eq. (15.13) gives $\overline{\alpha}_{12} = 985$ W/m^2 K. From Eq. (15.12), the heat transfer coefficient is calculated to be

$$k \cdot F = \frac{\pi \cdot 0.32}{\frac{1}{985 \cdot 4.8 \times 10^{-3}} + \frac{\ln(7.2/4.8)}{2 \cdot 3.5} + \frac{1}{985 \cdot 7.2 \times 10^{-3}}} = 2.45 \,[\text{W/K}].$$

The mass flow density $\dot{m}$, which is allowed if the flowing solid mixture is to be approximated to the elevated ambient temperature of the furnace, must not exceed, according to Eq. (15.11):

$$\dot{m}_1 = \frac{k \cdot F}{c_P} = \frac{2.45}{1100} = 2.2 \times 10^{-3} \text{ kg/s}. \tag{15.14}$$

Both mass flows $\dot{m}_0 = 3.5 \times 10^{-3}$ kg/s and $\dot{m}_1 = 2.2 \times 10^{-3}$ kg/s are almost equal, so that the batch flow impinges well heated on the heated specimen surface.

In the process, the carbide particles are well heated. The solution of the complementary error integral according to Gauss can be used for estimation (refer to Sect. 2.2.2):

$$\varphi = 1 - \text{erf}\left(\frac{x}{2 \cdot \sqrt{a \cdot t}}\right) \equiv \text{erfc}(A_P). \tag{15.15}$$

The spherical radius r_K [μm] of the particle is calculated, which is heated to 93% in its flight time ($t = 4 \times 10^{-3}$ s). For this purpose, (Eq. 2.2.27, Sect. 2.2.2) is used. Since $r_K = x$ and the factor for converting the argument of the error function from the plate (skillet) to the sphere is $A_P : A_K = 2.1$, the thermal conductivity of the silicon carbide gives $a_{\text{SiC}} = 0.045$ cm/s^2:

$$r_K = 2 \cdot 2.1 \times 10^4 \,\mu\text{m/cm} \cdot \sqrt{a \cdot t} = 563 \,\mu\text{m}.$$

The grain size used ($d_K = 300 - 600$ μm°) is practically heated all the way through.

The calculation of the wear is based on a semi-empirically derived equation for impact wear [3] which has not yet been tested on refractory materials. The equation formally resembles that of the kinetic crack progression (refer to Eq. 12.16, Sect. 12.2):

$$\dot{m}_V = 10^{-6} \cdot \rho_V \cdot F_V \cdot \frac{\dot{m}_P \cdot u_P^2 \cdot E_V}{d_P^2 \cdot \sigma_B^2} \,[\text{kg/s}]. \tag{15.16}$$

The factor 10^{-6} is empirical and can be improved by statistical comparison of a large number of individual measurements. Typical values for refractory materials, e.g. corundum, are $E_V = 35 \times 10^9$ kg/m s^2, $\sigma_B = 15 \times 10^6$ kg/m s^2 and $\rho_V = 3300$ kg/m^3 [6]. With a guided air jet, the impact area is similar to that of the tube $F = \pi \cdot d^2/4 = 1.8 \times 10^{-5}$ m^2 (refer to Remark 1). However, the carbide particles disperse in all directions after impact, resulting in a larger crater due to furrowing. The particle flux is $\dot{m} = 2.3 \times 10^{-3}$ kg/s, the impact velocity at room temperature $u_P = 19$ m/s and the mean diameter of SiC particles is $d_P = 4.5 \times 10^{-4}$ m. One obtains $\dot{m}_V = 3.8 \times 10^{-5}$ kg/s. This is 17 g wear in 7.5 min which is close to what is seen in practice [1].

The calculation shows that the laboratory test must be carried out very carefully. The main factors influencing wear are the diameter of the SiC particles, in other words their mass and their velocity. In addition, care must be taken to ensure that the batch jet does not disintegrate, to which, for example, the internal wear of the guide tube contributes. The gas flow must be carefully controlled.

Figure 15.4 [4] shows as a measurement result taking SiC-containing bauxite—LCC that the wear decreases with increasing temperature. This is also observed on other refractory materials and is the result of the plasticity of the materials increasing with temperature [2, 4]. In addition, wear increases with increasing SiC-content because the SiC grains are less well integrated into the microstructure than the oxidic wear carriers.

Remark 1 The measurement result is strongly dependent on the quality of the batch jet. As soon as the cylindrical jet becomes a truncated cone with an opening angle of $\beta°$, the impact surface changes as follows:

$$\frac{F_2}{F_1} = \left(\frac{2 \cdot tg(\beta/2) \cdot L}{d_1} + 1 \right)^2 . \tag{15.17}$$

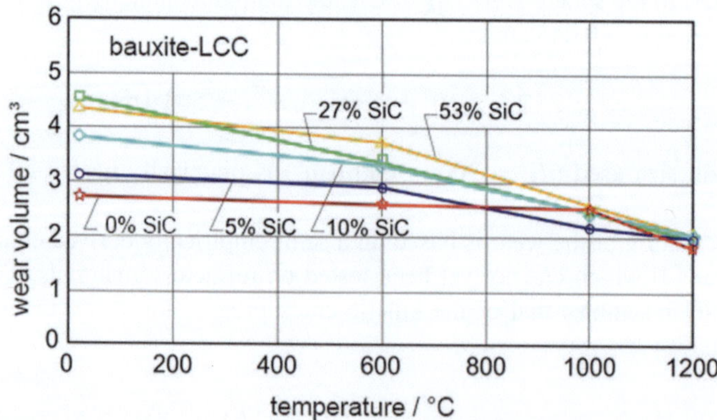

Fig. 15.4 Measurement of the wear volume of bauxite—LCC with different contents of silicon carbide at varying temperatures [4]

The cylinder has the diameter $d_1 = 4.8$ mm, the total length of the non-guided flow is $L = 203$ mm and the opening angle of the cone $\beta = 1.35°$ (on each side of the cone section $\beta/2 = 0.675°$). From Eq. (15.17), this calculates four times the impact area which would have to be taken into account for the calculation of the impact velocity according to Eq. (15.9).

Remark 2 As the tube length L increases, the pressure drop increases and the mean flow velocity decreases (15.3). An example is the pressure drop ΔP_{23} which is calculated at RT and $P_3 = 1$ atm from the impact velocity $u_3 = 53$ m/s calculated with Eq. (15.9) [3]:

$$\Delta P_{23} = 0.24 \cdot \frac{\eta^{1/4} \cdot L_3 \cdot \rho^{3/4} \cdot \left(\frac{\pi}{4}d_3^2 \cdot u_3\right)^{7/4}}{d_3^{4.75}} [\text{N/m}^2]$$

$$\Delta P_{23} = 0.24 \cdot \frac{\left(1.38 \times 10^{-5}\right)^{1/4} \cdot 0.32 \cdot 1.3^{3/4} \cdot \left(\frac{\pi}{4}\left(4.8 \times 10^{-3}\right)^2 \cdot 53\right)^{7/4}}{\left(4.8 \times 10^{-3}\right)^{4.75}}$$

$$= 5.6\,[\text{N/m}^2]. \tag{15.18}$$

The required pressure drop is given by $\Delta P_{23} = 5.6\,\text{N/m}^2$.

References

1. Krause, O., Urbanek, G., Körber, H.: Determining resistance to abrasion at ambient temperature—Improving comparability between laboratories. RHI Bull. **2**, 44–49 (2013)
2. Brüggmann, C., Krause, O., Pötschke, J., Simmat, R.: High Temperature Abrasion Resistance of Cement-Based Refractories; 48th International Colloquium on Refractories, pp. 35–39. Aachen, Germany (2005)
3. Brauer, H.: Basics of Single and Multiple Phase Flows. Sauerländer, Aarau a. Frankfurt/Main, Germany (1971)
4. Simmat, R., Brüggmann, C., Krause, O., Pötschke, J.: High Temperature Abrasion Resistance of Refractory Products; Proceedings UNITECR (2007), 514–517
5. Grigull, U.: (Groeber, Erk, Grigull) Basics of Heat Transfer, Springer, Berlin (1963)
6. Brochen, E., Pötschke, J.: A Contribution to the Measurement und Modelling of the Thermal Shock Resistance of Refractories Investigation of Al_2O_3- and MgO-Based Materials; 54th International Colloquium on Refractories, pp. 75–79. Aachen, Germany (2011)

Chapter 16
Sintering

16.1 Introduction

Together with pottery, the human being developed the sintering of powders. After drying, the powder particles, which are only loosely connected to each other by van der Waals forces, are firmly sintered together at high temperature, for which purpose a mass transfer takes place between them. The driving force is the creation of a surface with the lowest possible energy. Decisive for success are the size distribution of the powder particles (grain size distribution), the temperature and the time. In particular, a sufficiently large fine fraction (share) must be present. It forms, for example, the so-called binder phase in the refractory material.

The aim of sintering is to produce a dimensionally stable body. In the process, the porosity decreases and, to the same extent, the bulk density increases. Practice teaches us that two sintering mechanisms can be distinguished from one another.

Solid-phase sintering (single-phase/multiphase) and liquid-phase sintering. Reactions can occur and the process can be accelerated by the additional application of external pressure.

The theory distinguishes three separate in time phases of the sintering process: initial, shrinkage and final stages. Shrinkage is low in the initial and final stages. It occurs mainly in the second stage by grain boundary (interface) and volume diffusion. A detailed overview of this is given by Telle [1].

Grains have a convex surface. During sintering, contact points very quickly form a concavely curved sintering neck, the (negative) radius of which increases continuously, i.e. its volume increases (refer to Fig. 16.1). Kuczynski [2] assigns the temporal increase of the diameter of the sintering neck x with time to the sintering mechanism as follows: $X^n \sim t$. Let $n = 3$ (evaporation/recondensation), $n = 7$ (surface diffusion), $n = 5$ (volume diffusion), $n = 2$ (viscous/plastic flow) apply. To this should be added the experience made particularly frequently in the sintering of metal powders, according to which the sintering process depends strongly on the type of atmosphere and can be greatly accelerated by its skillful selection (activated sintering) [3].

© The Author(s), under exclusive license to Springer Nature Switzerland AG 2024 365
J. Pötschke, *Refractory Fundamentals in Metallurgical Practice*,
https://doi.org/10.1007/978-3-031-63709-4_16

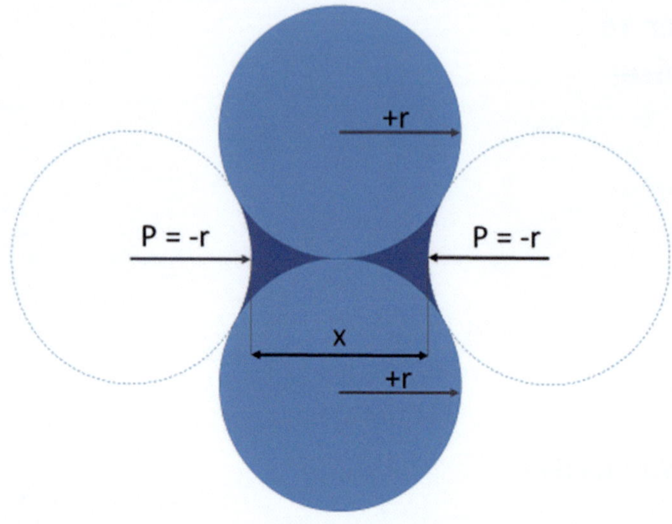

Fig. 16.1 Two-particle model

The direction of mass transfer between areas with different curvature of the surface can be calculated using the Gibbs–Thomson equation:

$$\ln \frac{P}{P_\infty} = \frac{2\sigma \cdot V}{RT \cdot r}. \tag{16.1.1}$$

P [Pa] is the vapor (steam) pressure of the surface curved with radius r [m] of the observed condensed phase and P_∞ is present above the plane spread phase. T [K] is the temperature, σ [N/m] is the specific free surface energy (surface tension) and V [m^3/mol] is the molar volume of the condensed phase, e.g. Al$_2$O$_3$.

If one compares a place of convex surface of radius $+ r_{\text{konvex}}$ with such a place of concave curvature $- r_{\text{konkav}}$, one immediately obtains a statement about the driving pressure ratio:

$$\frac{\ln \frac{P_{\text{konvex}}}{P_\infty}}{\ln \frac{P_{\text{konkav}}}{P_\infty}} = \frac{-r_{\text{konkav}}}{+r_{\text{konvex}}}. \tag{16.2}$$

Here, $P_{\text{convex}} > P_\infty > P_{\text{concave}}$. Consequently, the mass (material) flow is toward the concave neck.

But mass (material) transport via volume or surface diffusion also follows the gradient of the chemical potential caused by the curvature of the interface:

$$(\mu_r - \mu_\infty)_{\text{cond.}} = 2\sigma \cdot \frac{V}{r}. \tag{16.1.2}$$

The same applies to viscous flow.

The growth rate of a grain clearly follows the relationship

$$\frac{dr}{dt} = \frac{K}{r},$$
(16.3.1)

from which follows after integration in the limits r to r_0:

$$r^2 - r_0^2 = K \cdot t.$$
(16.3.2)

Usually, after a relatively short time, $r \gg r_o$, so that the well-known $\sqrt{t}$-law can be used for simplification.

The dependence of the transport processes on temperature can be described in a limited temperature range by the well-known equation of Arrhenius [1]:

$$K = K_0 \cdot \exp(-A_k / R \cdot T).$$
(16.4)

K [μm/$\sqrt{\text{min}}$] is the rate constant (refer to Eq. 16.3.2), K_0 [μm/$\sqrt{\text{min}}$] is the frequency factor, A_k [J/mol] is the apparent activation energy, $R = 8.31$ [J/mol·K] is the general gas constant, and T [K] is the absolute temperature. Upon sintering oxides, the apparent activation energy is usually $A_k \sim 500$ kJ/mol.

16.2 Single-Phase Solid Sintering

16.2.1 Influence of the Grain Size

The chosen example is a special case in two respects: The mass transfer takes place exclusively via the gas phase. The convex curvature $+\,\mathbf{r}$ [μm] of the two grains in contact, which are imagined as spheres, and the concave curvature ρ [μm] of the sintering neck connecting them have the same radius in magnitude but differ in sign (refer to Fig. 16.1). Then with $\rho = -\,\mathbf{r}$ it follows from Eq. (16.2)

$$P_{\text{konkav}} \cdot P_{\text{konvex}} = \exp(2 \cdot \ln P_\infty).$$
(16.5)

Corundum (Al_2O_3) reacts at high temperature to form gaseous components, according to $\langle Al_2O_3 \rangle \rightarrow \{AlO\} + \{AlO_2\}$. At 1700 °C, the partial pressure of both components is $P_\infty = 3.5 \times 10^{-12}$ atm $= 3.4 \times 10^{-6}$ Pa (FactSage). The surface free energy of the corundum, averaged over all surfaces, is 0.9 N/m [4]. The molar volume is $V_{Al_2O_3} = 2.58 \cdot 10^{-5}$ m^3/mol and the temperature $T = 1973$ K. With the gas constant $R = 8.31$ N · m/mol · K, Eq. (16.1) gives $\ln(P/P_\infty) = 2.8 \cdot 10^{-9}/r$ [m]. Choosing $r = 0.05$ μm $= 5 \times 10^{-8}$ m as the average grain size, we obtain $\ln(P/P_\infty) = 5.7 \cdot 10^{-2}$. Using $P_\infty = 3.4 \cdot 10^{-6}$Pa, it follows $P_{\text{konvex}} = 3.6 \cdot 10^{-6}$Pa. From Eq. (16.5), it is calculated $\exp(2 \cdot \ln P_\infty) = 1.2 \cdot 10^{-11}$, which is why $P_{\text{concave}} = 3.3 \cdot 10^{-6}$ Pa. Thus, mass (material) transfer occurs in the range of this grain size at very small pressure differences. The higher the vapor (steam) pressure and the finer the grain, the stronger the driving force (see Eq. (16.1)).

16.2.2 Influence of Temperature and Time

The initial mean grain radius of the powder of pure corundum (Al_2O_3) is $r_o = 1$ μm. The material is sintered isothermally and without pressure for up to 10 h at 1550, 1625 and 1700 °C; refer to Coble [5]. The result extrapolated to 1700° C is shown in Fig. 16.2. Converted to the $\sqrt{t}$—law

$$r = K \cdot \sqrt{t}\,[\mu m] \tag{16.6}$$

results in straight lines whose slope corresponds to the rate constant K [μm/$\sqrt{}$min] assigned to each temperature (Fig. 16.3). The resulting values for K are: K (1550 °C) = 0.22 μm/$\sqrt{}$min, K (1625 °C) = 0.81 μm/$\sqrt{}$min, and K (1700 °C) = 2.58 μm/$\sqrt{}$min. The general temperature dependence is calculated from two straight lines using Eq. (16.4): For example, if one forms the difference of the logarithmized Eq. (16.4) at 1700 °C and 1550 °C, one obtains for the activation energy

$$A_K = -8.31 \cdot \ln \frac{2.58}{0.22} / \{(1/1973) - (1/1823)\} = 4.90 \cdot 10^5 \,\text{J/mol}. \tag{16.7}$$

The frequency factor follows by direct application of Eq. (16.4) at a temperature, e.g. at 1.625 °C:

$$K_0 = 0.81 \cdot \exp(\frac{+4.9 \cdot 10^5}{8.31 \cdot 1898}) = 2.51 \cdot 10^{13}\,\mu m/\sqrt{\min}. \tag{16.8}$$

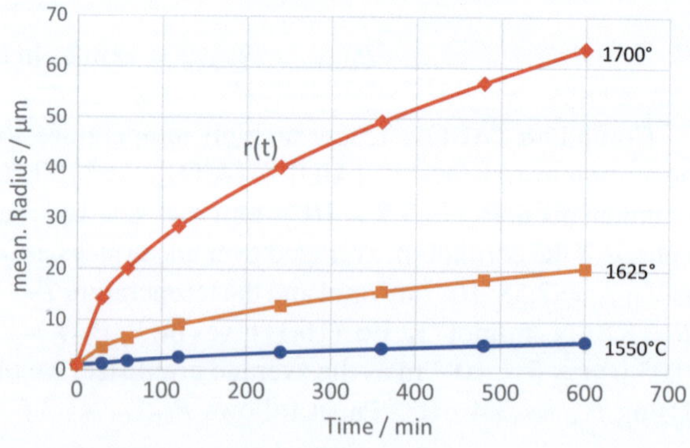

Fig. 16.2 Grain growth of Al_2O_3 powder dependent on temperature and time

Fig. 16.3 $\sqrt{t}$—illustration of the grain growth of Al_2O_3 powder

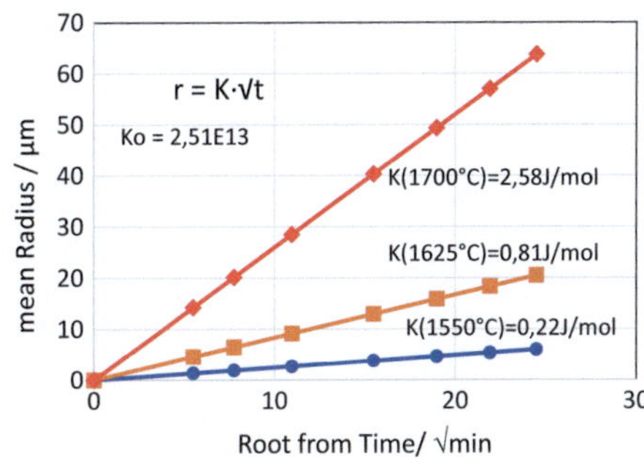

16.3 Two-Phase Solid Sintering

16.3.1 Tests and Results

The influence of the addition of powdered zirconium diboride on the behavior of corundum powder during depressurized, non-isothermal sintering [6] is examined. The goal is to determine the potential of this composite material in terms of its wear resistance as a refractory material while at the same time making it inexpensive to produce without the usual hot-pressing.

The most important properties of the powders in their initial state are
Al_2O_3: $T_m(2050\,°C)$, $\rho_A(3.95\,g/ccm)$, $d_{50}(\sim 0.6\,\mu m)$.

ZrB_2: $T_m(2990\,°C)$, $\rho_Z(6.10\,g/ccm)$, $d_{50}(\sim 10\,\mu m)$.
The mixture ratios used are in % by volume ZrB/Al_2O_3:
$F = 0$; 2.5; 5; 10; 20 % (refer to Table 16.1).
The carefully blended powders are pressed cold isostatically (150 MPa). The freshly pressed items (20 mm$^\varnothing$ × 10 mm), which contain 0% and 20% ZrB_2, are then heated to 1900 °C under argon (4.8) in a dilatometer at 10 K/min. The change in density over time is calculated from the length measured in each case:

$$\rho(t) = \rho_E \cdot (L_E/L(t)). \tag{16.9}$$

ρ_E and L_E are density and length at the end of the experiment.

All remaining freshly pressed items (2.5; 5.0; 10% ZrB_2) are heated to 1900 °C in a graphite resistance furnace also under argon at 10 K/min. After firing, the density of each sample is determined by the water absorption method (EN993.1) [7].

The experimental results and other information required for the evaluation are summarized in Table 16.1 and Fig. 16.4.

The theoretical density of the composite Al_2O_3/ZrB_2 is calculated as follows:

$$\rho_{cot} = (F \cdot 6.1 + (100 - F) \cdot 3.95)/100\,[g/cm^3]. \tag{16.10}$$

This is a true density, i.e. the porosity is zero. (2nd line, Table 16.1)

Table 16.1 Measurement results on the sintering behavior of Al_2O_3 /ZrB_2 powder (bulk densities of the composites)

F: Vol-share	ZrB_2 (vol%)	0	2.5	5	10	20
Theoret. density	Pcot (g/ccm)	3.95	4	4.06	4.17	4.38
ρco/"RT" →	1100	2.31	2.35	2.39	2.48	2.67
ρc → /$T[°C]$↓	1200	2.34	2.38	2.41	2.49	2.68
	1300	2.61	2.48	2.51	2.57	2.72
	1400	3.04	2.62	2.65	2.72	2.81
	1500	3.51	2.86	2.89	2.91	2.95
	1550	3.69	3	3.02	3.01	3.05
	1600	3.79	3.15	3.14	3.11	3.09
	1650	3.83	3.28	3.24	3.18	3.11
	1700	3.86	3.39	3.33	3.25	3.13
	1750	3.88	3.61	3.54	3.46	3.17
	1800	3.89	3.76	3.74	3.73	3.72
	1900	3.91	3.94	3.96	3.99	4.1
		$\rho Al_2O_3 = 3.95$ g/ccm		$\rho ZrB_2 = 6.1$ g/ccm		

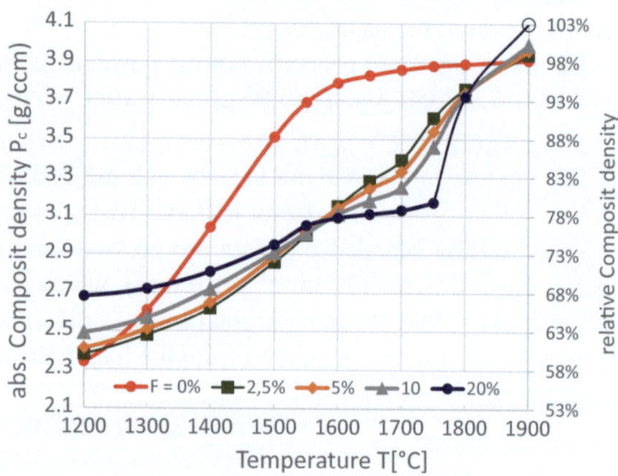

Fig. 16.4 Measured bulk density of the composites dependent on volume share F_{ZrB2} and the sintering temperature. Right ordinate based on $\rho_{cot} = 3.95$ g/ccm

Four more densities are distinguished from each other:

- ρ_c [g/cm^3] is the measured bulk density of the composite as a function of its composition F [% by volume] and temperature T [°C] (refer to Fig. 16.4).
- ρ_m [g/cm^3] is the density of the alumina matrix in the composite calculated from ρ_c. ρ_m determines the sintering behavior of the composite.
- At room temperature (25 °C), a zero is added to the index. Consequently, ρ_{co}, and ρ_{mo} are added as density values.

Figure 16.4 shows that in the experiment, the bulk density of the composites does not change significantly until above 1200 °C, i.e. only then does sintering begin noticeably. For the room temperature "RT," therefore, the values at 1100 °C are used in the later calculation. The following experimental result should be added as information: An additional dwelling time of one hour below 1700 °C changes the reported result only insignificantly [7].

Figure 16.4 further shows that ZrB_2 strongly impedes sintering and increasingly so with increasing volume share. However, above 1700 °C, sintering is strongly accelerated. The explanation follows upon giving the calculation of the matrix density.

16.3.2 Evaluation and Discussion

16.3.2.1 Calculation of the Matrix Density ρ_m of the Composite

The volume share of inert powder F [% by volume] used to prepare the specimens does not take into account the resulting porosity. Consequently, the pure density determined according to Eq. (16.10) must be converted into the bulk density for future observation. Then the volume content f [−] of the ZrB_2 is calculated at room temperature,

$$f = F \cdot \rho_{co}/(\rho_{cot} \cdot 100) = 5.8 \cdot 10^{-3} \cdot F. \tag{16.11}$$

Using the figures in Table 16.1, we then calculate $\mathbf{f} = 0.0145$ for $F = 2.5$ %.

The measured bulk density of the composite ρ_{c0} at room temperature "RT" (green (unfired) density) can be taken from Table 16.1 in the third row. It is

$$\rho_{c0} = 1.81 \cdot 10^{-2} \cdot F + 2.31. \tag{16.12}$$

If the fresh green (unfired) item contains an increasing volume content $\mathbf{f}$ of foreign phase $\mathbf{Z}$, the green density ρ_{c0} increases if $\rho_Z > \rho_A$. If there is no foreign phase in the cold compacted fresh green item ($\mathbf{f} = 0$), at "RT" the densities of the composite ρ_{c0} and its matrix ρ_{m0} are identical (refer to Fig. 16.5). In contrast, the density of the matrix ρ_{m0} calculated for "RT" decreases with increasing $\mathbf{f}$ (refer to Eq. 16.15 for $\rho_{c0} = \rho_c$ and Table 16.2, third row):

$$\rho_{m0} = (\rho_{c0} - f \cdot \rho_Z)/(1 - f) = -5.8 \cdot 10^{-3} \cdot F + 2.31. \tag{16.13}$$

The graphical illustration of both green densities is shown in Fig. 16.5 green density ZrB_2 share F (% by volume) relative density.

With the volume content $f \cdot V_{c0}$ of the inert foreign phase $\mathbf{Z}$ in the green item, the masses $m_C = m_A + m_Z$ and the corresponding volumes $V_C = V_A + V_Z$, the balance [8] is obtained.

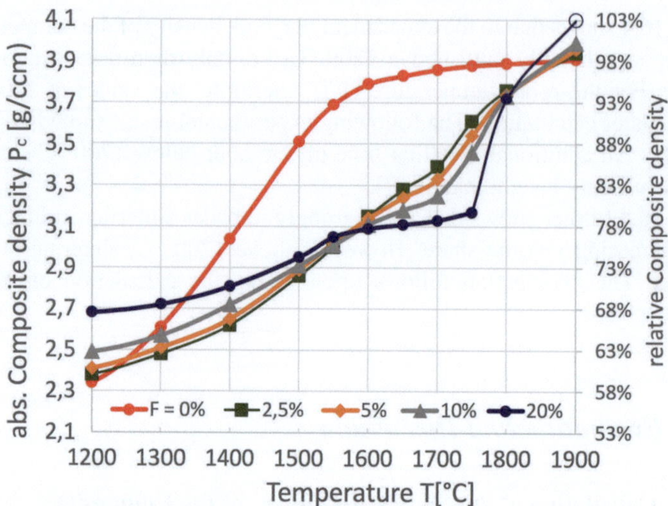

Fig. 16.5 Densities of the composite ρ_{co} and its matrix ρ_{mo} at room temperature dependent on the volume share F of ZrB_2

Table 16.2 Calculated Al_2O_3 matrix density ρ_m of the composite Al_2O_3/ZrB_2

F (vol%) →	0.0	2.5	5.0	10.0	20.0
T (°C) ↓	ρ_m (g/ccm)	ρ_m (g/ccm)	ρ_m (g/ccm)	ρ_m (g/ccm)	ρ_m (g/ccm)
"RT"	2.31	2.29	2.28	2.25	2.19
1200	2.34	2.32	2.30	2.26	2.20
1300	2.45	2.42	2.40	2.34	2.24
1400	2.61	2.56	2.53	2.48	2.33
1500	2.86	2.80	2.77	2.67	2.46
1550	2.99	2.94	2.90	2.77	2.56
1600	3.14	3.09	3.02	2.87	2.60
1650	3.29	3.22	3.12	2.94	2.62
1700	3.47	3.33	3.21	3.01	2.63
1750	3.68	3.55	3.42	3.22	2.67
1800	3.79	3.70	3.63	3.50	3.23
1900	3.9	3.89	3.85	3.77	3.64

$$\rho_m = (m_c - m_z)/(V_c - V_z) = (\rho_{c0} \cdot V_{c0} - \rho_z \cdot f \cdot V_{c0})/(V_c - f \cdot V_{c0}). \quad (16.14)$$

Since the mass m_C does not change, i.e. is always $\rho_{c0} \cdot V_{c0} = \rho_c \cdot V_c$, the bulk density of the Al_2O_3 matrix of the composite is obtained as follows:

$$\rho_m = (\rho_{c0} - f \cdot \rho_Z) \cdot \rho_c/(\rho_{c0} - f \cdot \rho_c). \quad (16.15)$$

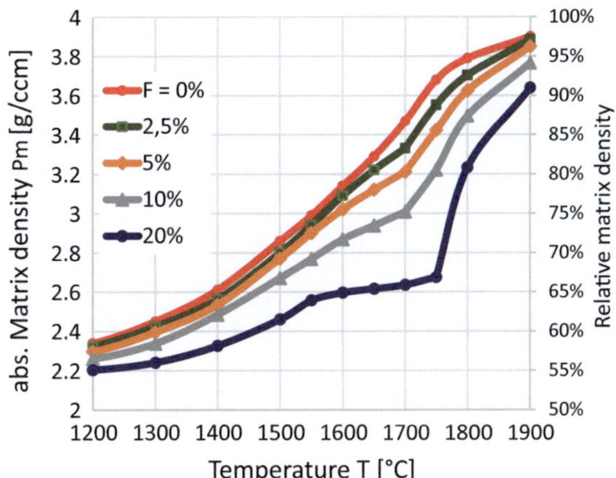

Fig. 16.6 Calculated densities of Al$_2$O$_3$ matrix ρ_m of composite

For 1600 °C and $F = 2.5\%$ (**f** $= 0.015$), the values $\rho_Z = 6.1\,\text{g/ccm}$, $\rho_{c0} = 2.35\,\text{g/ccm}$ and $\rho_c = 3.15\,\text{g/ccm}$ give the matrix density $\rho_m = 3.09\,\text{g/ccm}$, corresponding to the relative density $\rho_{c0} = 78\%$. All matrix densities calculated in this way have been summarized in Table 16.2 and shown graphically in Fig. 16.6.

The change in matrix density with time is the sintering rate,

$$\dot{\varepsilon} = \frac{1}{\rho_m} \cdot \frac{\mathrm{d}\rho_m}{\mathrm{d}T} \cdot \frac{\mathrm{d}T}{\mathrm{d}t} \left[\text{min}^{-1}\right]. \tag{16.16}$$

$\mathrm{d}T/\mathrm{d}t$ is the heating rate, here 10 K/min. The calculation and graphical illustration of the sintering rate is taken from the dissertation of Postrach [7].

It can be seen in Fig. 16.6, in comparison with $F = 0$, as the volume share ZrB$_2$ increases, the steric impairment of the sintering increases. For this purpose, the calculated values of the matrix density are extrapolated to the volume share $F = 0$, (Table 16.2). By comparison with Fig. 16.4, it becomes clear that there is an influence in addition to the steric impairment of sintering which has not been considered so far. (The same result is obtained by extrapolating the measured composite densities to $F = 0$; compare Eq. 16.15: for $f = 0$, $\rho_m = \rho_c$.) Figure 16.7 compares the curves of both densities, which depend only on temperature. As a cause for this obstruction of the sintering process, it is assumed that the contamination of the ZrB$_2$ by way of boron oxide (B$_2$O$_3$) alters the sintering of the corundum by "chemical influence." However, no new phase could be detected in the samples [7].

Above 1700 °C the sintering process accelerates, cf. Fig. 16.6 this is caused by the formation of liquid aluminum borate (9Al$_2$O$_3$·2ZrB$_2$). As the proportion of ZrB$_2$ increases, the resulting liquid-phase sintering is favored (refer to Sect. 16.4)

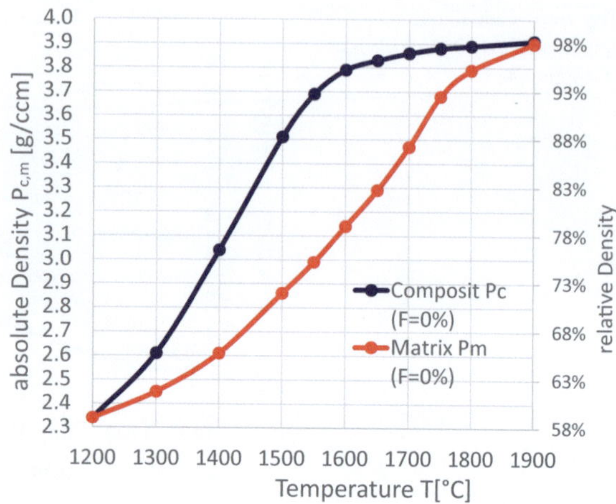

Fig. 16.7 Comparison of the density of the composite ρ_c with the density of its matrix ρ_m, both for the volume share $F = 0\%$

16.3.2.2 Steric Influence of the Foreign Phase

Zirconium diboride impedes the sintering of the corundum not only by "chemical" interaction but to the same extent sterically by initiating internal stress and, consequently, restricting the freedom of movement of the Al_2O_3 powder particles. F. F. Lange describes this effect with a model by comparing the behavior of the composite with the sintering behavior of its matrix material, in this case corundum [9].

Model-wise, the grains of the inert foreign phase ZrB_2 occupy locations of the Al_2O_3 matrix (as substitution). It determines the porosity **P**, which becomes

$$P_m\,[\%] = 100 \cdot (1 - \rho_m/\rho_A). \tag{16.17}$$

$\rho_A = 3.95$ g/ccm is the pure density of the matrix material Al_2O_3. ρ_m is the matrix density shown in Fig. 16.6. With time, the porosity decreases and, consequently, **f** increases as it takes the pore share into account (refer to Eq. 16.11).

Below 1700 °C, ZrB_2 is now considered an inert second phase. Consequently, the densification of the composite is solely attributable to the sintering behavior of its Al_2O_3 matrix. Its shrinkage is the result of acting internal stress and should be isotropic.

A rigid framework of inert foreign particles, to which a periodic crystal lattice is assigned, is embedded in the sinter-active powder matrix. In the simplest case, it is cubic primitive. The lattice constant is $\mathbf{l_0}$. The distance between any two lattice points is $\mathbf{l_n = \alpha\text{-}l_0}$, where the lattice factor α is $1 \leq \alpha \leq \sqrt{3}$ (space diagonal) in the cubic primitive lattice, $\sqrt{2}/2 \leq \alpha \leq 1$ (half space diagonal) in the face-centered lattice, and $\sqrt{3}/2 \leq \alpha \leq 1$ (half space diagonal) in the space-centered lattice. The volume content of the embedded phase is $2r/l_0 = (f/s)^{1/3}$. By using **s**, F.F. Lange denotes the volume content of particles of radius **r** embedded in a cell. If the particles touch each other (e.g. $2r = l_0$), $\mathbf{s = f}$ and no further shrinkage is possible. For the cubic primitive lattice, $s = \pi/6$, for the face-centered $s = 0.24\pi$ and the space-centered $s = 0.22\,\pi$.

According to F.F. Lange, the most suitable cell is the tetracaidecahedron from the cubic system with 14 surfaces, $1 \leq \alpha \leq 3.16$ and $s = 0.277$. The connection between the shrinkage of the composite ε_c and their Al_2O_3 matrix ε_m is calculated from the relation:

$$\varepsilon_c = \varepsilon_m \cdot [1 - 2r/l_n] = \varepsilon_m \cdot [1 - (1/\alpha) \cdot (f/s)^{1/3}]. \tag{16.18}$$

According to Lange [9], the shrinkages of the composite and its matrix during sintering are as follows:

$$\varepsilon_c = 1 - (\rho_{c0}/\rho_c)^{1/3} \text{ and } \varepsilon_m = 1 - (\rho_{m0}/\rho_m)^{1/3}. \tag{16.19}$$

At 1600 °C and $F = 2.5$, $\rho_{c0} = 2.35$ g/ccm (green density of composite), $\rho_c = 3.12$ g/ccm (density of composite), $\rho_{m0} = 2.29$ g/ccm (green density of matrix) and $\rho_m = 3.09$ g/ccm (density of matrix) (refer to Tables 16.1 and 16.2). From this, the values $\varepsilon_c = 0.090$ and $\varepsilon_m = 0.095$ are calculated. For $F = 10\%$, the figures at 1600 °C are $\rho_{c0} = 2.48$ g/ccm, $\rho_c = 3.11$ g/ccm, $\rho_{m0} = 2.25$ g/ccm and $\rho_m = 2.87$ g/ccm, from which $\varepsilon_c = 0.073$ and $\varepsilon_m = 0.078$ are calculated. Shrinkage decreases comparatively with increasing ZrB_2 content.

From Equations. (16.18) and (16.19), the densities of the matrix ρ_m and the composite ρ_c are obtained directly in the course of sintering:

$$\rho_m = \rho_{m0} \cdot \left[1 - \left(1 - (\rho_{c0}/\rho_c)^{1/3}\right)/\left(1 - (1/\alpha) \cdot (f/s)^{1/3}\right)\right]^{-3}. \tag{16.20.1}$$

and

$$\rho_c = \rho_{c0} \cdot \left[1 - \left(1 - (\rho_{m0}/\rho_m)^{1/3}\right) \cdot \left(1 - (1/\alpha) \cdot (f/s)^{1/3}\right)\right]^{-3}. \tag{16.20.2}$$

A prerequisite for the quantitative calculation of experimental test results is the (correct) choice of the lattice model and its parameters f, s and α.

To verify the theoretical approach of F. F. Lange, Eq. 16.20.1 is used to calculate the matrix densities of the composite ρ_m which are important for the mechanical behavior of the sintered material and to compare them with those shown in Fig. 16.6. The calculation is performed separately for all four ZrB_2 shares in the particularly sinter-active temperature range 1400–1700 °C. The required values are taken from Tables. 16.1 and 16.2.

The cubic face-centered lattice is chosen as the model. For the volume share of the embedded particles, $s = 0.75$ is used. f is calculated with Eq. 16.11 from the volume share F. The adjusted values for the lattice factor α reach high values which decrease with increasing volume share F because the space lattice is formed increasingly better. Figure 16.8 shows the course of the lattice factor:

Fig. 16.8 Lattice factor α
dependent on volume share
F [%]

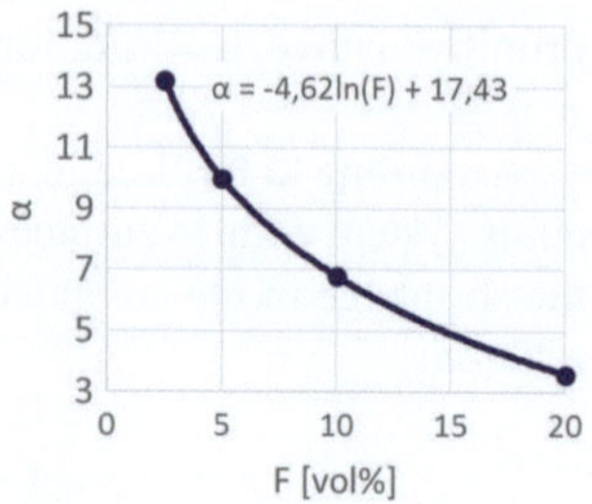

$$\alpha = -4.62 \ln(F) + 17.43. \tag{16.21}$$

Extrapolation to $\alpha = 1$ leads to $F = 33\%$.

Figure 16.9 compares both calculations, according to Eqs. 16.15 and 16.20.1, graphically. The solid lines result from the conventional calculation according to Eq. (16.15). The red points are obtained from the model given by F.F. Lange from Eq. (16.20.1). The correspondence is very good.

F.F. Lange's model allows the sintering conditions to be calculated for unknown volume shares F at a given temperature of a series of measurements. As an example, at 1600 °C, the matrix density for $F = 15\%$ is calculated according to Eq. 16.20.1. From Eq. 16.11, $f = 0.087$. For the cubic face-centered lattice, $s = 0.75$. The adjusted value of the lattice factor can be seen in Fig. 16.21 at $\alpha = 4.8$. The required density values are given in Figs. 16.5 and 16.6 at $\rho_{c0} = 2.58$ g/ccm, $\rho_{m0} = 2.22$ g/ccm, and $\rho_c = 3.1$ g/ccm. Equation 16.20.1 thus gives $\rho_m = 2.73$ g/ccm. Equation 16.15 gives $\rho_m = 2.74$ g/ccm with a slight deviation. Conversely, using this information from Eq. 16.20.2, the bulk density of the composite can be calculated to $\rho_c = 3.1$ g/ccm and compared with Fig. 16.4.

Fig. 16.9 Comparison of
the experimental course of
the matrix densities of the
composites (colored lines)
with their individual values
calculated according to
Eq. 16.20.1 (F.F. Lange) (red
markings)

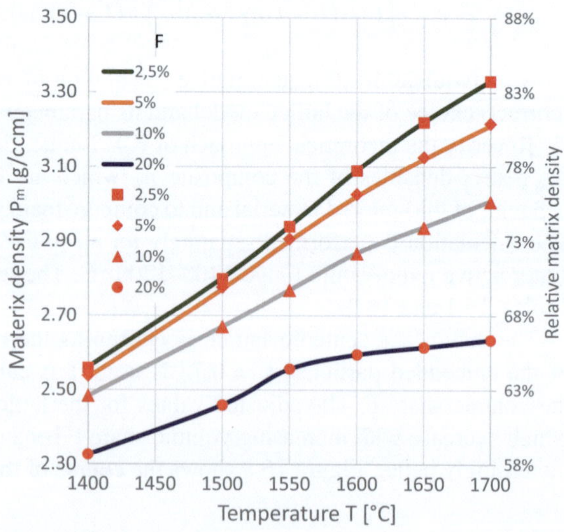

In principle, the model provided by F.F. Lange is suitable to describe the behavior of depressurized sintering of corundum powder in the company of inert powder shares of ZrB_2. However, the necessary adjustment of the lattice factor α to experimental results precludes the application of this theory without any conditions. This would be possible, for example, for a fixed value of the lattice factor independent of F, e.g. $\alpha = 1$, but does not apply to model calculations of F.F. Lange in the system ZnO/SiC [9].

Not considered in the model is the uneven stress distribution in the matrix which can lead to lattice defects and cracks in the microstructure. It can be calculated numerically but the result depends strongly on the conditions introduced [10]. The experimental result, on the other hand, is reliable and the best condition for extrapolation.

16.4 Liquid-Phase Sintering

Above 1700 °C, densification accelerates considerably during sintering of Al_2O_3/ZrB_2 powder. The previously observed sintering impairment is lifted. This is particularly true for the material doped with 20% by volume ZrB_2. It is obvious to look for the cause being the formation of a liquid phase and, consequently, to consider a change from dry to liquid-phase sintering. Then the following partial mechanisms act in chronological order:

1. Particle rearrangement.
2. Dissolution and renewed segregation processes.
3. Skeleton sintering.

The prerequisite for the first two partial mechanisms is the wetting of both types of particles by the melt and solubility of Al_2O_3 in the melt. After sintering of highly doped samples, glassy solidified phases, consisting of the components aluminum, boron and oxygen, are found in spandrels of the microstructure which point to melts based on aluminum borate. However, aluminum borate has not been experimentally proven to be an independent phase [7]. In any case, the microstructure shows that both mechanisms are at work. The typical shape-accommodative Ostwald ripening takes place [1].

The densification of the microstructure takes place predominantly with the aid of the first two partial mechanisms mentioned. The physical objective of the sintering process is to minimize the total surface energy for a given volume energy. The process is, therefore, proportional to the interfacial energy σ_{sl} between the alumina and the melt and inversely proportional to the viscosity η_l of the melt and the mean diameter of the grains $\overline{d}$. The ratio (σ/η_{sll}) [cm/min] has the dimension of a transport quantity. The driving potential of densification is expressed by the relative distance of the current density to the achievable maximum value. The equation [11] is obtained.

$$\dot{\varepsilon} = K \cdot (\sigma_{sl}/\eta_l) \cdot (1 - (\rho/\rho_0))/\overline{d}\,[1/\min]. \tag{16.22}$$

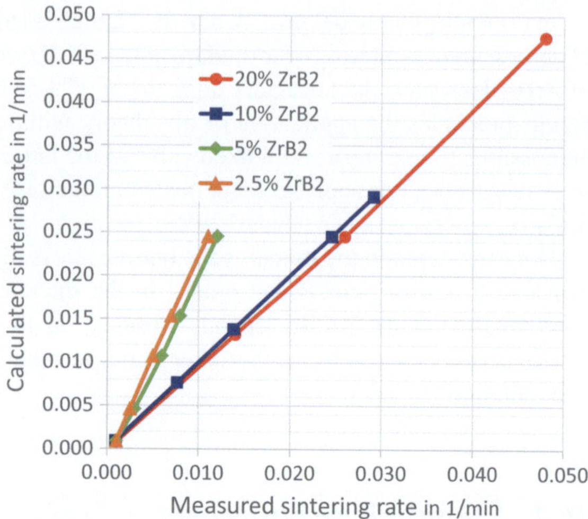

Fig. 16.10 Comparison of the measured sintering rate with the sintering rate calculated according to Eq. 16.22 in the temperature range of liquid-phase sintering

By extrapolating the values for boron oxide measured at lower temperatures in place of aluminum borate, however, the ratio σ_{sl}/η_l to ~ 6000 cm/min can be estimated [12]. With $\overline{d} = 1.1 \times 10^{-4}$ cm, the best fit for the time–temperature profile of the material doped with 20% by volume ZrB_2 is calculated to be $K = 2.7 \times 10^{-9}$ min^{-1}. The result of the calculation is shown as a red line in Fig. 16.10.

If the adjustment $K = 2.7 \times 10^{-9}$ min^{-1} determined for high doping is used unchanged, it becomes clear that there is already a small deviation at 10% doping but a large deviation at lower doping rates: The measured sintering rate lags behind that calculated for liquid-phase sintering. There is a lack of liquid phase. Skeleton sintering is the result.

The transport size (σ_{sl}/η_l) should increase with the temperature. This is not clear in the experiment. Obviously, the large distance away from equilibrium camouflages this influence.

References

1. Salmang, H., Scholze, H., Telle, R.: Ceramics, pp. 313–380. Springer, Berlin (2007)
2. Kuczynski, G.C.: Sintering Processes. Springer, NY (1980)
3. Dhingra, D., Förster, E., Pötschke, J.: Mechanism of the active sintering of technical iron powder. Archiv. Eisenhüttenwesen **45**, 1–7 (1974)
4. Kingery, D.: Surface energy of oxides. Am. Ceram. So. **37**, 42 (1954)
5. Coble, R.L.: Sintering crystalline solids. J. Appl. Phys. **32**, 787–799 (1961)
6. Postrach, S., Pötschke, J.: Pressureless sintering of Al_2O_3 containing up to 20 % by volume ZrB_2. J. Eur. Ceram. Soc. **20**, 1459–1468 (2000)
7. Postrach, S.: Manufacture and Characterization of Pressureless Sintered Al_2O_3-ZrO_2. Dr.-Ing. Dissertation. RWTH-Aachen, Germany (1999)
8. Rahamann, M.N., De Jonghe, L.C.: Effect of rigid inclusions on the sintering of glass powder compacts. J. Am. Ceram. Soc. **70**, 348–351 (1987)

9. Lange, F.F: Constrained network model for predicting densification behavior of composite powders. J. Mater. Res. **2**, 59–65 (1987)
10. Sudre, O., Bao, G., Fan, B., Lange, F.F.: Effect of inclusions on densification II, Numerical model. J. Am. Soc. (1992), Wiley Online Library
11. Mackencie J.K., Shuttleworth R.: A phenomenological theory of sintering. Proc. Phys. Soc. B **62**, 833–852 (1949)
12. Atlas, S.: Verlag Stahl-Eisen, Düsseldorf, Germany (1995)

Chapter 17
Porous Gas Purging Bricks Calculation

17.1 Introduction

To improve the homogeneity of molten metals, e.g. steel, gases, specifically argon, are injected through the bottom into the filled ladle with the aid of a so-called purging brick. In principle, a distinction is made between two types of purging bricks:

- Porous purging bricks,
  with randomly distributed pores, comparable to a uniform-grain fill:
  This also includes the labyrinth purging bricks with net-shaped pore distribution.
- Purging bricks with directed porosity:
  These are the channel purging bricks with straight continuous, round pore channels and the slit purging bricks with straight, rectangular channels.

The cross-sectional area of a channel is only a few square millimeters to prevent the penetration of the melt into the purging brick. The gas pressure can be up to five bar in continuous operation and significantly higher in the event of a malfunction. The steel mill operates at temperatures of up to 1750 °C.

17.2 Basics

17.2.1 Hagen-Poiseuille Equation

The starting point for a theoretical description of the gas flow is the equation of Hagen-Poiseuille [1]. According to this equation, the volume flow of an incompressible fluid in a long pipe L with a relatively narrow, circular cross-section is calculated as follows $\frac{\pi}{4}D_o^2$ to:

J. Pötschke, *Refractory Fundamentals in Metallurgical Practice*,
https://doi.org/10.1007/978-3-031-63709-4_17

$$\dot{V} = -\frac{\pi}{128} \cdot \frac{D_o^4(P_1 - P_2)}{\eta \cdot L}. \qquad (17.1.1)$$

$(P_1 - P_2)$ is the driving pressure difference and η the dynamic rigidity.
The average flow velocity is directly proportional D_o^2:

$$u = \frac{4\dot{V}}{\pi \cdot D_o^2} = -\frac{D_o^2 \Delta p}{32\eta L} \qquad (17.1.2)$$

Additional requirements for validity are:

- The flow must be laminar, i.e. the Reynolds number [2]

$$Re = (\rho \cdot u \cdot D_o)/\eta \qquad (17.2)$$

must not exceed the value 2.300. ρ is the density of the gas and u its velocity. D_o stands for the pipe diameter as a characteristic dimension.
- The flow profile must be parabolic.
- The fluid must not change its viscosity as a result of the flow (Newtonian behavior).
  These conditions are fulfilled insufficiently in a gas flow because gas is compressible and the flow is turbulent in many cases.

17.2.2 Darcy's Equation

Alternatively, it can be suggested to use the empirical equation, already known earlier and determined by Henry d'Arcy (Darcy) [1] on the flow of groundwater:

$$\dot{V} = -\mu \cdot \frac{F}{\eta} \cdot \frac{(P_1 - P_2)}{L} \cdot \frac{2P}{(P_1 + P_2)}. \qquad (17.3)$$

In it, the permeability of the material flowing through is described with the aid of the empirically measured permeability μ [m^2].

However, Darcy's equation also applies to incompressible fluids and, therefore, describes a gas flow only inadequately. Nonetheless, gas permeability traditionally plays a role in the characterization of refractory materials, as standard EN 993-4 proves. It describes the "determination of gas permeability" in detail. For small pressure differences ($\leq$ 1000 Pa), the value of the third fraction (Eq. 17.3) is very close to 1 and can be neglected. Use is made of this here.

Looking at Eq. (17.3), it becomes clear that only the dimension [m^2] of the empirical "permeability μ" formally enables the volumetric flow rate $\dot{V}$ [m^3/s] because according to Darcy, the volumetric flow depends on the flow area F [m^2] and not on its square.

Permeability is considered to be a material constant dependent on temperature. In fact, this is a consequence of the temperature dependence of the viscosity of the measuring gas used. The permeability μ should, therefore, be measured with the gas used later in practice. Furthermore, its temperature dependence should be known quantitatively, if one does not want to perform the laboratory measurement at service temperature.

17.3   Flow in Cylindrical Pores

The aim is to quantitatively describe the laminar flow of a gas in a uniform-grain bed. For this purpose, it is expedient to start with a simple model as first step:

A concrete block of length L and cross-sectional area F is considered. Its volume is $V = F \cdot L$. This block is perforated by N_z parallel cylinders of length l and cross-sectional area $\frac{\pi}{4}D_0^2$. The total void volume is $V_p = N_z \cdot \frac{\pi}{4}D_0^2 \cdot l$. The ratio of void volume to total volume is defined as porosity:

$$\varepsilon \equiv \frac{V_p}{V} = N_z \frac{\pi D_o^2}{4F} \cdot \frac{l}{L} \; [-] \tag{17.4}$$

If the holes are parallel to the center axis, $l = L$. If they form an angle θ to the a·is, they are longer than L:

$$\tau \equiv l/L = \cos^{-1}\theta = \frac{D_o \cdot A_o}{4\varepsilon} \; [-] \tag{17.5}$$

τ is designated tortuosity [2] according to Kozeny and describes the extension of the flow path. The reciprocal value is called labyrinth factor $\xi = 1/\tau$. A_o stands for the specific surface area of the entire pore space and is defined as the ratio of the wetted area F_p to the solid volume V. For $\tau = 1$, it follows from (17.5):

$$A_o = \frac{F_P}{V} = \frac{4\varepsilon}{D_o} \left[\frac{\text{m}^2}{\text{m}^3} \right]. \tag{17.6}$$

If the equations of Hagen-Poiseuille (17.1.1) and Darcy (17.3) describe the same volume flow, the following applies

$$-N_z \cdot \frac{\pi D_o^4}{128} = -\mu \cdot F. \tag{17.7}$$

The number of holes results for $\tau = 1$ from (17.4) to

$$N_z = \frac{4F \cdot \varepsilon}{\pi D_o^2}. \tag{17.8}$$

For a laminar pipe flow, the correlation between the permeability μ and the porosity ε follows from (17.7):

$$\mu_z = \frac{D_0^2}{32} \cdot \varepsilon \ [\mathrm{m}^2].$$
(17.9)

17.4 Flow in a Disorderly Spherical Bed

If the perforated concrete block is replaced by a body of the same size V consisting of n_k pebbles of the same diameter D_k, the volume V_k of all spheres (solid volume) is calculated to be

$$V_k = n_k \cdot \frac{\pi}{6} D_k^3 = (1 - \varepsilon) V \ [\mathrm{m}^3].$$
(17.10)

The wetted surface of all spheres F_k equals the pore area F_p

$$F_k \equiv F_p = n_k \cdot \pi D_k^2 = A_o \cdot V \ [\mathrm{m}^2]$$
(17.11)

The specific surface area of all spheres results from (17.6), (17.10) and (17.11) to be

$$A_o \equiv \frac{F_K}{V} = \frac{F_k}{V_k}(1 - \varepsilon) = \frac{6(1 - \varepsilon)}{D_k} = \frac{4\varepsilon}{D_o} \ [\mathrm{m}^{-1}]$$
(17.12)

It equals that calculated above (17.6). The diameter of a flow channel is then

$$D_o = \frac{2\varepsilon}{3(1 - \varepsilon)} \cdot D_k \ [\mathrm{m}].$$
(17.13)

The number of the continuous pores is calculated from (17.8) and (17.13) to give

$$N_k = \frac{4F \cdot \varepsilon}{\pi \cdot D_o^2} \cdot \frac{L}{l} = \frac{F}{\tau} x \frac{9(1 - \varepsilon)^2}{\pi \cdot \epsilon \cdot D_k^2}.$$
(17.14)

F is the outlet area of the purging brick. Let the tortuosity be $\tau = 1$. The equations of Hagen-Poiseuilles (17.1.1) and Darcy (17.3) are to describe the same volumetric flow and the following applies, as shown above

$$-N_k \cdot \frac{\pi D_o^4}{128} = -\mu \cdot F.$$
(17.7)

From (17.7), (17.13), and (17.14), one directly obtains the correlation between the permeability μ_k and the porosity ε for a laminar flow in a randomly arranged uniform-grain bed:

$$\mu_k = \frac{\varepsilon^3 \cdot D_k^2}{72(1 - \varepsilon)^2} \ [\text{m}^2].\tag{17.15}$$

This result has already been obtained by Carmann and Kozeny [2]. Practice shows that the number 72 is better substituted by 150 [2].

17.5 Flow in an Orderly Spherical Bed

The densest arrangement of spheres having the same size can be face-centered (fcc) or body-centered (bcc). In the face-centered lattice (ABCA..), 6 spheres each form an octahedral gap or 4 pebbles form a tetrahedral gap. For hexagonal densest packing (hcp) with the layer sequence (ABAB..) and a cubic face-centered arrangement (fcc), the porosity is calculated as follows $\varepsilon_{fcc} = 1 - \frac{\sqrt{2}}{6}\pi \approx 0.26$. In the simple cubic lattice (cpr), the porosity is particularly large: $\varepsilon_{cpr} = \frac{\pi}{6} \approx 0.48$. The body-centered lattice (bcc) forms octahedral gaps. The porosity is $\varepsilon_{bcc} = 1 - \frac{\sqrt{3}}{8}\pi \approx 0.32$. In the hexagonally densest sphere packing (hcp), straight flow channels of minimum diameter D_0 run through the entire stack in the direction of the c-axis. This is the case considered here.

The narrowest diameter D_0 of a channel in the hexagonally densest sphere packing is calculated from the sphere diameter D_k as is known to be

$$D_0^{hcp} = 0.5\left(\sqrt{6} - 2\right)D_k = \delta \cdot D_k = 0.225 D_k.\tag{17.16.1}$$

δ designates the ratio of the narrowest gap diameter to the sphere diameter. An *octahedral gap* is formed by six spheres and occurs with face-centered and body-centered arrangement: $D_0^{fcc} = \left(\sqrt{2} - 1\right)D_k = 0.415 D_k$ and

$$D_0^{bcc} = \left(\frac{2}{3}\sqrt{3} - 1\right)D_k = 0.155 D_k\tag{17.16.2}$$

The ratio $\delta \equiv D_0/D_k$ thus depends on the lattice type. The porosity has no formal influence.

The situation is different if using a model of arbitrarily arranged spheres. Here, the porosity ε alone determines the diameter ratio $\delta_k \equiv \frac{D_0}{D_k}$ (17.13).

It is not the narrowest gap but the equivalent diameter of the observed cylinder pore with the same flow admissibility that is the decisive variable for the flow. Both values, δ and δ_k, therefore, differ from each other. For the hexagonally densest sphere

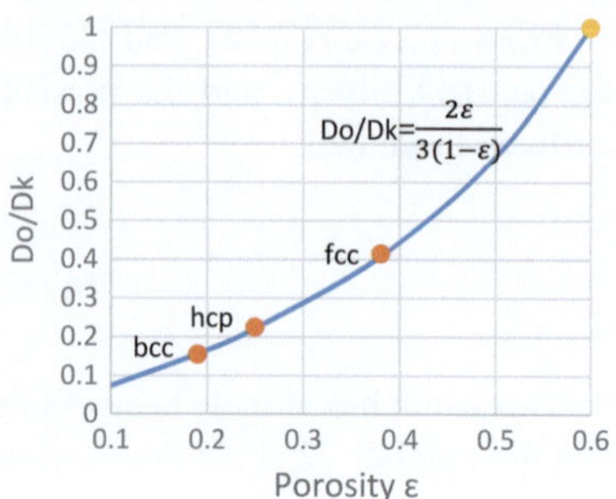

Fig. 17.1 Pore diameter D_0/sphere diameter D_k as a function of porosity ε

packing ($\varepsilon = 0.26$), however, the values almost coincide: $\frac{\delta}{\delta_k} = \frac{0.225}{0.234} = 0.962$. δ_k is calculated with (17.13). This makes sense because in the derivation of δ_k, the specific surface A_o (17.12), which is decisive for the flow, was taken into account.

The dependence of the diameter ratio $\delta_{zk} \equiv \frac{D_0}{D_k}$ on the porosity ε can be derived for the ordered spherical bed directly from the definition of the porosity $\varepsilon \equiv \frac{V_0}{V_{zk}+V_0}$ (17.4): With $V_0 = \frac{1}{4}D_0 F_0$, $F_0 \equiv F_{zk} = n_{zk}\pi D_k^2$ (17.11), one gets $V_0 = \frac{\pi}{4}n_{zk}D_0 D_k^2$ and $V_{zk} = \frac{\pi}{6}n_{zk}D_k^3$ (17.10). Due to $\frac{V_o}{V_{zk}} = \frac{\varepsilon \cdot V}{(1-\varepsilon)V}$ (compare (17.10)), there follows directly $\delta_{zk} = \frac{2}{3}\left(\frac{\varepsilon}{1-\varepsilon}\right)$ (17.13). The comparison with the corresponding expression in line (17.8) of the table for the disorderly spherical fill shows the conformance of the two. Preferably in both models δ_k and δ_{zk}, respectively, by means of (17.13) from the porosity ε is calculated. Figure 17.1 shows the result using the data of Eq. (17.16) as an example.

In contrast to the cylindrical flow channel considered in 17.3 above, the concave wall in the orderly sphere packing is not smooth but is formed from a sequence of convex sphere sections. If, despite this considerable difference in (17.9), the pore diameter D_0 is substituted by the sphere diameter D_k corresponding to it, it follows, for example, with the value calculated for the hexagonally densest sphere packing (hcp) $\delta = 0.225$ (17.16.1) the permeability

$$\mu = \frac{0.225^2}{32} \cdot \varepsilon \cdot D_k^2 = 1.58 \cdot 10^{-3} \cdot \varepsilon \cdot D_k^2. \qquad (17.17.1)$$

(The same can be done with the values from (17.16.2)).

However, from (17.13), with $\varepsilon = 0.26$, follows $\delta_{zk} = 0.234$ (see above) and from this

$$\mu_{zk} = \frac{0.234^2}{32} \cdot \varepsilon \cdot D_k^2 = \alpha \cdot \varepsilon \cdot D_k^2 = 1.706 \times 10^{-3} \cdot \varepsilon \cdot D_k^2. \qquad (17.17.2)$$

$\mu/\mu_{zk} = 0.962$. We refer to the number α as the void (gap) factor. It is calculated by setting (17.15) and (17.2) equal to

Table 17.1 Comparison of the models in Sects. 17.3, 17.4 and 17.5

		Z structure	K structure	ZK structure
1	Solid geometry	Concave	Convex	Concave/concave
2	Pore geometry	Convex	Concave	Concave/convex
3	Flow path	Linear	Nonlinear	Nonlinear
4	Flow cross-section	Constant	Not constant	Not constant
5	Solid number	1	$n_k = \frac{6V(1-\varepsilon)}{\pi D_k^3}$	$n_{zk} = n_k$
6	Pore count	$N_z = \frac{4\varepsilon F}{\pi D_0^2}$	$N_k = N_z$	$N_{zk} = \frac{4\varepsilon F}{\pi(\delta D_k)^2}$
7	Specific surface	$A_{0z} = \frac{4\varepsilon}{D_0}$	$A_{0k} = \frac{6(1-\varepsilon)}{D_k}$	$A_{0zk} = \frac{4\varepsilon}{\delta D_k}$
8	Pore Ø/Sphere Ø	$\delta \equiv \frac{D_0}{D_k}$	$\delta_k = \frac{2\varepsilon}{3(1-\varepsilon)}$	$\delta_{zk} = \frac{2\varepsilon}{3(1-\varepsilon)}$
9	Void (gap) factor	$\alpha \equiv \frac{\delta^2}{32}$	$\alpha = \frac{\delta_k}{32}$	$\alpha = \frac{\delta_{zk}}{32} = \frac{\varepsilon^2}{72(1-\varepsilon)^2}$
10	Calculated permeability	$\mu_z = \frac{\varepsilon D_0^2}{32}$	$\mu_k = \frac{\varepsilon^3 \cdot D_k^2}{72(1-\varepsilon)^2}$	$\mu_{zk} = \alpha\varepsilon D_k^2$
11	Prerequisite for 10	$A_{0z} \equiv A_{0k}$, e.g. $\varepsilon = 0,6$		$A_{0z} \equiv A_{0zk}$, for all ε

$$\alpha \equiv \frac{\delta_{zk}^2}{32} = \frac{\varepsilon^2}{72(1-\varepsilon)^2}. \tag{17.18}$$

When using the gap factor α, the permeabilities μ_k and μ_{zk} are identical calculated according to Eqs. (17.15) and (17.17.2) via ε. (refer to Table 17.1, line (10) and Fig. 17.1): Both models, 4 and 5, give the same result.

17.6 Discussion of the Models [3]

The models are compared with each other on the basis of Table 17.1 [3]:

Apart from the geometrical differences (lines 1–4), the comparison of lines (5) and (6) shows that, if all other parameters are kept constant, the number of spheres n decreases with increasing porosity ε, whereas the number of pores N increases. If the sphere diameter D_k increases, while the total volume remains unchanged, D_0 increases but the number of spheres n decreases as well as the number of pores N. The specific surface area A_0 decreases in the sphere model as the sphere diameter increases (line (7)). The pore to sphere diameter ratio increases rapidly with porosity (line (8)). The same is true for the void (gap) factor α (line (9)). It is calculated from the porosity according to (17.18). Similarly, the permeability increases with the size of the spheres (line (10)). The dependence of permeability on porosity, calculated according to Carmann and Kozeny [2], resembles the orderly spherical fill (line (10)). Line (11) shows that the permeabilities calculated according to the models are similar if their specific surfaces are similar. Then also the hydraulic diameters of the models $D_h \equiv 2\varepsilon/A_0$, which are often discussed in literature [2], are frequently the same.

As a result, it can be noted that the models lead to the same result and are equal in application. The decisive point is that the ratio D/D_{0k} (17.13) is always the same, regardless of the arrangement of the spheres.

Therefore, it is proposed to substitute the gas permeability μ (DIN EN 993-4), which is usually measured for the evaluation of purging bricks, by the more easily determined porosity ε (DIN EN 1094-4) and the "calculated gas permeability" μ_c from it. For the application of the equation of Hagen-Poiseuille for the calculation of the flow rate of a porous purging brick, from the characteristic grain diameter D_K [mm^2] and the porosity ε [$-$] with consideration of the gap factor $\alpha = \frac{\varepsilon^2}{72(1-\varepsilon)^2}$ (17.18), one gets the "calculated permeability" which becomes

$$\mu_c \ [\text{mm}^2] = \alpha \cdot \varepsilon \cdot D_K^2. \tag{17.15}$$

Deviating from the idealized models, practice shows that the permeability calculated according to Carmann and Kozeny [2] is not fulfilled that well even if in (17.18) the factor 72 is substituted by 150. Further influences, such as tortuosity, pore structure and friction of the flow are discussed in literature in order to achieve an improved alignment of theory and practice [2]. In any case, only practice will show to what extent the models can be applied to high gas velocities.

17.7 Applications

As an example, let us consider a porous purging brick constructed of ceramic spheres of the same size and diameter $D_K = 10$ mm. The outlet surface of the gas F has the dimension $D = 130$ mm$^\varnothing$ and its length is $L = 350$ mm. The temperature of the steel bath is 1600 °C. The inlet pressure is $P_2 = 2$ bar. The following applies to the dynamic resistance of argon at approx. 1 bar:

$$\eta[\mu\text{Pa s}] = 0.0556 \cdot T_{\circ C} + 21.21, \tag{17.19}$$

i.e. at 1600 °C: $\eta = 110\,\mu\text{Pa s}$. The material properties of the gas go into the calculation via the dynamic viscosity η.

For the conversion of the above dimensions to $\dot{V}$ [l/min] the factor $\mathbf{6 \times 10^6}$ is used.

17.7.1 Application of Hagen-Poiseuille's Equation to Channel Purging Brick

A typical channel purging brick with 20 channels, having a length of 350 mm and the cross-sectional area of a channel of approx. 4 mm^2 at a pressure of 2 bar, enables the gas transport of 280 l/min (personal communication with A. Eschner & RHI [4]).

Here, we consider a porous purging brick of the same length, consisting of densely packed ceramic spheres of diameter $D_K = 10$ mm. The "equivalent diameter" of one pore is calculated from the porosity $\varepsilon = 0.26$ and the diameter of the spheres using Eq. (17.13) to give $D_o = 2.34$ mm. This corresponds to the cross-sectional area of a pore $F_o = \frac{\pi}{4}D_o^2 = 4.3$ mm^2. With 20 channels of length $L = 350$ mm, Eq. (17.1.1) results in:

$$\dot{V}_N = 6 \times 10^6 \cdot 20 \cdot \frac{\pi}{128} \cdot \frac{2,34^4(P_1 - 2)}{110 \cdot 350} = 280\ \text{l/ min}. \tag{17.20}$$

The required pressure is $P_1 = 2.123$ bar: The porous purging brick with 20 channels of diameter $D_0 = 2.34$ mm allows the required gas flow of 280 l/min with an inlet pressure of 2 bar and a differential pressure of 0.12 bar.

The performance of the purging brick depends linearly on the pressure difference and increases proportionally with it. The purge capacity $\dot{V} = 140$ Nl/min, for example, is obtained at $P \approx 1[\text{bar}]$ with the pressure difference $\Delta P = 0.061$ bar. The equation for calculating this volume flow can, therefore, be written in the simple form for this example (refer to Fig. 17.2):

$$\dot{V}[\text{Nl/min}] = 140 \cdot P[\text{bar}].$$

The narrower the grain size distribution, the more accurate the result. It is recommended to use the most frequent distribution as the characterizing grain size.

Fig. 17.2 Purging performance as a function of gas pressure

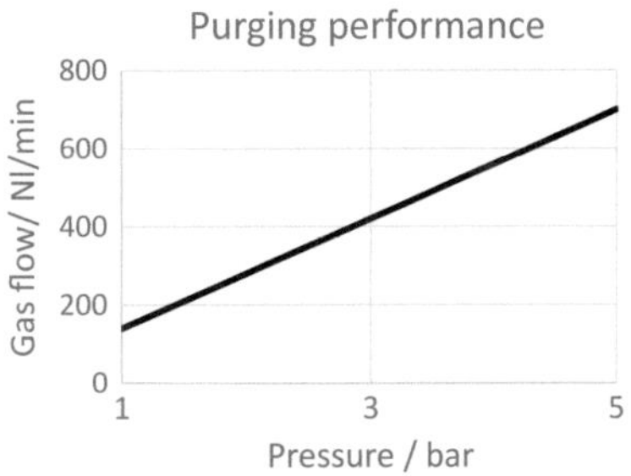

17.7.2 Application of Permeability μ (Darcy) on Spherical Bed

The arrangement of the spheres having a diameter $D_k = 10$ mm corresponds to a hexagonally extremely dense sphere packing. The porosity is $\varepsilon = 0.026$. According to (17.18), the gap factor is $\alpha = 1.71 \cdot 10^{-3}$. Equation (17.2) then gives for the permeability $\mu_{zk} = 4.43 \cdot 10^{-2}$ mm^2.

Here, 1 mm^2 is $\equiv 10^6$ Darcy $= 10^6$ μm$^2 = 10^7$ nPm. Darcy is an older term for permeability.

If Eq. (17.15), derived from Carman and Kozeny [2], is used to calculate permeability, the same value $\mu_k = 4.43 \times 10^{-2}$ mm^2 is obtained because the gap factor α links Eqs. (17.15) and (17.17) with one another (refer to Table 17.1, line (9) and Fig. 17.1). Thereby $\mu_k = \mu_{zk}$.

From Darcy's Eq. (17.3), it follows that with $\mu_{zk} = 4.43 \cdot 10^{-2}$ mm^2, the purging brick format $F \equiv D^2 = (130 \text{ mm})^2$ and $L = 350$ mm under the driving pressure difference $(P_1 - P_2) = 0.123$ bar:

$$\dot{V}_\mu = 6 \times 10^6 \cdot 0.0443 \cdot \frac{130^2}{350} \cdot \frac{(2.123 - 2)}{110} = 1.44 \times 10^4 \, 1/\min. \tag{17.21}$$

17.7.3 Application of Porosity ε (Hagen-Poiseuille) on Spherical Bed

The hexagonal most dense packed lattice (hcp) with unchanged large spheres is used. The porosity is unchanged $\varepsilon = 0.26$ and the number of channels is obtained from (17.14):

$$N_{zk} = \frac{4\varepsilon F}{\pi (\delta_{zk} D_K)^2} = \frac{4 \cdot 0.26 \cdot 130^2}{\pi \cdot (0.234 \cdot 10)^2} = 1022. \tag{17.22}$$

Inserted in (17.20) according to Hagen-Poiseuille

$$\dot{V}_N = \frac{6 \cdot 10^6}{32} \cdot \frac{D^2(P_1 - P_2)}{\eta \cdot L} \cdot \varepsilon \cdot (\delta_{zk} \cdot D_k)^2 \tag{17.23}$$

$$\dot{V}_N = \frac{6 \cdot 10^6}{32} \cdot \frac{130^2(2.123 - 2)}{110 \cdot 350} \cdot 0.26 \cdot (0.234 \cdot 10)^2 = 1.44 \times 10^4 \, 1/\min. \tag{17.24}$$

The pressure difference calculated in (17.20) for 2 bar $(P_1 - P_2) = 0.123$ bar is used.

17.8 Requirements for the Application

In practice, the flow rate $\dot{V}\,[Nl/\min]$ is measured as a function of the applied total pressure P [bar] of the gas. However, the decisive factor for the gas flow rate in the purging unit is the comparatively small pressure difference $\Delta P = (P_1 - P_2)$. Consequently, the minimum required gas pressure is determined by the inlet pressure P_1. It is initially calculated from the height of the metal bath and the capillary pressure because no metal should penetrate the purging brick.

The pressure calculated from the metallostatic height $h\,[\mathrm{m}]$ is

$$P_{\mathrm{Fe}} = \rho \cdot g \cdot h. \tag{17.25}$$

With the values for steel, density $\rho = 7000\,\mathrm{kg/m^3}$, bath height $h = 1.46$ m and acceleration due to gravity $g = 981\,\mathrm{m/s^2}$, the ferrostatic pressure is calculated as $P_{\mathrm{Fe}} = 10^5\,\mathrm{kg/m\ s^2}$ [1bar]. At a typical bath height of 3 m, the minimum gas pressure required for purging is, therefore, approximately 2 bar.

The capillary pressure P_K impairs the penetration of the melt if the wetting angle θ is greater than $90°$. It is calculated as [1]

$$P_K = -\frac{4\sigma \cdot \cos\theta}{D_o}. \tag{17.26}$$

With the typical values for liquid steel, surface tension $\sigma = 1.600\,\mathrm{dyn/c}$ [1.6 N/m], wetting angle to the ceramic $\theta = 120°$ and pore diameter in the mm range, e.g. $D_o = 1$ mm [0.001 m], one obtains as capillary pressure $P_K = 3.2 \times 10^3\,\mathrm{N/m^2}$ [0.032bar] which contributes only marginally to the prevention of infiltration by liquid steel. The gas pressure of ≥ 2 bar, on the other hand, reliably prevents infiltration of the purging brick and its failure at bath height ≤ 3 m.

In addition, the pressure loss in the pipes and in the purging brick must be considered in practice. Due to the friction of the flowing gas, pressure losses ΔP_{DV} occur in the piping and in the purging brick which are compensated for by the gas pressure but can increase the inlet pressure P_1 considerably [1]:

$$\Delta P_{DV} = \Psi \cdot \frac{l}{D_o} \cdot \frac{\rho}{2} u^2. \tag{17.27}$$

The pipe friction coefficient Ψ [1] is calculated as

$$\Psi = 64/\mathrm{Re} \tag{17.28.1}$$

with laminar flow and

$$\Psi = 0.32/\mathrm{Re}^{0.25} \tag{17.28.2}$$

in the turbulent flow range.

The density of the gas changes with temperature according to the ideal gas law:

$$\frac{\rho_2}{\rho_1} = \frac{T_1}{T_2}.$$
(17.29)

For one channel of the considered purging brick the following values are valid with (17.20):

$T_1 = 30\,°C$, $\rho_1 = 1.78\,kg/m^3$, $T_2 = 1600\,°C$ $\rho_2 = 0.28\,kg/m^3$, $\dot{V} = 280\,l/\,min$, $N = 20$, $D_o = 2.34$ mm. From (17.1.2), the maximum flow velocity in a channel is obtained:

$$u = \frac{4 \cdot \dot{V}}{N \cdot \pi \cdot D_o^2} = 16.7 \cdot \frac{4 \cdot 280}{20 \cdot \pi \cdot 2.34^2} = 54.5\,m/s.$$
(17.30)

The Reynolds number is calculated from (17.3) to be

$$\text{Re} = (\rho \cdot u \cdot D_o)/\eta = 10^3 \cdot \left(0.28 \cdot 54.5 \cdot \frac{2.34}{110}\right) = 325.$$
(17.31)

The flow is laminar. The coefficient of pipe friction is, therefore, calculated from Eq. (28.1) to $\Psi = 0.2$.

For the pressure drop in the purging brick, Eq. (17.27) gives the result:

$$\Delta P_{DV} = 10^{-5}\left(0.2 \cdot 350 \cdot \frac{0.28}{2.34} \cdot 54.5^2\right) = 0.25\,bar$$
(17.32)

With increasing transport capacity, however, the pressure loss increases rapidly which is why it should be estimated with foresight. To set the operating pressure P_1, the capillary pressure $P_K = 2\sigma/R_o$ and the pressure difference required for the gas transport ΔP must be added to the pressure loss ΔP_{DV}. If, for example, the grain size is halved ($D_K = 5$ mm), to transport $\dot{V} = 280\,l/min$ the pressure difference $\Delta P = 2, 3$ bar is required. The pressure loss in the purging brick is $\Delta P_{DV} = 4.6\,bar$. Consequently, the operating pressure P_1 must be $> 7\,bar$.

The practical applicability of this "calculated permeability μ_c, derived from the porosity ε, has not yet been verified experimentally.

Note: It is expected that the theory presented describes the flow of a molten metal in a ceramic filter material well, as this occurs more slowly and at a lower Reynolds number. This also needs to be verified experimentally.

References

1. Grassmann, P.: Basics of Chemical Technique, Salle + Sauerländer (1983)
2. Bird, R.B., Steward, W.E., Lightfoot, E.N.: Transport Phenomena. Wiley, NY, London, Sidney (1960)

3. Harmuth, H.: Personal communication as part of an discussion regarding the models and their importance for purging with gas
4. Excerpt from: R & D "One Week Training: Functional Products", RHI—Refractories 14/76–16/ 76 (2010)

Index